ENVIRONMENTAL PLANNING AND DECISION MAKING

ENVIRONMENTAL PLANNING AND DECISION MAKING

Leonard Ortolano
Stanford University

JOHN WILEY & SONS
NEW YORK CHICHESTER BRISBANE TORONTO SINGAPORE

Library of Congress Cataloging in Publication Data:

Ortolano, Leonard.
Environmental planning and decision making.

Includes index.
1. Environmental policy. 2. Environmental protection.
I. Title.

HC79.E5078 1984 363.7 83-19820
ISBN 0-471-87071-4

Printed in the United States of America

10 9 8 7 6 5 4 3 2 1

TO MY PARENTS

PREFACE

Environmental planning is a field made up of contributions from several academic disciplines including biology, engineering, geography, geology, and landscape architecture among others. The contents of courses in environmental planning typically reflect the professional biases of individual instructors. When such courses are taught by landscape architects they have one orientation, and when they are taught by ecologists (or engineers or geographers) they have rather different orientations. Because it emerged as a field only in the 1970s, ambiguity in what constitutes environmental planning is to be expected.

My motivation for writing this book was to provide a text reflecting the multidisciplinary nature of environmental planning as a field. As the writing progressed, it became clear that the principal challenge in the work consisted of integrating and synthesizing a disparate collection of information and perspectives. There are literally thousands of books that relate closely to the subjects treated in this work. For example, the specialized texts concerned with technologic issues in air and water pollution number in the hundreds. Since 1970, scores of books have been published on environmental economics and environmental impact assessment. The list of closely related subjects includes: environmental biology, environmental geology, environmental law, environmental policy, and environmental ethics. This book synthesizes information from all these areas. It is an introductory survey, not a comprehensive handbook.

A note about the level of presentation is in order. Elementary algebra is used to illustrate points in many chapters. Occasionally calculus is employed, but only in footnotes and appendixes. The main prerequisite for understanding the material is a desire to learn about what can be done by both technical specialists and individual citizens to improve the quality of the environment.

This book may be useful in several different contexts. At Stanford University, preliminary versions have been used in teaching undergraduates majoring in applied earth sciences, civil engineering, and human biology and graduate students specializing in environmental engineering and infrastructure planning. The text was conceived for courses concerned with environmental policy and regulation and environmental impact analysis. Instructors in the following program areas may find it suitable for professionally oriented courses: city and regional planning, landscape architecture, environmental and urban studies, and natural resource management.

Professional planners and engineers may find the text useful in introducing approaches employed by different specialists in treating environmental problems. The book may also be valuable to individual citizens who seek to influence public and private actions so that environmental quality can be preserved.

Leonard Ortolano
Cedro Cottage
Stanford, California

ACKNOWLEDGMENTS

I owe thanks to the many people who commented on sections of the book in its various preliminary forms. Although it is infeasible to mention all individuals by name, I would like to acknowledge the following Stanford students for reviewing substantial portions of the penultimate draft: Laura K. Charlton, Cheryl K. Contant, Mary Jane Delahunt, Timothy P. Duane, Melanie Julian, Alison S. Kneisl, James A. Levy, Laura R. Rines, and Gordon R. Shaw. Others who critiqued major portions of that draft include Professor Charles M. Brendecke of the University of Colorado and two of his students, Cindy Bishop and Douglas Kemper.

Several persons made suggestions on the entire manuscript. John J. Meersman of Bechtel Group, Inc. read the penultimate draft. The final version was reviewed by three Stanford students: William J. Beyda, Mary H. Masters, and Kenneth A. Wilke.

The following academic colleagues offered useful advice at key places: Jarir S. Dajani, Douglas A. Haith, Gilbert M. Masters, Perry L. McCarty, and Jery R. Stedinger. Many other persons helped in preparing sections of individual chapters and they are acknowledged in footnotes.

Duc Wong typed the several drafts of the text with the care and efficiency that characterizes all of her work. I consider myself very fortunate to have had her assistance.

My sons, Chris and Alex, and my wife, Patti Walters, provided emotional support that was absolutely essential to the undertaking. In addition, Chris commented on parts of the manuscript and Alex helped prepare the index. Patti encouraged me from the outset of the project and provided helpful suggestions on style and format. My debt to her is the greatest and the least simple to describe.

L. O.

CONTENTS

PART ONE

INTRODUCTION

CHAPTER 1

INFLUENCING ENVIRONMENTAL QUALITY

The impacts of people on the natural environment have been observed for centuries. However, for the most part, these observations were not used systematically to minimize the unintended environmental disruptions caused by public and private decisions. The evidence supporting this assertion is plentiful. Numerous case studies of inadvertent environmental degradation caused by human actions throughout the world are given by Farvar and Milton (1972).

During the 1960s, public concern over environmental degradation increased, and systematic planning to maintain environmental quality intensified in many countries. Past efforts to control water and air pollution were greatly expanded. New laws and administrative regulations were established requiring government agencies to account for the environmental impacts of their decisions. The increased attention to the environmental effects of human actions led to the development of a new field, environmental planning. Those practicing in this field are referred to as *environmental planners* or *environmental professionals.* Several new environmental planning journals and professional organizations were established in the late 1970s.[1] Although some aspects of the field are well established, the nature of environmental planning as a whole is still evolving.

People who view themselves as environmental planners generally have a speciality in a single discipline that is closely related to the environment. These include environmental engineering and a host of sciences such as biology, geology, hydrology, and meteorology. Some environmental planners have backgrounds in the social sciences, such as economics and sociology, and the design professions, especially landscape architecture. These disciplines are relevant to environmental planning because the term *environment* is interpreted broadly to include the "human environment" as well as the physical environment.

[1]Examples of new professional organizations are the National Association of Environmental Professionals (U.S.) and the International Association for Impact Assessment. New journals include *The Environmental Professional* and *Environmental Impact Assessment Review.*

Environmental planners commonly work in one of three occupational contexts: pollution control, environmental impact assessment, and land use planning. In pollution control, these professionals predict how environmental quality will be improved if various pollution control measures are implemented. They also design and implement programs requiring polluters to reduce waste discharges. Impact assessment involves environmental professionals in forecasting the consequences of various projects such as highways, airports, and dams. When they work as land use planners, environmental professionals predict how the air, land, and water will change in response to land use changes. They also design and implement strategies to manage the way land is used.

Because of the wide range of disciplines involved, there is no single orientation or approach used by environmental planners. The methods and biases of each of the contributing disciplines influence the overall outcome of environmental planning efforts. For example, an economist's view of a water pollution problem will not be the same as that of an engineer or biologist. Each approaches the problem differently, and this provides opportunities for pooling diverse perspectives. Efforts are often made to take advantage of these opportunities by having environmental planners work on teams formed to deal with particular problems.

The remaining parts of this book introduce the subject matter included in the field of environmental planning. Part Two, "Residuals Management," gives a framework for analyzing problems of air quality, water quality, noise, and other aspects of the environment. Adopting the terminology of Freeman, Haveman, and Kneese (1973), *residuals* refer to what are more commonly known as air pollutants, water pollutants, noise, and so forth. Emphasis is placed on developing a single management approach to deal with solid, liquid, and air-borne wastes, as well as energy residuals such as noise.

The third part of the book considers environmental impact assessment, a subject that has received considerable attention since the passage of the U.S. National Environmental Policy Act of 1969. This law, which has counterparts in other countries, requires federal agencies to give serious consideration to the environmental impacts of their proposed projects.

Part Four examines relationships between land use and environment. It is based on the view that environmental quality is not likely to improve significantly unless more attention is given to the unintended side effects of land use decisions. New procedures have been developed to integrate environmental considerations into land use decision making.

The fifth and final part of the book presents techniques to forecast specific environmental impacts. The types of effects considered are those that affect biological systems and visual (or scenic) resources. Procedures used in forecasting impacts on air quality, water quality, and noise are also introduced.

This book makes occasional mention of "social impact assessment," a new field that is related to environmental impact assessment. Examples of social impacts include the disruption of a neighborhood caused by a new highway and the job opportunities created by construction of a petroleum refinery. Social impact assessment rests heavily on contributions from psychology, sociology, economics, and other social sciences. It is beyond the scope of this work.[2]

Before proceeding with Part Two, a few general obsevations about attitudes and values are in order. People who consider themselves "environmentalists" often have diverse attitudes toward nature and the use of natural resources. The implications of these differences in attitudes and values are explored below.

HISTORICAL PERSPECTIVES ON "ENVIRONMENTALISM"

For many years scholars were mainly interested in how the environment influences people rather than how people affect the environment. At the turn of the century, an environmentalist was a person concerned with how the physical environment influenced the way societies functioned and developed. It is only during the past few decades that the term *environmentalist* has come to be associated with the converse view, namely, a concern with how human actions affect the natural environment.

There are several bases for concern about person–environment relationships, and they lead to ambiguities in the contemporary use of the term *environmentalist.* One view holds that nature's resources are there to be used, and this *use* should be characterized by efficiency and the absence of waste. A second basis for concern about the environment is that the absence of caution in human undertakings can lead to irreversible and sometimes disastrous impacts on the ability of natural systems to function. A third collection of views are those in which aesthetic, religious, and ethical concerns motivate a call for restraint on human actions affecting the environment.[3] Although the discussion below is presented in three distinct parts, many environmentalists possess a combination of these different views.

Conservation as the Efficient Use of Resources

Many people who consider themselves environmentalists do not question the idea of using the natural environment to satisfy material needs. Rather, their main concern is to avoid wasting natural resources. This outlook is epitomized by the conservation movement of Theodore Roosevelt's presidential administration, and especially by the views of Gifford Pinchot, Roosevelt's principal advisor on conservation issues.

Pinchot was the first head of the U.S. Forest Service and a leading advocate for the scientific management of forests. His attitudes toward the natural environment are reflected in his "three principles." For Pinchot, the "first principle

[2]In the impact assessment literature, the words "social" and "environmental" sometimes have blurry meanings. As an example, consider the title of the work by McEvoy and Dietz (1977): *Handbook for Environmental Planning, The Social Consequences of Environmental Change.* This book and the one by Finsterbusch and Wolf (1981) show how social science methods are used to assess the impacts of various projects.

[3]The use of three groups of perspectives relies on a discussion by Petulla (1980).

of conservation is development, the use of natural resources now existing on this continent for the benefit of the people who live here now."[4] Pinchot's emphasis on using resources *now* is, in part, a response to critics who argued that Roosevelt and Pinchot intended to withhold resources from development. The second of Pinchot's principles is that "conservation stands for the prevention of waste." Thus, although resources are to be used, this use is to be characterized by wise management and careful stewardship. The third of Pinchot's principles reflects the ideological overtones of the conservation movement of the turn of the century: "The natural resources must be developed and preserved for the benefit of the many, and not merely for the profit of a few." This is consistent with the antimonopoly sentiment that was widespread in America during the Theodore Roosevelt administration.

Leaders of the conservation movement in America argued that technicians trained in fields such as geology, forestry, and hydrology should play the key role in the planning and management of natural resources.[5] Highly trained specialists were needed to carry out resource management activities in a scientific and efficient manner. During the first half of the twentieth century, many resource development agencies in the United States relied on technically trained managers to accommodate multiple demands for land and water efficiently.

Many contemporary economists hold views that are consistent with Pinchot's concerns over the efficient use of natural resources. They are trained to identify combinations of labor, capital, and natural resources that can be used to produce various goods and services "efficiently." The economist's objective of maximizing the difference between benefits and costs (measured in terms of dollars) is termed the *economic efficiency objective*. For many years, economists involved in managing natural resources have been associated with various specialties such as agricultural economics, water resources economics, and so forth. In the past few decades a new specialty has evolved: environmental economics. One of its assumptions is that the acceptance of wastes from industries and municipalities is an appropriate use of the environment. Many environmental economists believe that the proper quantities of waste discharge should be determined by analyzing the economic benefits and costs involved. This view is consistent with Pinchot's ideas about how the environment should be used.

Maintenance of Harmony between People and Nature

The increased scientific understanding of natural processes provides another basis for concern about the influence of human actions on the environment. In modern times, the importance of maintaining the integrity of natural systems is commonly associated with ecologists who argue for the maintenance of harmony between people and nature.

[4]The quotations in this paragraph are from portions of Pinchot's book, *The Fight for Conservation*, reprinted in Nash (1968, pp. 58–62).

[5]The relationship between the conservation movement and the scientific management of resources is elaborated by Hays (1959).

Decades before ecology became an established science, Marsh (1864) argued persuasively for the need to keep "harmonies of nature" from being turned into discords. His widely cited book, *Man and Nature; or, Physical Geography as Modified by Human Action,* is a synthesis of numerous scientific works and personal observations. Marsh examined the ability of natural systems to withstand disturbances and to restore themselves following major perturbations such as those caused by "geologic convulsions." He was distressed with the way human actions could "interfere with the spontaneous arrangements of the organic or inorganic world" and thereby cause instabilities and irreversible changes in nature. Based on his studies of the adverse environmental effects of human actions, Marsh warned that continued "human improvidence" could threaten the earth's ability to support human life.

As the science of ecology developed in the early twentieth century, Marsh's concerns began to be articulated using a more rigorous scientific basis. Leopold's (1933, p. 635) ecological studies led him to the following position:

> *A harmonious relation to land is more intricate, and of more consequence to civilization, than the historians of its progress seem to realize. Civilization is not, as they often assume, the enslavement of a stable and constant earth. It is a state of* mutual and interdependent cooperation *between human animals, other animals, plants, and soils which may be disrupted at any moment by the failure of any of them. Land despoliation has evicted nations, and can on occasion do it again.*

The ability of humans to inadvertently destroy natural systems increased dramatically in the 1940s. The post-World War II period witnessed the emergence of a variety of new substances, including radioactive materials and synthetic organic chemicals. Many of these new substances are persistent. They do not decay or decompose rapidly into simpler, less harmful materials. Some scientists have responded to this dramatic increase in people's ability to disturb natural systems with demands for additional controls on human actions.

The contemporary scientists' calls for restraint are often based on the same concerns that motivated Marsh and Leopold: the human ability to irreversibly destroy natural systems. The sense of urgency that often accompanies these calls for restraint is exemplified by Rachel Carson's *Silent Spring* (1962). The title of this widely read book refers to a hypothetical future springtime in which birds and other animals are silenced inadvertently. They are destroyed by man-made chemicals that are transmitted from one life form to another as part of the normal production and consumption activities in nature. Pesticides, such as *n*[dichlorodiphenyltrichloroethane] (DDT), are the chemicals of principal interest in *Silent Spring.* Carson (1962, pp. 7–8) described the unintended effects of using such pesticides:

> *These sprays, dusts, and aerosols are now applied almost universally to farms, gardens, forests and homes—nonselective chemicals that have the*

power to kill every insect, the 'good' and the 'bad', to still the song of birds and the leaping of fish in the streams, to coat the leaves with a deadly film, and to linger on in soil—all this though the intended target may be only a few weeds or insects. Can anyone believe it is possible to lay down such a barrage of poisons on the earth without making it unfit for all life?

Silent Spring is one of many books written since World War II to alert the public to the seriousness of unintentional environmental disruptions.

Religion, Renewal, and Rights

The two perspectives above are linked to well-known disciplines: efficiency in natural resource use is associated with economics, and the need to maintain "harmonies of nature" is tied to ecology. Each of these viewpoints is pragmatic. The former calls for the elimination of waste in resource use, and the latter urges the maintenance of the earth as habitat that can support human life. In contrast, the ideas introduced below concerning religion, spiritual renewal, and ethics have a more ethereal quality. These ideas, however, are no less important than the more practical concerns of economists and ecologists. Indeed, some would argue that these spiritual and philosophic matters are of the utmost significance to the long-term survival of the planet.

Several of the issues examined below were initially raised during efforts to preserve portions of the American wilderness. Some observations about this "preservationist movement" are therefore in order. Until the eighteenth century, the untamed wilderness was generally considered to be mysterious and dangerous. It was a place to be conquered and put to "productive use." Romantic poets, such as William Wordsworth and Samuel Coleridge, played an important early role in putting a positive value on the remote, mysterious, and solitary nature of wild places. The cause of wilderness preservation gained momentum in America during the late nineteenth century, when only a small fraction of the original wilderness remained.[6]

Many reasons for preserving wilderness have economic and ecologic bases. For example, wilderness should be preserved to provide scientists with an opportunity to study natural processes, or because the numerous species present provide a pool of diverse genetic patterns that may eventually prove useful in agriculture or medicine. Alternatively, wilderness should be preserved because of the opportunities it provides to hunters, fishermen, and others engaged in outdoor recreation. Such practical propositions are important, but they are not considered below. Rather, the concern is with spiritual and philosophic arguments for wilderness preservation.

American Transcendentalism The religious fervor that has sometimes characterized the cause of wilderness preservation is frequently based on ideas of

[6]This discussion of the preservationist movement in America relies heavily on Nash (1968 and 1973).

the American transcendentalists of the nineteenth century, especially Emerson and Thoreau. A central tenet in their philosophy is that in possessing a soul, a person has the potential to transcend beyond the material portion of the universe and thereby perceive higher, spiritual truths. For the transcendentalists, "one's chances of attaining moral perfection and knowing God were *maximized* by entering wilderness."[7]

Emerson is generally acknowledged as the intellectual leader of the American transcendentalists. His view on the communion with God obtainable in wilderness are summarized in a passage from his essay, *Nature:*

> *In the woods, we return to reason and faith. . . . Standing on the bare ground – my head bathed by the blithe air and uplifted into infinite space – all mean egotism vanishes. I become a transparent eyeball; I am nothing; I see all; the currents of the Universal Being circulate through me; I am part or parcel of God.*[8]

The transcendentalists' views on how spiritual values are reflected in nature were used frequently in arguments for wilderness preservation at the turn of the century. This is illustrated by John Muir's valiant but unsuccessful effort to prevent the damming of Hetch Hetchy Valley in the Sierra Nevada mountains of California. Muir had studied Emerson's work and had similar, strong feelings about the presence of God in nature. Nash, quoting from several of Muir's writings, observes: "At one point Muir described nature as 'a window opening into heaven, a mirror reflecting the Creator.' Leaves, rocks, and bodies of water became 'sparks of the Divine Soul.' "[9] In defending Hetch Hetchy Valley from those who would inundate it with a reservoir, Muir asserted: "Dam Hetch Hetchy! As well dam for water-tanks the people's cathedrals and churches, for no holier temple has ever been consecrated by the heart of man."

Although religious sentiment influenced Muir's attempts to preserve Hetch Hetchy Valley and other wild areas, it was not his only motivation. A dominant theme in his work, and that of many other preservationists at the turn of the century, was the importance of providing people with opportunities for spiritual and emotional renewal. For Muir and others, the tribulations of the mechanized, technologic life ushered in by the industrial revolution required the kind of regeneration that only wilderness could provide.

Spiritual Renewal and Re-creation Those who view the wilderness as an important source of physical and spiritual "re-creation" often take their inspiration from Thoreau. For him, forest and wilderness provided "the tonics and barks

[7]This quotation is from Nash (1973, p. 86).

[8]The Emerson and Thoreau quotations in this discussion are from the volume of readings edited by Nash (1968).

[9]This quotation is from Nash (1973, p. 125); the famous quote below in defense of Hetch Hetchy Valley appeared originally in Muir's *The Yosemite* and is cited by Nash (1973, p. 168).

which brace mankind." In joining the political cause for preservation, Thoreau argued that the loss of wilderness would lead people to become weak and dull and lacking in creativity.

The importance of wilderness as a source of spiritual renewal is a theme that has been sounded many times since Thoreau wrote in the nineteenth century. For some, the very *idea* that wild places still exist is a source of strength. The American novelist Wallace Stegner articulated the importance of the wilderness idea:

> *The reminder and the reassurance that [wilderness] is still there is good for our spiritual health even if we never once in ten years set foot in it. It is good for us when we are young, because of the incomparable sanity it can bring briefly, as vacation and rest, into our insane lives. It is important to us when we are old simply because it is there—important, that is, simply as idea.*[10]

Sometimes the spectacular beauty of a wild place is what makes it a "tonic." The aesthetic features of wild areas are often cited in arguments for wilderness preservation, especially when these features are unique. This was the case in the 1960s when citizens' groups opposed federal attempts to inundate portions of the Grand Canyon in Arizona as part of a large water project. In this instance, the preservationists prevailed.

Rights for Non-human Beings and Natural Objects Although it has not been in the mainstream of western moral philosophy or law, the question of rights for nonhuman animals, and plants (and even rocks and soils) has come up from time to time. As far back as the sixteenth century, there is evidence of a concern for minimizing the cruelty suffered by animals.[11] The existence of societies for the prevention of cruelty to animals attests to the continuing strength of this sentiment.

There is, however, a broader view of the rights of nonhuman living things that has been used by preservationists. Consider, for example, John Muir's answer to the question, What are rattlesnakes good for? His response was that rattlesnakes are "good for themselves, and we need not begrudge them their share of life." In a different context, Muir indicated that "the universe would be incomplete without man; but it would also be incomplete without the smallest transmicroscopic creature that dwells beyond our conceitful eyes and knowledge." According to Muir, in the wilderness people learn a need for respecting "the rights of all the rest of creation."[12]

[10]The quotation is from Stegner's essay, "The Wilderness Idea," reprinted in Nash (1968, pp. 192–97).

[11]Passmore (1974) cites evidence for this in his philosophic treatise on *Man's Responsibility for Nature, Ecologic Problems and Western Traditions.*

[12]The quotations of Muir in this paragraph are cited in Nash (1973, pp. 128–129).

In recent times, the view that plants and animals have a right to exist has been closely linked to an ethical perspective advocated by the ecologist Aldo Leopold. The last portion of his book, *A Sand County Almanac,* contains the basis for what Leopold referred to as a "land ethic." For Leopold, the community to which ethics applies should not be restricted to people. It needs to be enlarged to include other members of the "biotic community," the soils, waters, plants, and animals. Leopold felt that decisions about the use of land should consider, in addition to economic factors, "questions of what is ethically and esthetically right." Leopold's land ethic provides a criterion for answering these questions. For him, a "thing is right when it tends to preserve the integrity, stability and beauty of the biotic community. It is wrong when it tends otherwise."[13]

Not surprisingly, some have argued that granting rights to nonhuman species, when followed to its logical end, leads to an absurdity. Passmore (1974, p. 126) put it this way: "If men were ever to decide that they ought to treat plants, animals, landscapes precisely as if they were *persons,* if they were to think of them as forming with men a moral community in the strict sense, that would make it impossible to civilize the world—or, one might add, to act at all or even to continue living."

Despite the criticism directed at Leopold's call for a land ethic, there has been much discussion of the proposition that plants, animals, and natural objects have rights that are legally defensible. This position was advanced in Stone's (1974) *Should Trees Have Standing?* a widely cited essay that played a role in the U.S. Supreme Court's deliberations on whether the U.S. Forest Service should permit the development of a ski resort in Mineral King Valley, a wilderness area in California. Stone's position was not used to stop the Forest Service's action. However, it was discussed in a minority opinion of Supreme Court Justice William O. Douglas, who pointed out that inanimate objects such as ships are sometimes parties in litigation. "So it should be" for natural bodies such as rivers and swamplands that feel "destructive pressures of modern technology and modern life."[14] In considering an example of a natural system under pressure, Douglas referred to a river that, he said, "as plaintiff speaks for the ecological unit of life that is part of it. Those people who have a meaningful relation to that body of water—whether it be a fishermen, a canoeist, a zoologist, or a logger—must be able to speak for the values which the river represents and which are threatened with destruction." Neither Justice Douglas' opinion nor the Stone essay were decisive in this instance. They nonetheless reflect an interest in applying the land ethic introduced by Leopold.

The question of whether plants, animals, and other natural objects have rights must be faced by those making decisions influencing the environment. It is part of a larger set of questions concerning the bases for judging whether one action

[13]This quotation is from Leopold (1949, p. 262).

[14]Justice Douglas' opinion in this case, *Sierra Club v. Morton,* is reprinted in Stone (1974).

affecting the environment is preferred to another. Criteria commonly used in making these judgments are introduced below.

BASES FOR MAKING DECISIONS AFFECTING ENVIRONMENTAL QUALITY

Reasons for caring about how human actions influence the environment include a desire to use resources efficiently, the need to maintain the earth as a human habitat, and a variety of religious and philosophic beliefs. Several of these concerns have been translated into ethical norms and government policies that guide the way decisions affecting the environment are made. Commonly used norms and policies rest on the idea that human welfare is diminished when natural resources are wasted, air and water is made unhealthy, and so forth. This approach to setting policies and norms is said to be *anthropocentric* in that the concern for the natural environment is based ultimately on the welfare of people.

In contemporary Western cultures, decisions affecting the natural environment are generally made using an anthropocentric standpoint, and this orientation is adopted here. Although it may appear that an anthropocentric basis for decision making cannot reflect the value of maintaining stable ecological systems or preserving rare species, this is not the case. To accommodate such considerations, it is only necessary that *some people* place values on stable ecosystems and rare species. The underlying reasons for these values may be purely intuitive or religious or based on a sense of obligation toward nonhuman living things. The reasons need not have a "practical" basis at all.[15]

There are many criteria for choosing among alternative actions affecting the environment that are consistent with an anthropocentric perspective. One that is commonly used is an adaptation of utilitarian philosophy known as the benefit–cost criterion.

Utilitarianism as a Framework for Making Decisions

Utilitarianism's central tenets are commonly ascribed to Jeremy Bentham, an eighteenth-century British philosopher concerned with the basis for developing social policy and legislation. Bentham and his followers argued that when considering alternative policies, lawmakers should estimate the beneficial and harmful consequences of each policy to the society as a whole. They should select the policy that produces the greatest net balance of beneficial over harmful consequences to society.

[15]There are some who argue that an anthropocentric framework is too narrow because it does not recognize that nonhuman species have a right to life independent of any value to humans. Further discussion of the philosophic arguments for and against an anthropocentric approach to decision making is given by Passmore (1974) and Tribe (1976), respectively.

For those who adopt a utilitarian view, an action such as a government project or regulation is judged entirely by its outcome to society as a whole. Traditional utilitarianism assumes that the harmful and beneficial consequences of an action to society can be predicted and evaluated quantitatively in units that can be added together to yield a net effect. Critics of the utilitarian approach often attack these assumptions.

"Benefit–cost analysis" is an adaptation of traditional utilitarianism that is widely advocated, but controversial. In this approach, the beneficial and harmful consequences of an action (termed *social benefits* and *social costs,* respectively) are described in monetary units. The social benefits are measured as the amount of money that people would be willing to pay to obtain the beneficial consequences. The social costs are measured in terms of the opportunities that a society gives up when its resources are used to implement the proposed action. The benefit–cost criterion for choosing among alternative proposals involves selecting the one that has the greatest numerical difference between monetary benefits and costs. It is always supposed, when using this type of analysis, that a proposed action should not be undertaken unless its total benefit exceeds its total cost.

The benefit–cost criterion is of limited usefulness because many effects of human actions cannot be evaluated in units of money. This is often the case when environmental consequences are involved. Suppose, for example, that a proposed action involves eliminating completely the known cancer-causing ("carcinogenic") substances in a community's drinking-water supply. It is hard to attach a meaningful economic value to the benefits of this proposal. One reason is the difficulty in estimating how many lives would be saved by removing all of the carcinogens from the water supply. Also, there are no widely accepted procedures for placing a dollar value on the estimated lives saved. Some people even object to the idea of evaluating lives in terms of money. For these reasons, many have argued against using a quantitative benefit–cost analysis as the principal basis for making decisions that affect environmental quality.

Even though a benefit–cost analysis in which all effects are described in monetary units is frequently impractical, it is often possible to make qualitative comparisons of the social costs and gains of a proposal. Adopting this broad conception of utilitarianism, it is only necessary that the consequences of a proposed action be described as clearly and accurately as possible. The information on benefits and costs, although not presented entirely in monetary units, is displayed so that people can make systematic comparisons among alternative proposals. As elaborated by Swartzman, Liroff, and Croke (1982), this qualitative conception of a benefit–cost analysis is sometimes advocated as a basis for establishing environmental policies and regulations.

Although a broad conception of utilitarianism seems like nothing more than common sense, it is often criticized as being insensitive to issues related to fairness and equity. Utilitarianism focuses on the aggregate welfare of a social unit rather than the welfare of particular individuals or groups. It is not concerned with who gains and who loses. Only the *net* effect is considered. Because issues related to the fairness of the distribution of gains and losses are ignored, many have urged that utilitarianism not be used as the sole basis for making decisions that influence the environment.

Equity and the Distribution of Benefits and Costs

The concept of fairness (or justice) is relevant to decisions affecting environmental quality because the individuals enjoying the benefits from such decisions are often different from those who pay the costs. Consider, for example, the case of a steel mill that closes because of strict government requirements for the reduction of its wastewater discharges. Beneficiaries of this action are the water users downstream of the steel mill. The costs, however, include the dislocation of the people unemployed as a result of closing the mill. The gainers are different from the losers. As a second example, consider a flood control reservoir that would inundate several homes. In this instance the benefits accrue to people protected from floods downstream of the dam, but the costs include the social disruption experienced by families displaced by the reservoir.

Although many people agree that equity issues should be considered in making decisions that affect the environment, these issues are not easily analyzed. The following are among the difficult questions that must often be treated in examining the equity of alternative proposals: How should fairness be defined and measured? If a decision leads to an unfair distribution of benefits and costs, should the inequity be tolerated if the social gains far outweigh the social costs? Suppose, in the flood control illustration above, that the benefits from controlling floods were enormous and only a few families had to be relocated. Would relocating the families be justified in those circumstances? Consider a decision that leads to inequities that at the same time corrects previous injustices. In the steel mill example above, the original injustice involved damages experienced by water users downstream of the mill. Should this injustice be remedied by imposing a wastewater reduction requirement that causes the mill to close and results in massive unemployment? If the steel mill had originally located at the site with the understanding that strict wastewater controls would not be required, would the mill closure still be justified?

The analysis of equity issues is often confounded by practical measurement problems. It is only since the 1970s that many government programs have been analyzed in terms of how much is gained and lost *by particular groups* when actions affecting environmental quality are undertaken. Before that time, systematic analyses of government proposals usually measured only aggregate social benefits and costs. In comparison to these aggregate measures, it is much more difficult to estimate how much is gained and lost by particular classes of individuals.

The concept of fairness has received increasing attention in discussions of government programs related to environmental quality. In recent years, several hypotheses about the distribution of costs and benefits from various environmental programs have been posed.[16] Although some data have been analyzed to explore these hypotheses, the results have not been definitive. For example, one hypothesis is that a disproportionate share of benefits from many environmental programs is derived by the middle- and upper-income classes. The reasoning used here is that, in proportion to their total numbers, middle- and upper-income families use national parks, wilderness areas, and nonurban public recreational areas more frequently than lower-income families. This occurs, in part, because it is necessary to have a car and a good income to travel to these areas and take advantage of the opportunities they provide for camping, fishing, and the like. The hypothesis that the poor are treated unfairly is not often advanced in the context of environmental programs that reduce ambient noise levels and improve air quality. Rather, it is argued that the urban poor are frequently burdened disproportionately by low air quality and high noise levels.

Another hypothesis related to the equity of environmental programs is that the poor often carry a disproportionate share of the cost of reducing pollution. The data relevant to this hypothesis are difficult to interpret because it is often hard to determine how much of the cost of pollution abatement results in higher prices and how much results in lower profits. In addition, many environmental programs are funded using federal and state tax revenues, and these tax programs take a smaller percentage of income from the poor in comparison to middle- and upper-income classes. One instance in which distributional impacts appear unambiguous concerns the cost of reducing emissions from auto exhausts. These costs are believed to fall disproportionately on the poor since lower-income families with autos spend a relatively high fraction of their disposable incomes on auto maintenance. It is sometimes argued that the poor carry a substantial burden when government environmental regulations cause plant closures and unemployment. However, it is not clear that unemployment is a major consequence of waste reduction programs, since many new jobs have been created by pollution abatement activities.

Indications that the costs and benefits of some environmental programs are distributed unfairly have led to calls for the systematic analysis of distributional issues in formulating new programs. As Baumol (1974, p. 266) put it, this would provide "some assurance that environmental policy does not become yet another influence that makes the rich richer and the poor poorer."

Rights to a Habitable Environment

Another basis for making decisions affecting environmental quality concerns the moral rights of humans and other living things to a habitable environment.[17] In this context, the term *right* refers to an individual's entitlement to something.

[16]The illustrative hypotheses introduced below are discussed in detail by Barbour (1980).

[17]The discussion of moral rights to a habitable environment benefits greatly from Velasquez's (1982) treatment of the subject. His presentation of utilitarianism and equity issues is also instructive.

A moral right is one that is independent of any particular legal system. It is based on "moral norms and principles that specify that all human beings are permitted or empowered to do something or are entitled to have something done for them" (Valasquez, 1982, p. 59). Some people feel that individuals have a moral right to a livable environment and that this right needs to be preserved at the expense of limiting people's legal freedom to engage in activities that destroy the environment. Such restraints on freedom are considered necessary to preserve the human habitability of the planet.

In some states, the moral right to a healthful environment has been made into a *legal* right. For example, Article 1 of the Constitution of the State of Pennsylvania was amended to read

> *The people have a right to clean air, pure water, and to the preservation of the natural, scenic, historic, and aesthetic values of the environment. Pennsylvania's natural resources . . . are the common property of all the people, including the generations yet to come. As trustee of these resources, the commonwealth shall preserve and maintain them for the benefit of all people.*

The notion that "each person has a fundamental and inalienable right to a healthful environment" almost made its way into federal law in the United States during the late 1960's. Such a right was specified in the Senate version of the bill that eventually became the National Environmental Policy Act of 1969. It was, however, deleted by the congressional conference committee that prepared the bill that was signed into law. Although it does not guarantee an individual's right to a healthful environment, the National Environmental Policy Act states that one of its purposes is to "assure for all Americans safe, healthful, productive and aesthetically and culturally pleasing surroundings."

Velasquez (1982) has reasoned that the moral right to a healthful environment was a key factor in determining the form of several of the U.S. environmental programs established during the 1970s. Some provisions of the Federal Water Pollution Control Act of 1972 illustrate his view. The act proclaims, as a national goal, the elimination of the "discharge of pollutants into navigable waters" by 1985. To meet this goal, the U.S. Environmental Protection Agency (EPA) was required to issue permits to all municipalities and firms that made significant wastewater discharges to streams and other surface waters. In deciding on levels of wastewater discharge to permit by 1983, Congress required the EPA to insist on the "best available technology economically achievable" for reducing waste loads at various *types* of facilities. According to this position, the water pollution control requirements for a particular firm or municipality would be based on the most stringent levels of wastewater reduction that, in the EPA's view, could be obtained economically with available technology. The utilitarian concern of whether the social benefits exceeded the costs of such stringent controls was not to affect EPA's decision making. The question of who would benefit and who would bear the costs was not to influence the EPA's decision process either.

Velasquez interprets this federal imposition of stringent controls as being based, implicitly at least, on a recognition of the rights of citizens to pure water.

The issue of rights to a habitable environment has been debated from many points of view, not just from the perspective of the present generation of humans. What, for example, are the moral rights of generations yet to come? Also, what are the rights of nonhuman animals? Of plants? Of inanimate natural objects? Philosophers have given serious attention to these questions. As is evident from the contrasting views of Feinberg (1974) and Passmore (1974), philosophers do not agree among themselves on these important questions. An example of a point of contention is the question of whether future generations have a moral right to an environment that is habitable and not depleted of natural resources. Although it is tempting to say that future generations should have such a right, many philosophers have argued against this view. They assert that since future generations do not exist, except in the imagination, they cannot possibly have rights. Moreover, because of high rates of change in technology and life-style, the present generation is not in a position to know which interests of future generations to protect. Consequently, it is not possible to say what rights future generations should have.

A concept of justice introduced by John Rawls, a Harvard professor of philosophy, is often considered in discussions of the rights of future generations to a habitable environment. To analyze the question of what is owed to future generations, Rawls asks that members of the present generation put themselves in the position of not knowing which generation they belong to. Then they should:

> *try to piece together a just savings schedule by balancing how much at each stage [of history] they would be willing to save for their immediate descendants against what they would feel entitled to claim of their immediate predecessors. Thus, imagining themselves to be fathers, say, they are to ascertain how much they would set aside for their sons by noting what they would believe themselves entitled to claim of their [own] fathers.*[18]

Although Rawls' position has been criticized by Passmore (1974) and others, it does provide a way of contemplating the moral rights of future generations. Rawls' analysis has been interpreted to mean that the next generation should receive the kind of environmental quality and natural resources that it could reasonably be assumed to need to make the types of choices available to the current generation. A concern for the kind of environment that is left for future generations has, since 1970, become a matter of national policy in the United States. The National Environmental Policy Act of 1969 indicates that each generation has responsibilities "as trustee of the environment for succeeding generations." Although the wording is general, it indicates an unmistakable

[18]This quotation is from Rawls (1971, p. 289).

concern for the kind of existence that generations to come will be able to have.

As illustrated by the above discussion relating to future generations, social decisions are generally not made on the basis of a single set of principles. Sometimes, a utilitarian benefit–cost analysis is the dominant basis for decision making. At other times, questions concerning individual rights or fairness may exert a significant influence. The chapters that follow demonstrate that utilitarianism, equity, and moral rights have each played a role in shaping environmental policies in the past several decades.

KEY CONCEPTS AND TERMS

ENVIRONMENTAL PLANNING AS A FIELD
- Specialties of environmental professionals
- Residuals management
- Environmental impact assessment
- Land use and environmental quality

HISTORICAL PERSPECTIVES ON ENVIRONMENTALISM
- American conservation movement
- Pinchot's principles of conservation
- Efficiency in natural resource use
- Ecological bases for restraint
- Preservationist's views on wilderness
- Leopold's land ethic
- American transcendentalism
- Legal rights of nonhuman species

BASES FOR MAKING DECISIONS AFFECTING ENVIRONMENTAL QUALITY
- Anthropocentrism
- Bentham's utilitarianism
- Benefit–cost criterion
- Equitable distribution of costs and benefits
- Moral rights to a habitable environment
- Rights of future generations

DISCUSSION QUESTIONS

1–1 Speculate on the advantages of having specialists from several disciplines work together on a team to solve environmental problems. What difficulties would you anticipate in using the team approach?

1–2 Examine your own reasons for caring about how people affect the environment. To what extent do your personal motives overlap with the economic, ecologic, spiritual, and philosophic viewpoints presented in the chapter?

1–3 Criticize the use of an anthropocentric framework for making decisions. In light of this critique, do you feel that anthropocentrism provides a reasonable approach to evaluating the environmental consequences of human actions?

1–4 Provide an example demonstrating that an exclusive reliance on utilitarianism could lead to an environmental program that either violates people's rights or distributes costs and benefits unfairly, or both.

REFERENCES

Barbour, I. G., 1980, *Technology Environment, and Human Values.* Praeger, New York.

Baumol, W. J., 1974, Environmental Protection and Income Redistribution, in R. Zeckhauser (ed.), *Benefit-Cost Analysis and Policy Analysis, 1974,* an Aldine Annual on forecasting decision making and evaluation, pp. 250–268. Aldine, Chicago.

Carson, R., 1962, *Silent Spring.* Houghton Mifflin, Boston.

Farvar, M. T., and J. P. Milton (eds.), 1972, *The Careless Technology: Ecology and International Development.* Natural History Press, Garden City, New York.

Feinberg, J., 1974, The Rights of Animals and Future Generations, in W. T. Blackstone (ed.), *Philosophy and Environmental Crisis,* pp. 43–68. University of Georgia Press, Athens.

Finsterbusch, K., and C. P. Wolf (eds.), 1981, *Methodology of Social Impact Assessment,* 2nd ed., Hutchinson Ross, Woods Hole, Mass.

Freeman, A. M., III, R. H. Haveman, and A. V. Kneese, 1973, *The Economics of Environmental Policy.* Wiley, New York.

Hays, S. P., 1959, *Conservation and the Gospel of Efficiency: The Progressive Conservation Movement, 1890–1920.* Harvard University Press, Cambridge, Mass.

Leopold, A., 1933, The Conservation Ethic. *Journal of Forestry* **31** (6), 634–643.

Leopold, A., 1949, *A Sand County Almanac.* Oxford University Press, Oxford. (Reissued in 1970 as a Sierra Club/Ballentine Book by Ballentine Books, New York.)

Marsh, G. P., 1864, *Man and Nature; or, Physical Geography as Modified by Human Action.* Charles Scribner, New York. (Reissued in 1965 by the Belknap Press of the Harvard University Press, Cambridge, Mass., and edited by D. Lowenthal.)

McEvoy, J., III, and T. Dietz (eds.), 1977, *Handbook for Environmental Planning, the Social Consequences of Environmental Change.* Wiley, New York.

Nash, R. (ed.), 1968, *The American Environment: Readings in the History of Conservation.* Addison–Wesley, Reading, Mass.

Nash, R., 1973, *Wilderness and the American Mind,* revised edition. Yale University Press, New Haven, Conn.

Passmore, J., 1974, *Man's Responsibility for Nature, Ecological Problems and Western Traditions.* Duckworth, London.

Petulla, J. M., 1980, *American Environmentalism.* Texas A&M University Press, College Station.

Rawls, J., 1971, *A Theory of Justice.* Harvard University Press, Cambridge, Mass.

Stone, C. D., 1974, *Should Trees Have Standing?* William Kaufmann, Los Altos, Calif.

Swartzman, D., R. A. Liroff, and K. G. Croke (eds.), 1982, *Cost-Benefit Analysis and Environmental Regulations: Politics, Ethics and Methods.* The Conservation Foundation, Washington, D.C.

Tribe, L. H., 1976, Ways Not to Think About Plastic Trees, *in* L. H. Tribe, C. S. Schelling, and J. Voss (eds.), *When Values Conflict,* pp. 61–92. Ballinger, Cambridge, Mass.

Velasquez, M. G., 1982, *Business Ethics: Concepts and Cases.* Prentice–Hall, Englewood Cliffs, N.J.

PART TWO

RESIDUALS MANAGEMENT

CHAPTER 2

CAUSES AND CONSEQUENCES OF RESIDUALS

This chapter explores concepts used by economists in analyzing environmental questions. Economists frequently rely on benefit–cost analysis to decide how much environmental degradation to tolerate or how much environmental improvement to strive for. The benefit–cost approach is easily criticized because it ignores the moral rights of individuals to a healthful environment. In addition, it does not consider whether the distribution of benefits and costs is equitable. Despite its narrow focus, benefit–cost analysis often plays a role in the political processes that yield environmental policies and regulations.

Economists did not take a major interest in environmental problems until the 1960s. Before then, they viewed what is commonly termed *pollution* as one of several types of "external effects," that is, effects that are external to ordinary markets. The early work of Kneese (1964) and other economists affiliated with Resources for the Future, Inc., in Washington, D.C., played a central role in expanding economic theories to include environmental quality issues.

THE ENVIRONMENT AS AN ECONOMIC ASSET

Economists have found it valuable to examine what different forms of pollution, such as noise and air pollution, have in common.[1] They use the term *residuals* to mean the material or energy "left over" from the various consumptive and productive activities carried out by individuals, firms, and governments. For example, the residuals generated by a ride on a city bus include the noise, heat,

[1]The material in this section draws heavily on Freeman, Haveman, and Kneese (1973).

hydrocarbons, particulates, and carbon monoxide that emanate from the bus during the ride. By analyzing residuals in general, instead of individual emissions from smokestacks, sewers, and so forth, it is possible to develop a more comprehensive framework for examining environmental problems.

The natural environment can be viewed as a shell within which traditional economic activities take place. A common representation is depicted in Figure 2.1. Firms and governments produce goods and services, which are sold to consumers. The production of goods and services utilizes labor, capital, and raw materials from the environment; it also yields residuals, which flow back to the environment. Consumers use the income they receive from firms and governments to purchase various goods and services. The process of "consuming" these goods and services yields additional residuals, which are ultimately released into the environment.

The generation of residuals is a normal consequence of virtually all productive and consumptive activities. Residuals consist of either materials or energy. *Materials residuals* are returned to the environment in solid, liquid, and gaseous forms. *Energy residuals* take the form of noise or waste heat, for example, the heat returned to the atmosphere when coal is burned to produce electricity. Flows of residuals can be recycled (in which case they become "inputs") or they can be treated using physical, chemical, and biological processes. However, residuals cannot be eliminated entirely. To do so would violate the laws of conservation of mass and energy.

A Materials Balance Perspective

A sensitivity to the law of conservation of mass, referred to as a *materials balance perspective,* is useful in determining government policies for controlling resid-

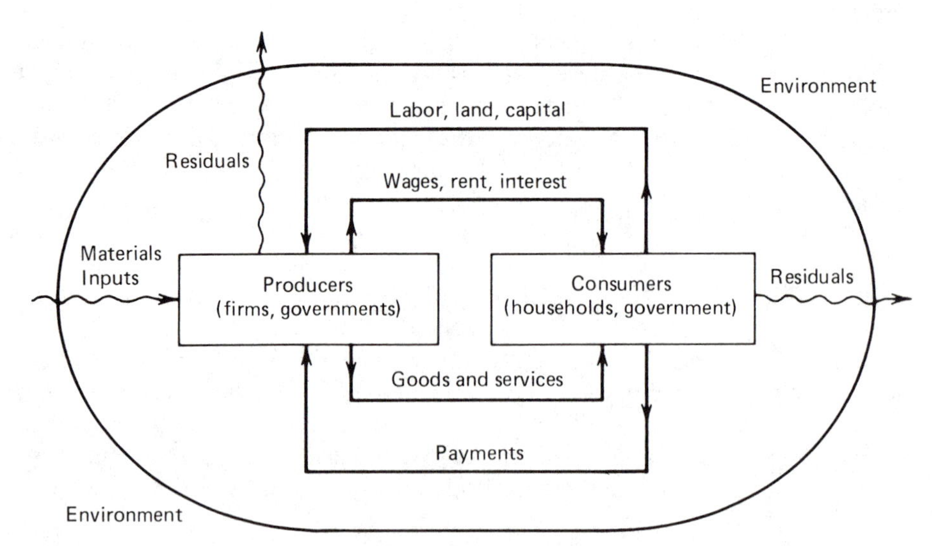

FIGURE 2.1 Traditional economic activities and the environment.

uals. Policies that fail to consider this physical law sometimes merely shift the problems caused by residuals among the "media" (air, land, and water) that receive them. For example, regulations aimed at improving water quality may end up causing solid waste and air quality problems. When additional matter is removed from wastewaters, more solids have to be disposed of in landfills or by incineration; the latter is a source of air-borne residuals.

A materials balance perspective recognizes that processes used to treat residuals do not eliminate them. Rather, treatment processes transform residuals into substances that may cause fewer objectionable effects when they are ultimately released into the environment. Consider, for example, the use of screens and other physical devices to remove solids from wastewater at a municipal treatment plant. Solids are taken out of wastewaters with the hope that when they are discharged to the environment as "dried sludge" or air-borne materials they will cause less of a problem than if they flowed directly into a watercourse.

Although the materials balance idea seems elementary, there are more than a few instances where residuals management policies have been adopted as if the law of conservation of mass did not exist. Such policies may result inadvertently because of the narrow focus of many environmental programs. Often there is one government program for air quality, one for water quality, one for noise, and so on, with little coordination among the programs. An example involving air quality management in New York City demonstrates what can happen if materials balance concepts are not adequately considered. The city adopted very strict controls on air-borne emissions from apartment house incinerators.[2] Requirements were so stringent that many apartment house owners found it more economical to stop using their incinerators than to bring them into compliance with the new regulations. The result was a substantial increase in solid waste left for curb-side pickup and ultimate disposal by the city's Sanitation Department. Necessary expansion in solid waste collection and disposal efforts was so great that the stringent incinerator regulations were not enforced.

Environmental Outputs

Figure 2.1 indicates two ways that the natural environment serves as what Freeman, Haveman, and Kneese (1973, p. 20) call a *nonreproducible capital good.* As shown in the figure, the environment supplies *materials inputs,* such as the fuel, lumber, and minerals used in production activities. The environment also provides *residuals (or waste) receptor services.* These services reflect the environment's ability to transform residuals into harmless substances and to transport and dilute residuals. In viewing the environment as a capital good, Freeman, Haveman, and Kneese identify two additional types of "environ-

[2]This example is from Freeman, Haveman, and Kneese (1973, p. 29). Other examples of unintentional transfers of residuals from one medium to another are given by Lowe, Lewis, and Atkins (1982).

mental outputs": *life support functions* that are necessary for human existence, such as the provision of clean air; and *amenity services* in the form of surroundings that people find pleasant, attractive, and renewing. Table 2.1 lists all four categories of outputs.

When viewed as a nonreproducible capital good, the natural environment is of concern only to the extent that it affects people. This deliberately anthropocentric perspective is not as restrictive as it seems at first. The destiny of nonhuman species can be considered when treating the environment as a capital good. All that is necessary is that someone be concerned about their fate. For example, many individuals feel it is important to preserve rare species such as the bald eagle. Therefore, the value of preserving these species is included in the economists' framework for analyzing environmental problems.

The environmental outputs in Table 2.1 can be viewed as economic goods and services in that people are *willing to pay* (1) to receive more of them or (2) to avoid a reduction in their quality or quantity. What is commonly termed *pollution* is defined as the adverse effects of one category of outputs, namely, waste receptor services, on one or more of the other outputs. Using this definition, it is possible to have waste discharges into the environment that would not be called *pollution.* An example is the disposal of human waste by a lone backpacker in the Alaskan wilderness. If backpacking in the now-isolated sections of Alaska became popular, the discharge of human wastes in these areas might adversely affect someone and satisfy the above definition of pollution.

Another concept related to waste receptor services is the "assimilative capacity of the environment." Societies often make judgments about which levels of environmental quality are satisfactory. For example, the quality of a stream may be judged acceptable if the concentration of oxygen dissolved in the stream is above a certain level, if the bacterial counts are below some number, and so on. When such judgments are made, it may be possible to calculate the amount of residuals that can be released into the environment without violating established levels of acceptability. This quantity of residuals is the environment's assimilative capacity.

TABLE 2.1 Categories of Environmental Outputs[a]

Life Support Services

A hospitable, healthful environment including clean air and pure water

Amenity Services

Pleasant spaces for recreation and personal renewal, such as the Grand Canyon in Arizona

Materials Inputs

Inputs, such as oil and lumber, to various economic activities

Waste Receptor Services

Acceptance of residuals such as wastewater and noise

[a]Based on information in Freeman, Haveman, and Kneese (1973, pp. 21–22).

THE MANAGEMENT OF RESIDUALS

Many parts of the environment are not privately owned. This makes them different from the goods and services, such as surf boards and haircuts, that are bought and sold in markets. An obvious illustration of an environmental feature that is not privately owned is the atmosphere. A more subtle example is a highly valued vista, which people cannot easily be prevented from seeing. The view of Mt. Rainier from Seattle on a clear day is widely appreciated. This sight cannot be sold like an ordinary commodity because it would be impractical for a private entrepeneur to keep people who did not purchase the view from enjoying it anyway.

Environmental outputs that are not privately owned are frequently overused unless social restraints are imposed. Forces leading to overuse are illustrated by Hardin's (1968) "The Tragedy of the Commons." He describes events in an imaginary cattle grazing area, "the commons," that is open to all. There are no constraints on how many cattle an individual member of the community may graze. If a limited number of cattle per year are grazed, the commons can renew itself and provide a satisfactory grazing area indefinitely. However, each herdsman seeks to maximize his own gain by adding additional animals to his herd. Since there are no restrictions on the size of an individual's herd, the commons are used excessively and eventually ruined for grazing.

"The Tragedy of the Commons" applies directly to the waste receptor services provided by the environment. Consider, for example, a stream on which standards of quality have been established. This stream can accept a certain quantity of water-borne residuals without violating the standards, and that quantity is its assimilative capacity. Like the grazing area, the assimilative capacity of the stream is "common property." In the absence of socially imposed controls, the stream's ability to carry residuals will be viewed by each discharger as a "free good." Thus, no discharger will have an incentive to invest money to reduce wastewater discharges. Without this incentive, the stream's waste receptor services are likely to be overused, causing a violation of the stream quality standards. The need for incentives to prevent overuse of common property is frequently cited in justifying government intervention to manage residuals.

There are many ways of structuring a residuals management program. For example, absolute limits can be imposed on the quantities of residuals released at various locations. Alternatively, individuals can be required to pay fees for use of the environment's waste receptor services. Many other possible strategies exist. Regardless of the approach, however, the law of conservation of mass indicates that residuals cannot be eliminated entirely. Thus, it is necessary to decide on what levels of environmental quality are acceptable and what discharges of residuals can be tolerated.

BENEFITS OF RESIDUALS REDUCTION

Economists frequently argue that what constitutes acceptable environmental quality should be determined in light of the benefits and costs of discharging different quantities of residuals. Assessing the costs of residuals reduction involves engineering and economic computations that are well established. However, this is not the case for benefits. The following questions must be answered in computing the economic benefits of residuals reduction:[3]

1 How does reducing the discharge of residuals influence the quality of the environment?
2 How does a change in environmental quality affect the outputs that the environment provides?
3 What are the monetary values of changes in environmental outputs?

Discharges of Residuals and Ambient Quality

A first step in characterizing the effects of residuals reduction is to define variables (or "indicators") describing the residuals and the ambient environment. Many different variables are used. *Discharges of materials* residuals are often measured as the mass of substance emitted per unit of time. For example, nitrogen dioxide released from a smokestack is commonly reported in tons emitted per day. Similarly, the solids suspended in domestic wastewater are often given in pounds per day. Discharges are also described by the amount of a substance released per unit of "activity." For example, hydrocarbons from an automobile exhaust system are often reported as the number of grams of hydrocarbons per mile traveled by the vehicle.

Materials residuals in the *ambient environment* are typically characterized using concentration, the mass of a substance per unit volume. For instance, one indicator of a stream's quality is the number of milligrams of suspended solids per liter of water (mg/l). The quality of ambient air is also described using concentrations, but the units are different. Commonly used measures include the number of micrograms (10^{-6} g) of a substance per cubic meter of air ($\mu g/m^3$) and the number of volume units of a gas per million volume units of air (parts per million, or ppm).

Often there are so many substances within a discharge that it is impractical to measure all of them. Assumptions are therefore made about which characteristics of a discharge are likely to be the most important in degrading the environment. For example, hundreds of different hydrocarbon compounds enter the atmosphere from automobile exhausts. Because it is not feasible to measure each of the compounds, a single aggregate measure of hydrocarbon concentration is often used. It is assumed that this single measure adequately reflects how the many compounds affect people.

For residuals that take the form of *energy,* completely different indicators are used to characterize discharges and ambient environmental quality. For example, the waste heat from the cooling system of a power plant is often described by the volume of water discharged per unit time *and* the temperature of that

[3]These three questions are based on a presentation by Freeman (1979, p. 17). The following discussion of residuals reduction benefits also relies on Freeman's work.

water. Both influence the temperature of the watercourse that receives the discharge. Another example involves noise. In this case, both the noise source and the ambient environment are frequently described in decibels, a measure of sound pressure level. Because noise sources such as airplanes and jackhammers affect people in distinctive ways, different sound pressure level indicators are used to characterize them.

Before the benefits of residuals reduction can be calculated, it is necessary to predict how the residuals measured in the ambient environment will change if the discharge is reduced by a particular amount. Equations can be used to make such predictions for some residuals.[4] Consider, for example, an emission of sulfur dioxide (SO_2) from a smokestack. The equations used in forecasting are based on physical laws and empirical observations. To use the equations, it is necessary to specify the rate of discharge of SO_2 as well as information concerning local meteorological conditions and topography (see Figure 2.2). The forecast consists of estimates of SO_2 concentration at selected locations downwind of the source under postulated conditions of discharge, meteorology, and topography.

Persons developing residuals management programs typically have less information then they want concerning relationships between discharges and ambient quality. A shortage of information often results because the physical, chemical, and biological processes governing the dissemination of residuals in the environment are poorly understood. Even when the underlying processes are understood well enough to provide a useful forecasting procedure, field data are generally needed to "calibrate" the procedure for local circumstances. Sometimes the cost of gathering and processing the required data is prohibitively high.

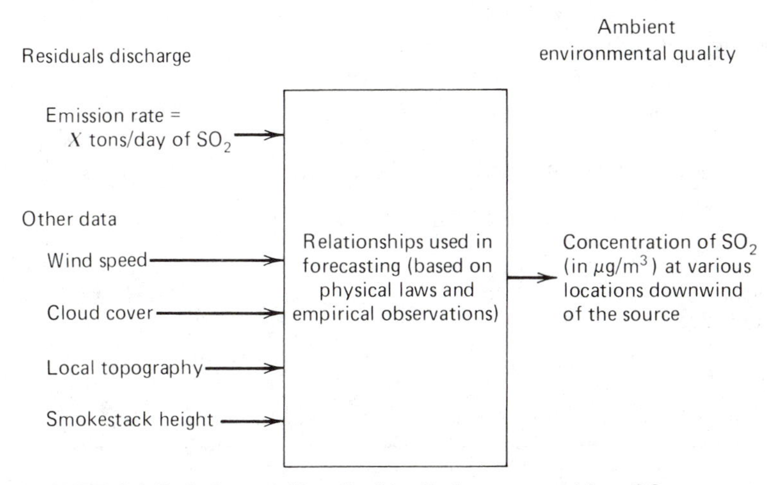

FIGURE 2.2 Relating a sulfur dioxide discharge to ambient SO_2 concentrations.

[4]Methods commonly used in forecasting how environmental quality responds to changes in residuals discharges are surveyed in Part Five.

Ambient Quality and Flows of Environmental Outputs

To determine the benefits of residuals reduction, it is necessary to know how changes in the ambient environment affect the goods and services that the environment provides.[5] Table 2.2 classifies the effects of changes in environmental quality according to their relationship with each of three environmental outputs: life support services, amenity services, and materials inputs. An excessive use of a fourth output, waste receptor services, is often the cause of adverse effects on the other three.

Of the "life support services" listed in Table 2.2, those concerning human health are of the greatest immediate concern. Information on health effects has been gathered since the turn of the century for water-borne contaminants, but more recently for other residuals. Early water quality studies concerned the transmission of infectious diseases by water-borne bacteria. As a result of modern disinfection techniques, problems caused by bacteria in water have diminished significantly. Recently, however, several organic compounds found in drinking-water supplies have been suspected of being carcinogenic at very low concentrations. Some of them are produced synthetically, and others are formed when naturally occurring organic compounds combine with chlorine used to disinfect water supplies. The identification of suspected carcinogens in drinking water has caused concern over the magnitude of potential problems. Only a small fraction of the organic compounds in drinking water has even been iden-

TABLE 2.2 Effects of Ambient Environmental Quality on Environmental Outputs[a]

Life Support Services

Human health—morbidity, mortality

Ecological diversity and stability

Weather, climate

Amenity Services

Recreational uses of ecosystems—swimming, fishing, and hunting

Aesthetics—odor, visibility, visual aesthetics

Materials Inputs

Economic productivity of ecological systems
- Agricultural productivity
- Forestry
- Fisheries

Damages to materials in the form of corrosion, soiling, and so forth.

[a]Based on information in Freeman (1979, p. 20).

[5]The discussion here only introduces the subject of how changes in environmental quality influence environmental outputs. For further details, see Freeman (1979), and the many references cited by him.

tified, and only a few of those identified have been tested for adverse health effects.[6]

Much effort has gone into examining the effects of air-borne residuals on human health. Although scientists have not reached agreement on the precise contaminants and physiological mechanisms involved, there is evidence that higher exposures to some air-borne residuals lead to increased human illness and mortality. An example is given by statistical analyses of relationships between human mortality and pollution due to sulfur oxides and suspended particulates.[7] Findings from 10 recent statistical studies have been interpreted by Freeman (1982, p. 17) to mean "that a decrease in sulfur and particulate pollution of, say 50 percent would be expected to lead to a reduction in mortality of between 0.5 percent and 5 percent." These results are controversial. Critics argue that the statistical studies often omit important variables and interactions among variables and are based on poor-quality data. Additional research is needed to clarify the associations between air quality and human health.

The other items listed in Table 2.2 under Life Support Services, ecological diversity and stability and weather and climate, are so complex and poorly understood that it is frequently difficult to predict how they will be influenced by human actions. An example is the effect of carbon dioxide released during the combustion of fossil fuels on average global temperatures. As carbon dioxide levels in the atmosphere increase, the balance between incoming solar energy and outgoing longer wavelength energy radiated from the earth is disrupted. The resulting "greenhouse effect" could lead to an increase in the earth's surface temperature.[8] Whereas many find the arguments in support of this theory persuasive, average global temperatures depend on many other factors besides carbon dioxide. For example, particulate matter in the air may reflect incoming solar energy, reducing the earth's temperature. The interactions among factors are so complex that it is difficult to predict their long-term effects on average temperature with confidence.

[6]For a discussion of organic compounds in drinking water, see Page, Harris, and Bruser (1981).

[7]A widely discussed statistical study of relations between sulfur oxides and particulates and human health is the one by Lave and Seskin (1977). More generally, information on the health effects of air-borne residuals is summarized in EPA publications on air quality criteria for specific contaminants; see, for example, Environmental Protection Agency (1971).

[8]The term *greenhouse effect* is based on the analogy between the roles of carbon dioxide in the atmosphere and the glass in a greenhouse. In a greenhouse, the glass easily allows the short-wavelength solar energy to pass through and be absorbed by objects in the interior. The warmed-up interior in turn radiates long-wavelength energy back toward the glass. The glass is, however, relatively opaque to this longer-wavelength energy, with the net effect that energy enters the greenhouse more easily than it can get back out. Therefore, it heats up. (It also heats up by reducing convection, but that is another matter.) Similarly, in the atmosphere, carbon dioxide is transparent to incoming solar energy and opaque to the long-wavelength reradiated energy from the earth. As carbon dioxide levels increase, the incoming solar energy is not affected, but the earth has a more difficult time reradiating energy back to space. The balance between the two is upset; more energy arrives than is lost, and the earth heats up. (This footnote and the discussion of carbon dioxide and global temperatures in the text were contributed by Professor Gilbert M. Masters of Stanford University.)

People's abilities to discern changes in their surroundings play a key role in determining how the amenity services in Table 2.2 are affected by changes in environmental quality. The importance of perceptual ability is demonstrated by considering a wastewater discharge in the vicinity of an ocean beach. Suppose the beach is closed to swimmers because of the discharge. The benefits of treating the wastewater so that the beach can be opened are clear: a new recreation area is provided. Alternatively, suppose the beach was open all along and that treatment serves largely to reduce the numbers of bacteria in the discharge. This situation is more complex because it is not clear that the improved water quality would even be noticed by the people using the beach. Some research on this topic suggests that it may be difficult to predict how people perceive environmental changes. Results reported by Dornbush (1975), for example, indicate a very low association between what people perceive as high-quality water and what objective measures such as bacterial counts indicate to be high-quality water.

The situation seems more promising as far as perceptions of air quality are concerned. For example, Flachsbart and Phillips (1980) found that objective measures of smog correlate well with what individuals perceive to be smoggy conditions. Additional research must be done before information concerning the effects of residuals on amenity services can be widely used for systematic analyses of environmental problems.[9]

A long list of examples concerning the economic productivity of biological systems can be provided: increased total dissolved solids concentrations in streams used for irrigation lead to decreased agricultural outputs, high concentrations of sulfur dioxide in the air cause damages to ornamental vegetation and crops, high concentrations of mercury and other heavy metals interfere with the yields of commercial fisheries, and so forth. Similarly, there are numerous cases in which materials and objects are adversely influenced by changes in environmental quality. Examples include the effects of air pollutants in soiling clothing and the corroding of metal parts. Government publications on environmental quality criteria document these types of effects.[10]

Economic Evaluation of Changes in Environmental Outputs

The changes caused by residuals discharges are described above in different units, such as mortality rates and indicators of smoggy conditions. In some cases, it is possible to put monetary values on these changes. By describing effects on environmental goods and services using a single unit such as dollars, the benefits of different residuals management programs can be compared. In addition, the dollar value of a change in environmental outputs provides an indication of how strongly people feel about the change.

[9]For reviews of the literature on people's abilities to distinguish differences in air quality, water quality, and noise levels, see Craik and Zube (1976).

[10]For example, many publications on air and water quality criteria have been issued by the EPA; see, for instance, Environmental Protection Agency (1976).

To measure the economic value of reducing residuals, it is necessary to estimate how much people are willing to pay for the resulting improvements in life support services, amenity services, and materials inputs. These estimates of willingness to pay (in dollars) are termed the *benefits* of reducing the discharges. Sometimes the word *damage* is used in economic evaluations of residuals management policies. Whereas benefits refer to gains associated with a decrease in discharges from the existing condition, damages are losses in moving from a hypothetical "pristine" environment to the *status quo*. Some economists, such as Freeman (1979), have chosen not to use the damages concept because it requires the specification of a pristine condition in which no residuals are discharged. This specification can be difficult to make in practice.

During the past few decades, many studies have estimated the monetary benefits of reducing residuals discharges. A number of these studies are mentioned below. A warning is in order, however. Although numerous monetary benefit estimates are reported in the literature, specialists often disagree about the proper theoretical foundation for computing benefits.[11]

Life Support Services Of the life support services listed in Table 2.2, only health effects have been described in terms of dollars. Many health benefit studies have been conducted for air quality. To estimate monetary benefits, it is first necessary to determine how a given reduction in residuals will improve ambient air quality and how this improvement will, in turn, reduce the rate of human illness or death. The previously mentioned statistical studies linking sulfur oxides and mortality rates illustrate how these reductions in illness or death can be calculated.

Many techniques exist for estimating the dollar value of a change in the rate of human illness or death. Consider, for example, procedures that use the dollar values for human life that are *implied* by the decisions of individuals and governments. Some implied values have been determined by analyzing the wages that people accept for jobs with high occupational risks. These analyses give values of "statistical" life ranging from $275,000 to $625 million.[12] Another benefit estimation technique employs questionnaires to determine people's willingness to pay for a decrease in the risk of death, for example, by paying higher air fares to ride on safer airplanes. This method yields values of statistical life ranging from approximately $30,000 to $5 million. Since the various numerical values span three orders of magnitude, not much confidence can be placed in dollar estimates of air quality health benefits.

[11]A perspective on the "state-of-the-art" is given by an excerpt from the Preface of Freeman's monograph on benefit estimation:

> *The notion of writing a layman's guide to benefit estimation was appealing. But I felt the need first to address a number of technical issues and to provide a unified theoretical treatment. A layman's guide might better come after the specialists have brought some order and cohesiveness to the subject. (Freeman, 1979, p. xiii)*

[12]The values of statistical life reported in this paragraph are from a review of the literature on air quality health benefits by Graves, Krumm, and Violette (1982, pp. 38–40).

Other effects on life support services listed in Table 2.2 involve changes in physical and biological systems that have not yet been evaluated in monetary terms. Some economists argue that such effects are beyond the domain of economics.

Amenity Services Of the amenity services in Table 2.2, much attention has been given to economic benefits associated with recreation, especially water-based outdoor recreation. Many procedures have been developed to compute the monetary value of new recreation facilities at reservoirs developed by U.S. water resource agencies. However, these techniques generally do not consider variations in water quality as a factor affecting the degree of recreational enjoyment. If an improvement in water quality makes a previously unusable area suitable for recreation, it is equivalent to providing a new recreational facility. In this instance techniques for estimating the value of a *new* outdoor-recreation area may be useful.[13]

Estimating the benefits of enhanced water quality at an existing facility requires information on how much additional use would result from the improved quality. Sometimes questionnaires are employed to estimate increased facility use. Such surveys are easily criticized because often people do not know the water quality levels they have been exposed to. Thus, it is difficult for them to indicate how they would react to improved quality. Feenberg and Mills (1980) developed a procedure to estimate recreation benefits associated with improved water quality that does not rely on questionnaires. Their research may stimulate additional work in this field. Such research is significant because many experts believe the principal economic benefits of high levels of wastewater treatment will be due to increased water-based recreation.

The amenity services labeled as "aesthetics" represent another area in which economic evaluation techniques are not highly developed. Recently, economists have introduced methods of estimating the value of preserving "atmospheric visibility" in U.S. national parks. For example, an experimental approach using survey questionnaires gauged the benefits of SO_2 emission controls to maintain high visibility in the southwestern parklands. Estimated benefits, as reported by Schulze et al. (1983), totaled several billion dollars per year. Because many Americans place a high value on outdoor leisure experiences, Lave (1978) has suggested that the monetary benefits of aesthetic air quality improvements may be even greater than those associated with improved health.

Materials Inputs The third category in Table 2.2 includes effects of residuals on biological systems whose yields are sold in markets; fisheries, forests, and agriculture are examples. This category also concerns effects of residuals on nonliving things. Examples here include the physical deterioration of machines and buildings caused by poor air quality.

[13]Many of the techniques for estimating the economic benefits of new outdoor-recreation facilities are given by Clawson and Knetsch (1966).

Several literature surveys have been made of economic benefits for the materials inputs category in the table. For instance, the National Academy of Sciences (1975) reviewed measurements of the economic impacts of air pollution on materials and vegetation. As another example, Heintz, Hershaft, and Horak (1976) estimated damages due to water pollution in several categories: water treatment costs for municipalities and industries, damages to agricultural irrigation systems caused by sedimentation, losses to commercial fisheries, damages to domestic plumbing and appliances, and damages to navigation systems due to sedimentation and corrosion.

Benefit Estimation Using Property Values Another procedure for putting a dollar value on improved environmental quality is to assess how it influences the value of real estate. This method relies on statistical analyses of the many variables thought to influence the selling price of real property. Examples are distance of property to transportation facilities, distance of property to the central business district, and average concentration of particulates in the ambient air. Assuming certain conditions governing the behavior of a real estate market are satisfied, the resulting statistical equations can be used to calculate how much people value certain environmental quality characteristics. Numerous property value studies have been conducted in connection with air quality. Suppose, for example, a study of a particular city shows real estate values increasing with improved air quality. If all other factors are equal, property in a neighborhood with clean air would sell for a higher price than an equivalent property in a neighborhood with less clean air. The difference in selling price reflects the value people place on the difference in air quality at the two locations.

The property value approach provides no insights into *why* the price of real estate varies with changes in environmental quality. Consider the above case in which values increase with improved air quality. This result does not indicate whether the enhanced property value results because people feel it is healthier to live in an environment having cleaner air, because they appreciate improved visibility, or because of some other reason. It is inappropriate to simply add monetary benefits computed using statistical analyses of real estate prices with the monetary benefits obtained using other techniques for health effects, decreased soiling, and so on. To do so would involve double counting of these effects.

Property value analyses have been used to estimate benefits of reducing ambient noise levels, especially noise associated with airports. In addition to being annoying, airport noise causes adverse psychological and physical health effects. As in the case of air quality, it is hard to identify the aspect of airport noise that is responsible for decreasing the price of real estate. Despite the difficulties in interpreting study results, the property value approach has been used in many airport location studies.[14]

[14]Examples of property value analyses to find the economic consequences of airport noise are given by James, Jansen, and Opschoor (1978). Property value studies to estimate benefits of improved air quality are discussed by Rubinfeld (1978).

BENEFIT–COST ANALYSIS AND RESIDUALS MANAGEMENT

Information on how residuals influence environmental goods and services can be used to perform a benefit–cost analysis for a proposed residuals management program. A benefit–cost analysis defines the appropriate discharge of residuals as one that yields a maximum difference between total social benefits and total social costs.

To illustrate how a benefit–cost analysis is conducted, consider a firm emitting residuals into a stream. Another firm, located downstream of the discharge, uses the stream for its water supply. The upstream firm's discharge imposes a downstream cost by increasing the treatment undertaken by the downstream firm to make the water suitable for its production processes.

In this situation, the key question is, What quantity of residuals should the upstream firm be allowed to release into the stream? The benefit–cost approach to this question requires that the quantity of residuals emitted be set at the value that maximizes the difference between total benefits and costs. The costs are the expenses incurred by the upstream firm in reducing its discharge.

The computation of benefits is more complex and involves bringing together information introduced in the previous section. It is essential to know how the stream's quality improves in response to reductions in the upstream firm's discharge. It is also necessary to estimate the monetary benefit the downstream firm receives when the upstream discharge is reduced. The benefit is computed by estimating how much the downstream firm's water treatment cost can be reduced if the upstream firm decreases its discharge by a particular amount. The next chapter explains how this net benefit maximization process is carried out.

KEY CONCEPTS AND TERMS

THE ENVIRONMENT AS AN ECONOMIC ASSET
- Materials and energy residuals
- Materials balance perspective
- Environment as a nonreproducible capital good
- Four types of environmental outputs
- Willingness to pay for environmental outputs
- Assimilative capacity

THE MANAGEMENT OF RESIDUALS
- Tragedy of the commons
- Waste receptor services
- Free goods and common property
- Justifications for government intervention

DISCHARGES OF RESIDUALS AND AMBIENT QUALITY
- Variables that characterize discharges
- Concentration as a measure of ambient quality
- Decibels as a measure of energy
- Relating residuals discharges to ambient quality

AMBIENT QUALITY AND FLOWS OF ENVIRONMENTAL OUTPUTS
- Carcinogens in drinking-water supply
- Statistical correlations between air quality and health
- Ability to perceive changes in ambient quality
- Effects on recreation and aesthetics
- Effects on productivity of forests, farms, and fisheries

DISCUSSION QUESTIONS

2–1 Provide an example that illustrates the economist's definition of pollution. How must the example be modified to demonstrate a case in which a residuals discharge does not constitute pollution? Your example should be different from the one presented in Chapter 2.

2–2 What is the relationship between the definitions of pollution and the assimilative capacity of the environment? Give an example in which a residuals discharge causes pollution without exceeding the assimilative capacity of the environment. Indicate factors that can cause the assimilative capacity to change.

2–3 The materials balance perspective can be useful to those developing environmental policies and programs. How would you advise a busy, new member of Congress about how materials balance considerations should be used in devising water quality management policies and programs?

2–4 Imagine that you are a consultant to a government that has no policies or standards regarding the discharge of "waste heat" effluents from steam electric power plants. These discharges consist of high-temperature water flows from cooling water systems. What information would you need in order to assist the government in devising a strategy for managing the waste heat discharge from a power plant to be located on the shore of a large lake?

2–5 Suppose you are asked to help assess the effects of noise from construction equipment on construction workers. In conducting your analysis you are to consider four items:

(i) the cost of reducing noise from construction equipment,
(ii) the transmission of noise from equipment to workers,
(iii) the effect of that noise on hearing, and

(iv) the economic cost of hearing loss.

Which of these items would be the most difficult to describe? Which would be the easiest? What kinds of experts would you call upon to help you develop a full understanding of each item?

REFERENCES

Clawson, M., and J. Knetsch, 1966, *Economics of Outdoor Recreation.* John Hopkins University Press for Resources for the Future, Inc., Baltimore, Md.

Craik, K. H., and E. H. Zube (eds.), 1976, *Perceiving Environmental Quality: Research and Applications.* Plenum, New York.

Crandall, R. W., and L. B. Lave (eds.), 1981, *The Scientific Basis of Health and Safety Regulation.* The Brookings Institution, Washington, D.C.

Dornbush, D. M., 1975, *The Impact of Water Quality Improvements on Residential Property Prices,* prepared for the National Commission on Water Quality, Washington, D.C.

Environmental Protection Agency, 1971, *Air Quality Criteria for Nitrogen Oxides* (January). EPA, Washington, D.C.

Environmental Protection Agency, 1976, *Quality Criteria for Water.* EPA, Washington, D.C.

Feenberg, D., and E. S. Mills, 1980, *Measuring the Benefits of Water Pollution Abatement.* Academic Press, New York.

Flachsbart, P. G., and S. Phillips, 1980, An Index and Model of Human Response to Air Quality. *Journal of the Air Pollution Control Association* **30** (7), 759–768.

Freeman, A. M., III, 1979, *The Benefits of Environmental Improvement, Theory and Practice.* Johns Hopkins University Press for Resources for the Future, Inc., Baltimore, Md.

Freeman, A. M., III, 1982, Benefits of Air Pollution Control Overview, *in* G. S. Tolley, P. E. Graves, and A. S. Cohon (eds.), *Environmental Policy, Air Quality,* Vol. II, pp. 15–26. Ballinger, Cambridge, Mass.

Freeman, A. M., III, R. H. Haveman, and A. V. Kneese, 1973, *The Economics of Environmental Policy.* Wiley, New York.

Graves, P. E., R. J. Krumm, and D. M. Violette, 1982, Issues in Health Benefit Measurement, *in* G. S. Tolley, P. E. Graves, and A. S. Cohon (eds.), *En-*

vironmental Policy, Air Quality, Vol. II, pp. 27–116. Ballinger, Cambridge, Mass.

Hardin, G., 1968, The Tragedy of the Commons. *Science* **162** (13 December), 1243–1248.

Heintz, H. T., Jr., A. Hershaft, and G. C. Horak, 1976, *National Damages of Air and Water Pollution,* report to the U.S. Environmental Protection Agency by Enviro Control Inc., Rockville, Md.

James, D. E., H. M. A. Jansen, and J. B. Opschoor, 1978, *Economic Approaches to Environmental Problems.* Elsevier/North Holland, New York.

Kneese, A. V., 1964, *The Economics of Regional Water Quality Management.* Johns Hopkins University Press for Resources for the Future, Inc., Baltimore, Md.

Lave, L. B., 1978, Comments on Market Approaches to the Measurement of the Benefits of Air Pollution Abatement, *in* A. F. Friedlaender (ed.), *Approaches to Controlling Air Pollution,* pp. 280–284. MIT Press, Cambridge, Mass.

Lave, L. B., and E. P. Seskin, 1977, *Air Pollution and Human Health.* Johns Hopkins University Press for Resources for the Future, Inc., Baltimore, Md.

Lowe, J., D. Lewis, and M. Atkins, 1982, *Total Environmental Control, the Economics of Cross-Media Pollution Transfers.* Pergamon Press, Oxford, England.

National Academy of Sciences, Commission on Natural Resources, 1975, *Air Quality and Stationary Source Emission Control,* Committee Print, Senate Committee on Public Works, 94th Congress, 1st session (March).

Page, T., R. Harris, and J. Bruser, 1981, Water-borne Carcinogens—An Economists View, *in* R. W. Crandall and L. B. Lave (eds.), *The Scientific Basis of Health and Safety Regulation,* pp. 197–228. The Brookings Institution, Washington, D.C.

Rubinfeld, D. L., 1978, Market Approaches to the Measurement of the Benefits of Air Pollution Abatement, *in* A. F. Friedlaender (ed.), *Approaches to Controlling Air Pollution,* pp. 240–273. MIT Press, Cambridge, Mass.

Schulze, W. D., D. S. Brookshire, E. G. Walther, K. K. MacFarland, M. A. Thayer, R. L. Whitworth, S. Ben-David, W. Malm, and J. Molenar, 1983, The Economic Benefits of Preserving Visibility in the National Parklands of the Southwest. *Natural Resources Journal* **23** (1), 149–173.

CHAPTER 3

MANAGING RESIDUALS ON THE BASIS OF ECONOMIC EFFICIENCY

Benefit–cost analysis provides one way of determining the appropriate discharges of residuals. Its premise is that economic efficiency should be used as the criterion for selecting a desirable level of residuals reduction. A numerical example is used below to demonstrate the benefit–cost procedure and to point out its limitations.

Before considering the example, types of actions to reduce or modify residuals are introduced. The number of possible actions is much greater than is commonly assumed.

ACTIONS TO MODIFY RESIDUALS

"Treatment" is often the first thing that comes to mind when people think about controlling residuals. Consider, for example, a firm that releases air-borne residuals to the atmosphere. The firm could reduce its emissions by intercepting residuals at the end of its production process and removing the residuals using physical devices and chemical processes. However, this "end-of-pipe treatment" is not always the most effective approach.

Table 3.1 lists a number of ways to control residuals. These options are relevant to firms, as opposed to individuals and governments, since they involve matters relating to plant design, materials inputs, and product outputs. As indicated by the first three items in the table, if the *type* of output is held constant, the residuals management options available to a firm include decreasing production, changing production processes, and changing the types of inputs. An example of a decrease in production is the common practice of having firms cut back their levels of output during various stages of a smog alert.

TABLE 3.1 Modifying a Firm's Residuals by Means Other Than "Treatment"[a]

Type of Action	Example
Reduce level of output	Firm cuts back production during a "smog alert"
Produce same outputs, but use *production processes* yielding less damaging residuals	Electric power plant with higher "thermal efficiency" produces lower waste heat per unit of power output
Produce same outputs, but use *inputs* yielding less damaging residuals	Paper company eliminates use of recycled paper containing PCBs to get rid of PCBs from its effluent
Produce *different outputs* and thereby generate less damaging residuals	Paper company produces same mix of outputs, but with all products having a lower level of brightness
Increase materials recovery and reuse (that is, "recycling")	Refinery decreases residuals by adding "skimmers" to remove crude oil components from refinery wastes

[a]This table is based on ideas presented by Kneese and Bower (1979, pp. 41–52).

Firms may reduce residuals by making the same outputs using different processes. This type of action often involves the selection of an engineering efficiency. Typically, there are many different engineering efficiencies that can be adopted when designing a production process. The choice of a particular value is generally based on financial considerations. High levels of efficiency are often possible, but they may require costly investments. Steam–electric power plants provide a good example of how increased efficiency can lead to decreased discharges. The higher the thermal efficiency, the greater the plant construction cost, but the lower the waste heat discharged from cooling water systems.[1]

Firms may also reduce residuals by using different inputs to produce the same outputs. An example is provided by a paper company in Los Angeles that was required, by U.S. Environmental Protection Agency (EPA) regulations, to diminish the concentration of polychlorinated biphenyls (PCBs) in its wastewater discharge. The company examined two treatment schemes and decided against both. Instead, it chose to get rid of the PCBs by changing the mix of recycled office paper used as inputs to its production process. By eliminating the recycled paper that contained PCBs (especially carbonless copy paper), the firm reduced

[1]Another example of how changing production processes can yield less damaging residuals is found in the timber industry. Timber harvesting practices may be shifted from "clear cutting," which often has adverse impacts on erosion and scenic resources, to "selective harvesting." The latter involves harvesting only trees that have reached a certain level of maturity.

its discharge of PCBs to zero. It was cheaper for the firm to change its inputs than to use either of the treatment schemes.[2]

Another alternative to end-of-pipe treatment is to change the firm's outputs so that the resulting residuals are less damaging. In considering this option, Kneese and Bower (1979) found it useful to distinguish between changing product mix and changing product specifications. An example of an alteration in product mix is an automobile manufacturer's decision to produce only low-weight "economy models" instead of a full range of cars that includes "luxury models." A change in product specification is illustrated by a paper company's decision to manufacture the same mix of bond paper, paper towels, and so forth, but with a reduced level of brightness. Producing a white paper of high brightness requires greater quantities of chemicals, water, and energy than an unbleached paper that is otherwise similar in quality. By making paper products with lower brightness, the discharge of some residuals can be diminished substantially.

Table 3.1 also indicates that residuals can be cut back by expanding the level of materials recovery and reuse. The use of "oil skimmers" in an oil refinery exemplifies materials recovery. Skimmers can remove selected crude oil components that normally might be left in the wastewater from a refinery.[3] Materials recovery and reuse often involve the development of "by-products." Among the many examples are the production of animal foods from distillery residuals, and the recovery of animal fats from slaughter houses for use in making soaps and glues.

How might firms be encouraged to take the steps listed in Table 3.1? In many cases, actions to reduce residuals will be adopted independent of any concern for environmental quality. There are instances where changing product mix, increasing engineering efficiency, and reusing materials are implemented because they make economic sense. For instance, before the 1960s, the relatively high levels of thermal efficiency used in steam–electric power plants were often decided upon in the absence of stringent controls on waste heat loads from power plant cooling systems. There are, however, a number of procedures to modify residuals that will not be undertaken unless the government intervenes to force firms to account for the harmful effects of their discharges. Alternative forms of government intervention are described later in this chapter.

In addition to end-of-pipe treatment and the approaches listed in Table 3.1, residuals management can also involve changes in the location and timing of discharges. These options can be used by individuals and governments as well

[2]The paper company example is from Bartos (1979). Another instance of changing inputs as an alternative to end-of-pipe treatment involves electric power production using coal-fired power plants. By shifting from high-sulfur coal to low-sulfur coal, electric utilities have been able to generate the same quantities of electricity with lower discharges of sulfur oxides. Other cases in which inputs were modified as a residuals management strategy are given by Kneese and Bower (1979); see, also, Stockholm's (1981) discussion of the 3M Company's "Pollution Prevention Pays" program.

[3]For examples of the use of skimmers by oil companies, see Ackerman et al. (1974).

as firms. Changing the *location* of a discharge can be beneficial since some places are better able than others to accommodate residuals without harmful results. A provision of the Clean Water Act of 1977 provides an example. In passing this act, the U.S. Congress specified conditions under which cities discharging wastewater to the ocean could provide lower treatment levels than inland municipalities. In providing this opportunity for reduced treatment, Congress recognized that some near-shore marine environments can assimilate and disperse residuals without significant negative effects.

The *timing* of discharges is important because the environment can accommodate more residuals in some periods than in others. For example, emissions of air-borne residuals on windy days are often much less objectionable, other things being equal, than the same discharges on days when the atmosphere is stagnant. This is why automobile travel is restricted under certain meteorological conditions in Los Angeles and other cities with smog problems. Curtailing airplane takeoffs and landings over residential neighborhoods during normal sleeping hours is another way timing is used to manage residuals.

Another management option is to augment the environment's ability to accept residuals without causing significant damages. An example is "low-flow augmentation," the release of water from a reservoir solely to provide enough flows to dilute wastewater entering the river below the reservoir. Generally, these flows do not substitute for minimum levels of wastewater treatment. Instead, they provide a supplement to treatment.

The assimilative capacity of the environment can also be increased by "artificial aeration," the use of special equipment to transfer oxygen into a watercourse. There are many ways to do this. One technique uses an air compressor attached to a perforated hose placed across the bottom of a stream. Bubbling air through the hose increases the concentration of oxygen dissolved in the stream. This, in turn, enhances the stream's ability to accept wastewaters containing organic matter. The bacterial decomposition of organic material frequently depletes a stream's dissolved oxygen. By using artificial aeration, a stream can accept greater quantities of organic waste without oxygen levels dropping so low that fish and other aquatic life are adversely affected.[4]

A thorough consideration of the laws of conservation of mass and energy is required in making good choices among the many alternative ways of modifying residuals. This point is emphasized in a warning by Kneese and Bower (1979, p. 51):

> *[The]* net *environmental effects of any of the alternatives discussed above — or any others — can be adequately assessed only in the context of the total system. Such a system must include the residuals associated with all of the material and energy inputs relating to the production of the good or service and all of the material and energy inputs, and related residuals, necessary to handle or dispose of the good subsequent to the end of its useful service.*

[4]Thackston and Speece (1966) present numerous applications of artificial aeration.

The need for a broad perspective in designing a residuals management program is reinforced by the 1968 U.S. regulations on emissions from automobiles. The government called for cutbacks in the emissions of hydrocarbons and carbon monoxide from the exhausts of new cars. Auto manufacturers met the federal regulations by modifying engine designs so that a lower fuel-to-air ratio could be used. With a lower ratio, the fuel burned more completely, and there were lower emissions of hydrocarbons and carbon monoxide. However, the change in engine design also increased the temperature of fuel combustion. As a result, the releases of oxides of nitrogen became even greater than they were before the emission control program was implemented.[5] Because emissions of nitrogen oxides cause numerous adverse effects, federal regulations were subsequently modified to control them. The 1968 regulations demonstrate that unless a residuals reduction strategy is evaluated in a "total systems" context, the strategy may yield only a shift in environmental degradation, not a net improvement in quality. Unfortunately, technological complexities often make it difficult to predict all consequences of a residuals reduction option before it is implemented, even when a total systems approach is adopted.

BENEFITS AND COSTS OF RESIDUALS REDUCTION: THE CEDRO RIVER EXAMPLE

The example below examines two important residuals management questions: what are the appropriate levels of residuals reduction, and how can dischargers be given incentives to attain those levels? The example reflects a utilitarian basis for decision making. It assumes that the appropriate levels of reduction are those that provide a maximum "net benefit" — the difference between the total economic benefits and the costs of reducing discharges.

To simplify the presentation, the example first determines the appropriate level of residuals reduction using the minimization of total cost as the basis for making decisions. This is done because matters related to cost are familiar and easy to grasp. Following this, a technique for estimating benefits is introduced. It is shown that the residuals reduction that minimizes cost in the example is identical to the reduction that maximizes net benefits. This level of residuals reduction is said to be "economically efficient." Any other level is wasteful of society's resources.[6] Readers familiar with calculus are encouraged to consult the footnotes to see how calculus simplifies the computations.

The example concerns two firms on the fictitious Cedro River in Figure 3.1. The upstream firm, the Margarita Salt Company, takes water from the river,

[5]This explanation of why the nitrogen oxides increased relies on Masters (1974, pp. 233–234).

[6]Economists define a situation as being economically efficient "if it is impossible to make one person better off except by making someone else worse off" (Layard and Walters, 1978, p. 7). This criterion for efficiency is based on ideas of the economist Pareto; economically efficient outcomes are sometimes said to be "Pareto optimal." For a discussion of economic efficiency concepts and how they are used in designing residuals management programs, see Dorfman and Dorfman (1977).

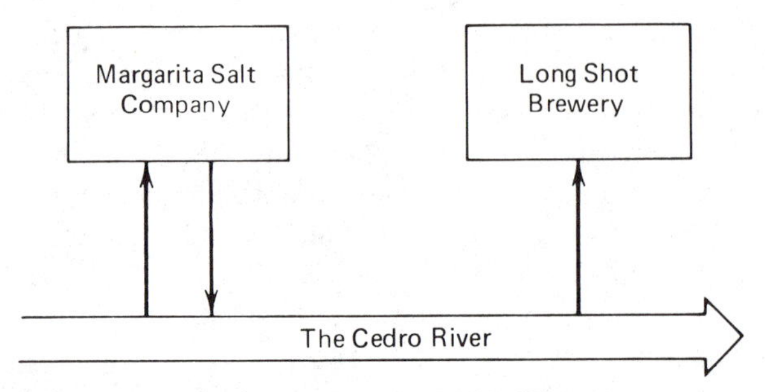

FIGURE 3.1 Two firms on the Cedro River.

uses the water in its production process, and then discharges it back to the river. The only change in the water that results from this usage is that the concentration of chlorides, in milligrams per liter(mg/l), is increased. Thus, when the Margarita Salt Company's wastewater is discharged, the level of chlorides in the river is increased. This adversely affects the downstream firm, the Long Shot Brewery, since the brewery must treat the water to remove chlorides before it uses the water to make beer.

Cost of Chloride Reduction by Margarita Salt Company

The Cedro River Agency to Control Quality (CRACQ) has been established to come up with a water quality management plan. Its aim is to get the Margarita Salt Company to account for the damage it is causing to the Long Shot Brewery as a result of discharging its chloride load. Each firm has a representative that is a liaison to CRACQ, and both firms are interested in providing CRACQ with all the data it needs in making its decisions. (This *is* a hypothetical example.) The chief engineer at CRACQ asks the representative of each firm for some information. From the person representing Margarita Salt, she asks how much it would cost to reduce the existing Margarita Salt Company discharge by 25, 50, 75, and 100%, respectively. (One hundred percent reduction is unrealistic in practice. It is used here only because it simplifies the arithmetic in the example.) The chloride reduction can be accomplished by means of end-of-pipe treatment or by a combination of the actions listed in Table 3.2. Economic efficiency dictates that any particular chloride reduction must be attained with the least costly combination of actions. Costs provided by the Margarita Salt Company are given in Table 3.2. The chief engineer at CRACQ plots these numbers (see Figure 3.2) and then uses her mathematical ability to get the equation of a curve that passes closely through all the points. That equation, plotted in Figure 3.2, takes the form

$$C_M(R) = 100(R)^{2.3} \tag{3-1}$$

where R represents the percentage of the "raw" (that is, before reduction) chloride load removed by Margarita Salt, and $C_M(R)$ is the cost of reducing the

TABLE 3.2 Cost of Chloride Reduction by Margarita Salt Company

Percent of Chloride Discharge Reduced	Cost of Reduction ($\$10^3$)
25	164
50	808
75	2050
100	3980

raw load by R percent. The raw load represents the quantity of chlorides that Margarita Salt releases in the absence of any government intervention to influence its discharge.

Relationship between Chloride Discharge and Cedro River Quality

The chief engineer at CRACQ then asks her own staff to analyze how different levels of chloride reduction by Margarita Salt translate into changes in the chloride concentration at the intake of the Long Shot Brewery. First, the staff examines the historic record of flows in the Cedro River. On the basis of these records they decide to perform their analysis for conditions when the river flow at Margarita Salt is approximately 1000 cubic feet per second (cfs), and it does not increase significantly between the Margarita Salt Company and the brewery. They also examine water quality records and find that under this flow condition, the chloride concentration just above Margarita Salt's point of discharge is about 25 mg/1. The discharge from Margarita Salt is of negligible volume (1 cfs), but

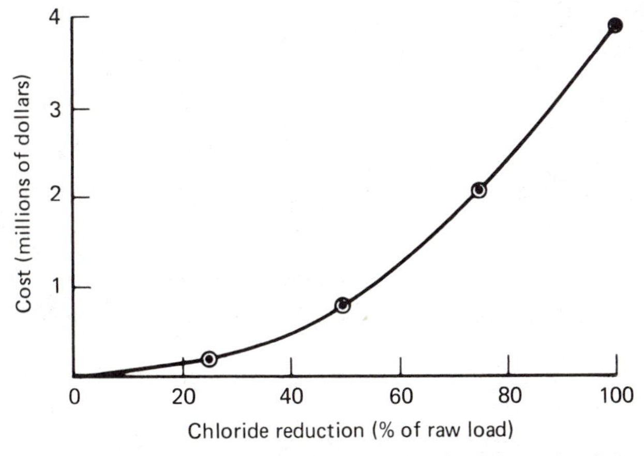

FIGURE 3.2 Cost of chloride reduction by Margarita Salt.

it is highly concentrated (100,000 mg/1). The load from Margarita Salt without any chloride reduction is 100,000 cfs · mg/1, the product of flow times concentration.[7]

Because chlorides do not degrade over time, the CRACQ staff has an easy job. To find the concentration of chlorides at Long Shot Brewery, they simply conduct a "mass balance" analysis at the Margarita Salt Company's point of discharge. The chloride load downstream of the point is set equal to the load upstream plus the load from the Margarita Salt Company (see Figure 3.3). Once the chloride load downstream from Margarita Salt is found, all that remains is to compute the concentration of chlorides. This is done by dividing the total downstream chloride load by the downstream flow. Thus, if Margarita Salt did not reduce its discharge, the chloride concentration at Long Shot Brewery would be

$$\frac{[(1000)(25) + (1)(100{,}000)]\ \text{cfs} \cdot \text{mg/l}}{1001\ \text{cfs}} = 125\ \text{mg/l}$$

Using the same reasoning, the staff at CRACQ derives a formula showing the relationship between chloride concentration in the river and discharges at Margarita Salt. If *R* equals the percentage of chloride load reduction, then the diagram for a chloride balance at Margarita Salt is as shown in Figure 3.4. To deduce the relationship between chloride load and concentration, let [C1] represent the concentration of chlorides (in mg/l) downstream from the discharge point. The value of [C1] can be found using a mass balance equation like the one above. In this instance,

$$[\text{C1}] = \frac{[(1000)(25) + 100{,}000 - 1000\,R]\ \text{cfs} \cdot \text{mg/l}}{1001\ \text{cfs}}$$

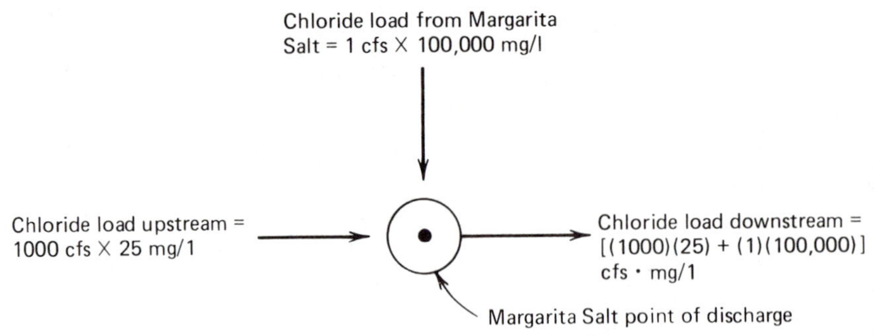

FIGURE 3.3 Chloride balance for Margarita Salt discharge with no chloride reduction.

[7]Although flow and concentration are not given in a consistent unit system, the units indicated are commonly used by water quality specialists in the United States. The product of flow (volume/time) and concentration (mass/volume) is a "mass flowrate" (mass/time).

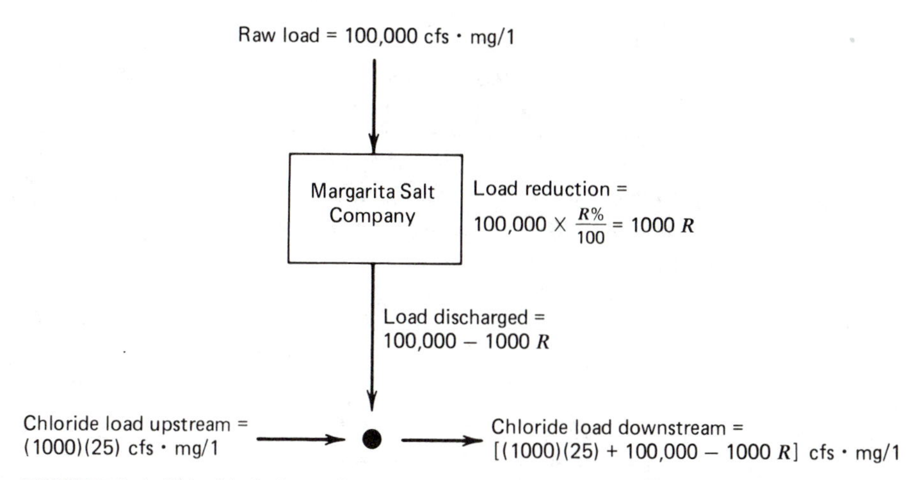

FIGURE 3.4 Chloride balance for Margarita Salt discharge with $R\%$ reduction.

or, after some simplifications,

$$[C1] = 125 - R \tag{3-2}$$

In equation 3-2, the result of dividing by 1001 cfs is rounded off to the nearest whole number.

Cost of Water Treatment at Long Shot Brewery

The chief engineer at CRACQ next turns to the representative from the Long Shot Brewery for information. She asks for an estimate of what the brewery's water treatment costs would be if its intake water had each of several chloride concentrations (see Table 3.3). The engineers at Long Shot Brewery perform the necessary analyses and provide the cost estimates in Table 3.4. Once more

TABLE 3.3 Relation between Chloride Reduction and Downstream Chloride Concentration

Percent of Chloride Reduced by Margarita Salt	Total Chloride Load from Margarita Salt (cfs · mg/l)	Concentration of Chlorides in Cedro River at Long Shot Brewery (mg/l)
0	100,000	125
25	75,000	100
50	50,000	75
75	25,000	50
100	0	25

TABLE 3.4 Cost of Water Treatment at Long Shot Brewery

Chloride Concentration of Intake Water (mg/l)	Cost of Water Treatment for Long Shot Brewery ($\$10^3$)
25	63
50	250
75	563
100	1000
125	1560

the chief engineer calls upon her analytical training, this time to derive the equation of a curve that links water treatment cost with the river's chloride concentration. The resulting equation is

$$C([\mathrm{C1}]) = 100[\mathrm{C1}]^2 \tag{3-3}$$

where $C([\mathrm{C1}])$ is the Long Shot Brewery water treatment cost for any particular value of chloride concentration at the point of water intake. If equations 3-2 and 3-3 are combined, the result can be expressed as

$$C_L(R) = 100(125 - R)^2 \tag{3-4}$$

where $C_L(R)$ is the Long Shot treatment cost in terms of the percentage of chloride reduction by Margarita Salt. The chief engineer at CRACQ plots this cost equation against both [C1] and R, and the result is given in Figure 3.5. The horizontal axis in this plot is based on the relationship between [C1] and R in equation 3-2.

Minimizing the Total Cost of Margarita Salt's Discharge

The chief engineer for CRACQ observes that an economically efficient approach to managing the chloride discharge involves minimizing the total cost associated with the discharge. It can be *imagined* that a single firm owns both the Margarita Salt Company and the Long Shot Brewery. For such a firm, it would make good economic sense to decrease the chloride discharge to the point where Margarita Salt's chloride reduction cost *plus* Long Shot Brewery's water treatment cost is at a minimum. Any other level of chloride reduction would be unnecessarily expensive to a single firm that owned both the salt company and the brewery.

The determination of the chloride reduction that minimizes total cost is based on Table 3.5. For any particular reduction, R, the cost to the Margarita Salt

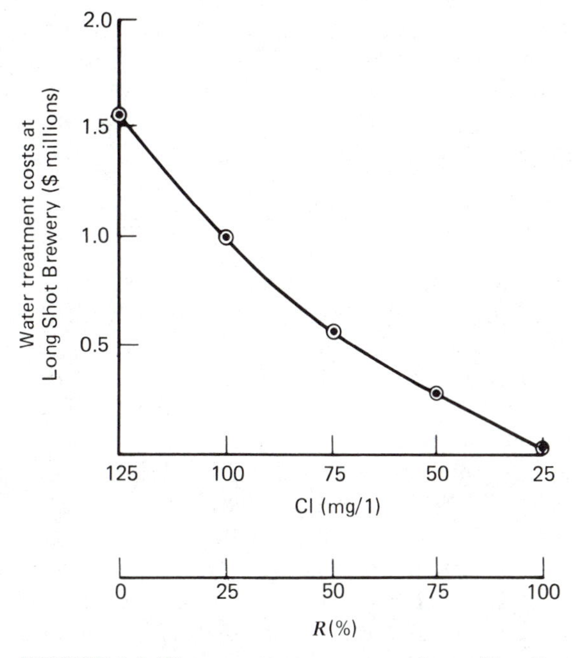

FIGURE 3.5 Water treatment costs at Long Shot Brewery.

TABLE 3.5 Cost of Chloride Reduction and Water Treatment ($$10^3$)

Percent of Chloride Reduced by Margarita Salt	Chloride Reduction Cost Incurred by Margarita Salt Company	Water Treatment Cost Incurred by Long Shot Brewery	Cost of Chloride Reduction Plus Water Treatment[a]
0	0	1560	1560
10	20	1320	1340
20	98	1100	1200
30	250	903	1150
40	484	723	1210
50	808	563	1370
60	1230	423	1650
70	1750	303	2050
80	2380	203	2580
90	3120	123	3240
100	3980	63	4040

[a]Entries in this column have been rounded off to three significant figures.

Company is computed using equation 3-1. The corresponding water treatment cost for Long Shot Brewery is determined from equation 3-4. These costs are shown for different values of R in Table 3.5. By inspection of the last column of the table, the cost of chloride reduction *and* water treatment is at a minimum when Margarita Salt decreases its chloride load by 30%.[8]

Benefits Estimated as Water Treatment Costs Avoided

People who feel that society's resources should be allocated efficiently would advocate that CRACQ set R to minimize the *total* cost associated with the chloride discharge. A closely related view is that the appropriate level of discharge is one that maximizes the total economic benefit minus the cost of chloride reduction. In this instance, the benefit is defined as the decrease in the brewery's water treatment cost due to chloride removal by the salt company.

The procedure for computing the economic benefit rests on the idea that when Margarita Salt reduces its chlorides by $R\%$, there are water treatment costs (or "damages") avoided by the Long Shot Brewery. If $R = 0$, no damages are avoided and thus the benefit is zero. Alternatively if $R = 100\%$, the damage avoided is the difference between Long Shot's water treatment cost with no upstream chloride reduction ($1,560,000) and its treatment cost with full chloride reduction ($62,500). Continuing with this reasoning, the benefits to Long Shot Brewery, in terms of water treatment cost avoided, can be expressed as $B_L(R)$, where

$$B_L(R) = 1{,}560{,}000 - C_L(R) \qquad \textbf{(3-5)}$$

Combining equation 3-5 with equation 3-4 gives

$$B_L(R) = 1{,}560{,}000 - 100(125 - R)^2 \qquad \textbf{(3-6)}$$

Table 3.6 presents benefits for several values of R.

[8]Calculus can also be used to determine the value of R that minimizes $C_T(R)$, the cost of chloride reduction plus water treatment. This total cost is found by adding equations 3-1 and 3-4:

$$C_T(R) = 100R^{2.3} + 100(125 - R)^2$$

A theorem from calculus requires that $\hat{R}$, the value of R that minimizes $C_T(R)$, satisfy the condition

$$\left.\frac{dC_T(R)}{dR}\right|_{R = \hat{R}} = 0$$

Solving

$$\frac{dC_T(R)}{dR} = 0 = 230R^{1.3} - 200(125 - R)$$

for R yields $\hat{R} = 30\%$. For this value to provide a minimum total cost, the second derivative of $C_T(R)$ evaluated at $\hat{R}$ must be positive. This can be shown to be the case.

TABLE 3.6 Computation of Benefits to Long Shot Brewery ($$10^3$)

Percent of Chloride Reduced by Margarita Salt	Cost to Long Shot Brewery if Margarita Salt Does Not Reduce Its Load	Cost to Long Shot Brewery if Margarita Salt Reduces *R*% of Its Load	Benefit to Long Shot Brewery, in Terms of Treatment Costs Avoided[a]
0	1560	1560	0
25	1560	1000	560
50	1560	563	997
75	1560	250	1310
100	1560	63	1500

[a]Entries in this column have been rounded off to three significant figures.

The chief engineer knows that with this definition of benefits, the value of R that minimizes total cost ($R = 30\%$) will also be the R that maximizes net benefits.[9] To demonstrate this to CRACQ's policy makers, she prepares a table that allows her to select R^*, the net benefit maximizing percentage of chloride reduction. For any particular R, the equations for $B_L(R)$ and $C_M(R)$ are used to compute the benefits to Long Shot Brewery and the costs to Margarita Salt, respectively. The results for different values of R are given in Table 3.7.

The last column in Table 3.7 indicates that the net benefit maximizing value of R is 30%. This is confirmed by making a graph of total costs and total benefits and inspecting the differences (see Figure 3.6). The 30% reduction yields a chloride concentration of 95 mg/l at the intake of the Long Shot Brewery. This value of concentration is obtained using equation 3-2 with $R = 30\%$.

[9]The chief engineer uses her knowledge of calculus to deduce this result. She observes that the net benefits, $N(R)$, are defined as

$$N(R) = B_L(R) - C_M(R)$$

or

$$N(R) = 1{,}560{,}000 - 100(125 - R)^2 - 100R^{2.3}$$

The value of R that maximizes $N(R)$ must satisfy

$$\left.\frac{dN(R)}{dR}\right|_{R = R^*} = 0$$

Solving

$$\frac{dN(R)}{dR} = 0 = 200(125 - R) - 230R^{1.3}$$

yields $R^* = 30\%$. Inspection of the first derivatives for $N(R)$ above and $C_T(R)$ in footnote 8 indicates that these derivatives are equal except for their signs. It follows that the R that minimizes the cost of chloride reduction plus water treatment must be identical to the R that maximizes net economic benefits.

TABLE 3.7 Summary of Benefits to Long Shot Brewery and Costs to Margarita Salt ($10^3)

Percent of Chloride Reduced by Margarita Salt	Total Benefits to Long Shot Brewery	Total Costs to Margarita Salt	Difference Between Total Benefits and Total Costs
0	0	0	0
10	240	20	220
20	460	98	362
30	657	250	407
40	838	484	354
50	997	808	189
60	1140	1230	−90
70	1260	1750	−490
80	1360	2380	−1020
90	1440	3120	−1680
100	1500	3980	−2480

The Cedro River example can be extended to treat a more general case that considers the linkage between Margarita Salt's chloride discharge and its level of output. This extension, which relies on calculus, is given in the appendix to Chapter 3. Readers familiar with calculus are encouraged to consult the appendix for a compact review and extension of the ideas presented here.

How the Cedro River Example Differs from Reality

The Cedro River case demonstrates the benefit–cost analysis approach to residuals management, but it does not provide a realistic representation of how decisions concerning appropriate levels of discharge are actually made. One reason is that the example considers only a single residual, chlorides. In real settings, several residuals are likely to be significant. Moreover, the Cedro River

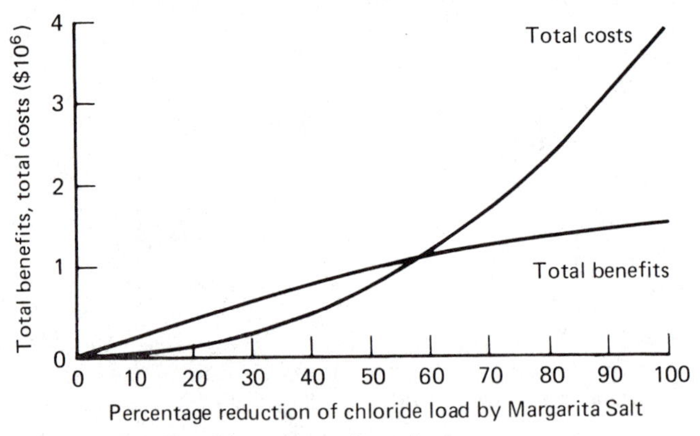

FIGURE 3.6 Total benefits and total costs.

has only a single source of residuals, the Margarita Salt Company. Usually there are many discharges involved.

The linkage between the residuals released by the Margarita Salt Company and the residuals concentration downstream is much simpler than corresponding relationships in real rivers. The simplicity in the example results because chlorides are stable and because both the discharge and the river flow are assumed constant. Not many residuals behave as simply as chlorides. Complex physical, chemical, and biological processes often transform residuals after they are discharged into the environment. In addition, the characteristics of the environment itself change over time. These complicating factors often make it difficult to predict accurately the effect that an emission will have on the quality of the ambient environment. When there are numerous discharges, the forecasting task is even more burdensome. In many settings, it is impossible to make reasonable predictions.

Another simplification concerns the procedure for estimating the benefits of residuals reduction (or its counterpart, the damages caused by the discharge). The benefit estimation procedure used in the example is straightforward. However, most real situations involve much more than damages in the form of additional treatment costs downstream. Typically, the benefits concern less easily quantified effects such as improvements in aesthetics and reductions in human illness. Many effects of residuals reduction are so difficult to evaluate in dollar terms that the computations are frequently not even attempted. Without such estimates, a net benefit maximization computation cannot be made.

Despite its many simplifications, the Cedro River case serves several purposes. It demonstrates a rationale that is often advocated in discussions of environmental quality: residuals reduction should only be undertaken if the resulting monetary benefits exceed the costs. In addition, the example illustrates the simplifications and assumptions that must be made to calculate the benefits and costs of a particular residuals management program.

The Cedro River example also assists in understanding the limits of the economic efficiency concept. Even if a complete benefit–cost analysis could be performed, the numerical results would not be used without considering other factors. The fairness of the distribution of the costs and benefits of residuals reduction is often an important consideration. Issues related to moral rights are also relevant. Here is a sample of questions related to moral rights and fairness in the Cedro River case:

- Should Margarita Salt Company be given the right to cause any damages at all?
- If damages are caused, should they be offset by payments from Margarita Salt to Long Shot Brewery?
- If forcing Margarita Salt Company to pay for either chloride reduction or downstream damages would cause it to close, should anything be expected of the company?
- Would strong action against Margarita Salt be justified if a plant closure would cause widespread unemployment?

Such questions are important, and they go beyond the boundaries of the economic efficiency orientation to residuals management.

NEED FOR GOVERNMENT INTERVENTION IN RESIDUALS MANAGEMENT

To what extent is government intervention required in managing residuals? In examining this question, consider Margarita Salt's expected behavior if there were no intervention by CRACQ or any other agency. In these circumstances, the Margarita Salt Company would probably discharge its entire raw load. From the salt company's perspective, waste receptor services are provided by the Cedro River free of charge. *Any* level of chloride reduction costs the company money, but it provides no economic return. Even though substantial downstream costs are associated with the chloride discharge, there is nothing to compel Margarita Salt to account for these costs in its decision making.

It might be argued that the salt company would provide some chloride decrease out of a sense of moral obligation. As a practical matter, the ethical perspectives of Margarita Salt cannot be relied on to yield substantial chloride reductions. This position is supported by evidence that Baumol and Oates (1979) gathered concerning the voluntary activities that firms engage in to eliminate environmental damages. They observed that "moral suasion" is effective in promoting voluntary actions only when brief, acute emergencies are involved. They cite, as an example, emergency conditions associated with intense smog. In such cases, voluntary residuals reductions may be the only way to deal with the emergency. The analysis by Baumol and Oates provides little support for the view that discharges will be cut back voluntarily in routine circumstances.

A cash payment to the Margarita Salt Company would probably provide a more effective incentive for chloride reduction than an appeal to the company's moral obligations. The benefits and costs in Table 3.7 suggest that a suitable payment might be offered without any government intercession. Consider, for example, the chloride reduction costs and the savings in water treatment costs when R = 10%. As shown in the table, Margarita Salt's chloride reduction costs are $20,000. At R = 10% the Long Shot Brewery can save $240,000. Clearly there is the potential for a private agreement in which Margarita Salt provides a 10% chloride reduction and receives a payment that more than covers its cost. Long Shot Brewery has an incentive to offer payments for chloride reductions up to R = 30%. Beyond 30%, it is cheaper for the brewery to treat its intake water than to pay for Margarita Salt's chloride reduction. For example, to attain reductions of 30 and 40%, the costs are $250,000 and $484,000, respectively (see Table 3.7). The $234,000 difference represents the cost of the 10% increment of chloride reduction. The benefits to Long Shot Brewery in going from R = 30% to R = 40% are $657,000 and $838,000, respectively. The difference, $181,000, is the additional benefit the brewery receives for the extra 10% reduction. Since this amount is lower than the corresponding increment in costs, Long Shot Brewery has no incentive to pay for the additional chloride decrease.

There are two important reasons why Margarita Salt and Long Shot Brewery might not strike a mutually beneficial bargain regarding the reduction of chlorides. One concerns "transaction costs," the costs that firms incur in identifying and implementing mutually advantageous exchanges. There are costs involved in generating information, such as the estimates in Table 3.5, and in negotiating and enforcing agreements. The negotiation process can be complex and lengthy since the firms generally would not know *each other's* treatment costs, and they might rely on tactics, such as stalling and bluffing, that are used commonly in negotiations.

A second impediment to a private agreement between the firms is an ambiguity in legal rights. Does Margarita Salt have a legally enforceable right to emit chlorides, or is its discharge occurring in the absence of laws controlling the use of the Cedro River? Alternatively, does Long Shot Brewery have a legal right to water that is uncontaminated by Margarita Salt's discharge? Long Shot Brewery would not agree to pay for chloride reduction by Margarita Salt if the latter did not have a right to release chlorides. If the legal rights are ambiguous, it might be impossible to enforce a private bargain between the two firms.

Arguing in a more general way, Coase (1960) has shown that the presence of transaction costs and the lack of well-defined legal rights are often barriers to mutually advantageous exchanges of the type described above.[10] Before such private agreements can be expected to materialize, governments must clarify who has the legal right to the services provided by the environment. Of course, even if legal rights are clearly specified, the existence of high transaction costs could prevent agreements from being reached. For example, imagine the costs involved in reaching agreements when the sulfur oxides discharged by a power plant adversely influence the health of an entire community. When many individuals are affected, the cost of bringing them together as a negotiating unit may be much greater than the resulting benefit.

FORMS OF GOVERNMENT INTERVENTION

The above discussion indicates that without government intervention, residuals reduction would be unlikely, except at levels that made economic sense to the firms generating the residuals. Three categories of government action to manage residuals can be distinguished: (1) direct regulation using standards to control discharges; (2) economic incentives such as subsidies; and (3) government construction and operation of facilities to collect, treat, and dispose of residuals. Illustrations of actions from each category are given below.

Direct regulation to limit discharges often involves effluent or emission standards. The term *effluent standard* applies to water-borne residuals, such as the

[10]The economist Coase is widely known for a theorem which bears his name: if costless negotiation is possible and rights are well specified, then the allocation of resources will be economically efficient (paraphrased from Layard and Walters, 1978, p. 192).

chlorides from the Margarita Salt Company. For noise and air-borne residuals, the regulations are often called *emission standards.* Penalties are commonly imposed on those discharging in excess of standards. Examples include monetary fines and legally enforceable administrative orders to halt operations which cause a standard to be violated.

Effluent (or emission) standards are often based on an "ambient standard," a target level of environmental quality. Effluent and ambient standards can be distinguished by considering the Cedro River example. Suppose CRACQ decides to impose an ambient standard corresponding to the chloride concentration in the Cedro River that maximizes net economic benefits. As indicated in the discussion of Figure 3.6, the net benefit maximizing chloride concentration is 95 mg/l, and this results when $R = 30\%$. Thus CRACQ's *ambient standard* for the Cedro River is a maximum chloride concentration of 95 mg/l. Assuming it has the needed authority, CRACQ can try to have this ambient standard met by promulgating the following *effluent standard:* the chloride load released by the Margarita Salt Company can be no greater than 70% of its raw load, which is equivalent to a flow of 1 cfs with a chloride concentration of 100,000 mg/l.

An alternative approach to setting an effluent standard relies on the technology available for residuals reduction. An explicit ambient standard need not exist. As an example of this "technology-based" approach, suppose CRACQ promulgated an effluent standard for the Margarita Salt Company that specified the highest technically feasible chloride reduction, say 98%. The standard might also indicate the devices and methods to be used in achieving the 98% decrease. Because the costs of $R = 98\%$ far exceed the monetary benefits, CRACQ could not argue that its standard is economically efficient. CRACQ's defense of the technology-based standard might be that those who generate residuals do not have a right to degrade the environment.

Economic incentives are frequently used to supplement standards. In the United States, for example, the federal government provides cash subsidies to municipalities for the construction of wastewater treatment plants. Another economic incentive is the deposit refund given to individuals for returning beverage containers.

Many economists have advocated use of an *effluent* (or *emission*) *charge,* a fee paid per unit of residuals discharge. The Cedro River case serves again as an illustration. Assume that CRACQ places no standard limiting the quantity of chlorides released by Margarita Salt. Instead, it imposes a fee of $\$Z$ per pound of chlorides discharged. Margarita Salt's response will depend on how the fee compares to the cost of chloride reduction. If Z is set much lower than the incremental cost of chloride reduction, Margarita Salt will have no financial motivation to reduce its discharge. It would release its raw load and pay the consequent effluent charge. If, on the other hand, Z is set very high compared to the cost of chloride reduction, Margarita Salt will try to minimize its discharge. If CRACQ knew Margarita Salt's basis for making decisions and its cost of chloride reduction, then the agency could set Z to induce Margarita Salt to decrease its discharge by any amount between 0 and 100%.

Another scheme advocated by economists involves "marketable pollution permits." A marketable permit system could be established for the Cedro River as follows: CRACQ first sets an ambient standard for chlorides. Using a mass balance analysis, it then computes the total chloride load the river can accept and still meet this standard. Let Y lb/day represent this total load. Next, CRACQ issues permits to discharge chlorides into the Cedro River. Collectively, the permits allow for only Y lb/day of chlorides. By requiring that each unit of discharge be covered by a permit, the ambient standard is always met. Suppose CRACQ initially allocates *all* of the permits to the Margarita Salt Company.[11] After the initial allocation, the individual permits can be bought and sold. Thus, if the price were right, Margarita Salt would sell some of its permits to Long Shot Brewery. As indicated by Table 3.7, the brewery enjoys substantial savings in water treatment when chloride discharges are reduced up to $R = 30\%$. Another possibility is for Margarita Salt to sell some permits to a new firm that wants to locate in the area and that needs to dispose of chlorides. This system would have to be expanded for the more realistic case in which the concentration of residuals in the river depends on the locations of the discharges. The necessary modifications do not change the main idea, which is to encourage private transactions that use the environment's waste receptor services efficiently and do not violate ambient standards.

A different class of interventions consists of the direct government provision of services to collect, store, modify, and dispose of residuals. In the United States, this type of intervention is commonly used to deal with water-borne residuals generated within municipalities. Another example involves the collection and disposal of solid wastes in urban areas. These services are often provided by local government.

This discussion of strategies of government intervention is intentionally terse. The next two chapters provide details on how residuals management programs are designed and implemented.

[11]An alternative is to give all the permits (totalling Y lb/day) to Long Shot Brewery. Another possibility is for CRACQ to auction the permits off to the highest bidders.

APPENDIX TO CHAPTER 3

AN EXTENSION OF THE CEDRO RIVER EXAMPLE

Textbooks in environmental economics sometimes treat a situation that is more general than the Cedro River example by allowing both the level of output and the level of residuals reduction of the upstream firm to be variable. The discussion below, which is adapted from Nijkamp (1977), uses calculus to present this more general case.

Consider a firm producing a single homogeneous output and discharging its liquid waste to a stream. The following notation is adopted:

Q = level of the firm's output

W = price per unit of output

$U(Q)$ = total amount of residuals (that is, the raw load) generated in producing Q, assuming no government intervention to influence the firm's discharge

R = percentage reduction of raw load that results from end-of-pipe treatment, in-plant process changes, and so forth

$C(Q,R)$ = total cost of producing Q and reducing $R\%$ of the raw load

$D(Q,R)$ = total damage to downstream water users as a result of producing Q and reducing raw load by $R\%$

Assume the firm operates in perfectly competitive markets, and thus the firm's behavior has no influence on the prices of its inputs and outputs.

The following aspects of the shapes of the various functions are given:

$\frac{dU}{dQ} > 0$ Total residuals load generated (before residuals reduction) increases with increasing output

$\frac{\partial C}{\partial Q} > 0$ Cost increases with increasing output

$\frac{\partial C}{\partial R} > 0$ Cost increases with increasing percentage of residuals reduction

$\frac{\partial D}{\partial Q} > 0$ Damage to downstream users increases with increasing output (for a given percentage of residuals reduction) because $dU/dQ > 0$

$\frac{\partial D}{\partial R} < 0$ Damage to downstream users decreases with increasing percentage of residuals reduction (for a given level of output)

PART 1: ECONOMIC EFFICIENCY PERSPECTIVE

Economic efficiency dictates that the firm should set its output and residuals reduction levels to maximize the difference between social benefits and costs. The firm's revenues provide a measure of benefit since they reflect people's willingness to pay for the firm's outputs. Social costs are the sum of the firm's own costs and the costs it imposes on downstream water users. Adopting this perspective, Q and R should be chosen to maximize $S(Q,R)$, where

$$S(Q, R) = WQ - C(Q, R) - D(Q, R)$$

The first term on the right-hand side represents the firm's revenues. The remaining terms are costs to the firm and to downstream water users, respectively.

Calculus can be employed to maximize $S(Q, R)$, as long as the function possesses first and second partial derivatives and is otherwise mathematically "well behaved." The procedure requires that the first partial derivative of $S(Q, R)$ with respect to each variable be set equal to zero. The resulting equations are then solved simultaneously. As elaborated by Gue and Thomas (1968), well-established conditions on various combinations of second derivatives can be used to test that the resulting solution is a maximum (and not a minimum or a saddle point). Applying this procedure requires finding values of Q and R that satisfy

$$\frac{\partial S}{\partial Q} = 0 = W - \frac{\partial C}{\partial Q} - \frac{\partial D}{\partial Q}$$

or

$$W = \frac{\partial C}{\partial Q} + \frac{\partial D}{\partial Q} \tag{3-7}$$

and

$$\frac{\partial S}{\partial R} = 0 = -\frac{\partial C}{\partial R} - \frac{\partial D}{\partial R}$$

or

$$\frac{\partial C}{\partial R} = -\frac{\partial D}{\partial R} \tag{3-8}$$

Equation 3-7 indicates that the unit price of the firm's output should equal the sum of the incremental (or "marginal") cost to the firm ($\partial C/\partial Q$) and the incremental damage to downstream users ($\partial D/\partial Q$) that occurs because of a unit increase in output. Equation 3-8 requires that the incremental cost of residuals reduction equal the incremental benefits of residuals reduction. For this interpretation to follow, the benefits of treatment at level R are defined as the damage *costs avoided.* With $R = 0$ the damage is $D(Q, 0)$. Thus, the damages avoided are the benefits of reduction level R

$$D(Q, 0) - D(Q, R)$$

Therefore, the incremental benefits of reduction level R are

$$\frac{\partial}{\partial R}\{D(Q, 0) - D(Q, R)\} = -\frac{\partial D(Q, R)}{\partial R}$$

Let the values of Q and R that satisfy equations 3-7 and 3-8 be referred to as $Q^*_{(1)}$ and $R^*_{(1)}$.

PART 2: THE FIRM'S PERSPECTIVE WITHOUT GOVERNMENT INTERVENTION

If there is no government intervention, assume that the firm's objective is to pick levels of Q and R to maximize profits, π, given by

$$\pi = WQ - C(Q, R) \qquad \textbf{(3-9)}$$

The following conditions must be satisfied by Q and R:

$$\frac{\partial \pi}{\partial Q} = 0 = W - \frac{\partial C}{\partial Q} \qquad \textbf{(3-10)}$$

and

$$\frac{\partial \pi}{\partial R} = 0 = -\frac{\partial C}{\partial R} \qquad \textbf{(3-11)}$$

Equation 3-10 is the decision rule for a perfectly competitive firm given in textbooks in economics, namely, "price equals marginal cost." Equation 3-11 is, strictly speaking, outside the domain of the function $C(Q, R)$, since it has been postulated that $\partial C/\partial R > 0$. However, it can be argued that a profit maximizing firm must choose $R = 0$, since having $R > 0$ results in a cost and does not increase revenues no matter what value of Q is selected.

PART 3: THE FIRM'S PERSPECTIVE WITH GOVERNMENT INTERVENTION

A key question concerns how the government can intercede to make the firm set its output and residuals reduction levels at the values of Q and R that were optimal in Part 1. One way to do this is to determine the values of $Q^*_{(1)}$ and $R^*_{(1)}$, and then have the government forbid the firm from operating unless it chooses these levels of output and residuals reduction. This approach recognizes the linkage between output and residuals discharge. However, it is unrealistic to consider using direct government controls on output levels in a capitalist economy.

An alternative intervention is to force the firm to pay an effluent charge, a monetary penalty that varies with the firm's discharge. Let $h(Q, R)$ represent this penalty. To determine the form of $h(Q, R)$, examine π_{new}, the firm's profits after the penalty is imposed:

$$\pi_{new} = WQ - C(Q, R) - h(Q, R) \tag{3-12}$$

By comparing π_{new} and $S(Q, R)$ in Part 1, it follows that the penalty function that the firm should face is precisely equal to the damage to downstream users. If this penalty is imposed, a firm maximizing π_{new} will set its output and residuals reduction at an economically efficient level.

KEY CONCEPTS AND TERMS

ACTIONS TO MODIFY RESIDUALS
- End-of-pipe treatment
- Changing type or quantity of outputs
- Changing inputs and production processes
- Increased materials recovery and reuse
- Changing timing and location of discharges
- Augmenting assimilative capacity

BENEFITS AND COSTS OF RESIDUALS REDUCTION
- Economic efficiency
- Mass balance analysis
- Linking discharges and ambient quality
- Costs of residuals reduction
- Benefits as damage costs avoided
- Minimizing residuals reduction plus damage costs
- Maximizing net benefits
- Limitations of benefit–cost approach

NEED FOR GOVERNMENT INTERVENTION IN RESIDUALS MANAGEMENT
- Voluntary residuals reductions
- Private agreements to reduce residuals
- Transaction costs
- Legal rights to discharge

FORMS OF GOVERNMENT INTERVENTION
- Direct regulation using standards
- Effluent versus ambient standards
- Technology-based standards
- Cash subsidies
- Effluent charges
- Marketable pollution permits

DISCUSSION QUESTIONS

3–1 Chapter 3 provides several illustrations of actions to control residuals that do not involve end-of-pipe treatment. Pick three types of actions and provide examples of each that are different from those in the chapter.

3–2 How does "engineering efficiency" differ from "economic efficiency"? Which type of efficiency would you expect to play a more dominant role in the decision making of firms?

3–3 Consider a market economy of the type that prevails in the United States. What arguments can be used to justify government interventions to influence the way individuals and firms make decisions regarding the generation and disposal of residuals? In framing your response, consider the differences between social costs and private costs.

3–4 Perform a "sensitivity analysis" for the Cedro River example by determining the net benefit maximizing levels of chloride reduction if the cost functions for Margarita Salt were as follows

(i) $C_M(R) = 100\ R^2$
(ii) $C_M(R) = 100\ R^3$

Suppose CRACQ sets effluent standards to attain economic efficiency. Describe the effects, in terms of differences in chloride reduction costs, if CRACQ imposed an effluent standard assuming that Margarita Salt's costs were those in (i) above, when in reality, its costs were those in (ii).

3–5 Consider a stream receiving the wastewater discharge from a cannery. A heavily used park lies downstream of the discharge. Many individuals would be willing to pay a significant amount to rid the park of odors caused by the cannery's waste. Assume the total amount is enough to pay for residuals modifications to minimize the odors. Discuss the barriers to having the odor reduction measures paid for by private cash transfers from the park users to the cannery.

REFERENCES

Ackerman, B. A., S. R. Ackerman, J. W. Sawyer, Jr., and D. W. Henderson, 1974, *The Uncertain Search for Environmental Quality.* Free Press, New York.

Bartos, M. J., Jr., 1979, EPA Goes to BAT against Toxic Industrial Wastewater. *Civil Engineering* **51** (9), 87–89.

Baumol, W. J., and W. E. Oates, 1979, *Economics, Environmental Policy and the Quality of Life.* Prentice–Hall, Englewood Cliffs. N.J.

Coase, R. H., 1960, The Problem of Social Cost. *Journal of Law and Economics* **3,** 1–44.

Dorfman, R., and N. S. Dorfman (eds.), 1977, *Economics of the Environment,* 2nd ed. Norton, New York.

Gue, R. L., and M. E. Thomas, 1968, *Mathematical Methods in Operations Research.* Macmillan, London.

Kneese, A. V., and B. T. Bower, 1979, *Environmental Quality and Residuals Management.* Johns Hopkins University Press for Resources for the Future, Baltimore, Md.

Layard, P. R. G., and A. A. Walters, 1978, *Microeconomic Theory.* McGraw–Hill, New York.

Masters, G. M., 1974, *Introduction to Environmental Science and Technology.* Wiley, New York.

Nijkamp, P., 1977, *Theory and Application of Environmental Economics.* Elsevier/North Holland, New York.

Stockholm, N., 1981, The Real World of Environmental Decision Making, *in* D. L. Brunner, W. Miller, and N. Stockholm (eds.), *Corporations and the Environment: How Should Decisions Be Made?* Committee on Corporate Responsibility, Stanford University, Stanford, Calif.

Thackston, E. L., and R. E. Speece, 1966, Review of Supplemental Reaeration of Flowing Streams. *Journal of the Water Pollution Control Federation* **38** (10), 1614–1622.

CHAPTER 4

STANDARDS, CHARGES, AND MARKETABLE PERMITS

No single form of government intervention to manage residuals is superior. This chapter considers several forms of intervention and points out the strengths and weaknesses of each. Traditional approaches based on standards are emphasized since they are used most widely. Two less commonly employed methods, effluent charges and marketable permits, are also assessed.

In designing residuals management programs, choices must be made about permissible discharges from each source. These choices lead to different distributions of costs among dischargers. This chapter examines alternative bases for determining appropriate discharge levels. The decision criteria most often considered are economic efficiency and equity. An economically efficient set of discharges is one that either minimizes the total social costs of the releases or maximizes net economic benefits. An equitable or fair decision is much more difficult to characterize.

To introduce equity considerations into residuals management decision making requires an answer to the question, What is a fair distribution of residuals reduction costs? A basic principle for determining what is fair is that "equals should be treated equally and unequals, unequally."[1] However, this concept is difficult to apply because the terms *equals* and *equal treatment* are so general. In the context of residuals management, Kneese and Bower (1968) propose a simple rule based on fairness: waste dischargers should bear the costs of the damages they impose. This idea is also difficult to apply, since it is often impossible to estimate damages in monetary terms. Other conceptions of an equitable distribution of residuals reduction costs consider the ability of dischargers

[1]This principle is elaborated in Velasquez's (1982) introduction to concepts of equity applied to decision making.

to pay. Is it fair to impose waste reduction requirements that are so costly as to cause a firm to close down and put people out of work? Questions like this do not have clear-cut answers.

Frequently, the design of a residual management program involves making compromises between achieving economic efficiency and providing an equitable distribution of costs. The examples below illustrate these compromises and highlight the difficulties in applying various principles of equity.

EFFLUENT STANDARDS BASED ON AMBIENT STANDARDS

Discharge standards are sometimes set to force cutbacks in residuals so that an ambient standard can be met. This approach is illustrated by a case involving effluent standards for water-borne residuals. The concepts apply equally well to *emission standards,* a term used for direct controls on air-borne residuals and noise.

The Three Waste Source Example

Consider a hypothetical stream that receives waste discharges ("effluents") from three sources. Assume that each discharge can be described adequately using "biochemical oxygen demand" (BOD), an aggregate measure of organic matter. A high concentration of BOD in wastewater indicates that the bacterial decomposition ("biodegradation") of organic matter will use up a large amount of the oxygen dissolved in the wastewater. After the waste is discharged, the oxygen required during biodegradation comes from the stream. Thus, if the BOD of the effluent increases, the dissolved oxygen (DO) concentration immediately downstream of the point of discharge decreases. The concentration of dissolved oxygen in a stream is important since if it drops below about 6 mg/l, some types of fish will not survive. If the DO drops all the way to zero, offensive odors will result from the biodegradation process.

Biochemical oxygen demand measures the strength of waste discharges and DO characterizes stream quality. These indicators are widely used. The analysis of BOD and DO relies on standard procedures for predicting how much a stream's dissolved oxygen will decrease if a BOD discharge is increased. These procedures make it convenient to use BOD and DO in illustrating how effluent standards can be derived to meet ambient standards. In real situations, however, many other aspects of water quality, such as turbidity and presence of disease-causing bacteria, are also considered in setting standards.

Assume that an ambient standard for the stream has been set at 6 mg/l of dissolved oxygen under specified low streamflow conditions.[2] Suppose, also,

[2]When setting water quality standards, it is customary to specify how much streamflow is available to dilute wastewater discharges. Typically, periods of low streamflow are selected. If standards can be met when there are low flows for diluting wastes, they will generally be met under other conditions. Velz (1976) describes ways to define low-flow conditions.

that if no wastes were released into the stream, the dissolved oxygen would be 7 mg/l under these flow conditions. The difference between the DO standard and the DO that would exist in the absence of discharges is 1 mg/l. It represents the portion of the stream's oxygen used to provide waste receptor services.

Figure 4.1 shows the three waste discharges. Sources labeled 1 and 2 are located opposite from each other, and source 3 is further downstream. The figure also includes a point "α" downstream of source 3. The 6 mg/l DO standard applies only at point α, a location valued for its good fishing. In a real case, standards would be set at many stream locations. However, the single point α is sufficient for illustrative purposes. Figure 4.1 also shows the "raw loads" of BOD generated at each source. The loads are in units of 1000 lb/day of BOD. Thus, for example, source 1 has a raw load of "100 units" or 100,000 lb/day of BOD.

There are numerous ways that the three waste sources can reduce their discharges so that only 1 mg/l of stream dissolved oxygen is used up. To analyze the possibilities, let X_1 represent the percentage reduction in the raw BOD load originating at source 1. Define X_2 and X_3 similarly for sources 2 and 3, respectively. As noted in Table 3.1, there are many options for residuals reduction. For simplicity, this example assumes that only end-of-pipe treatment is used.

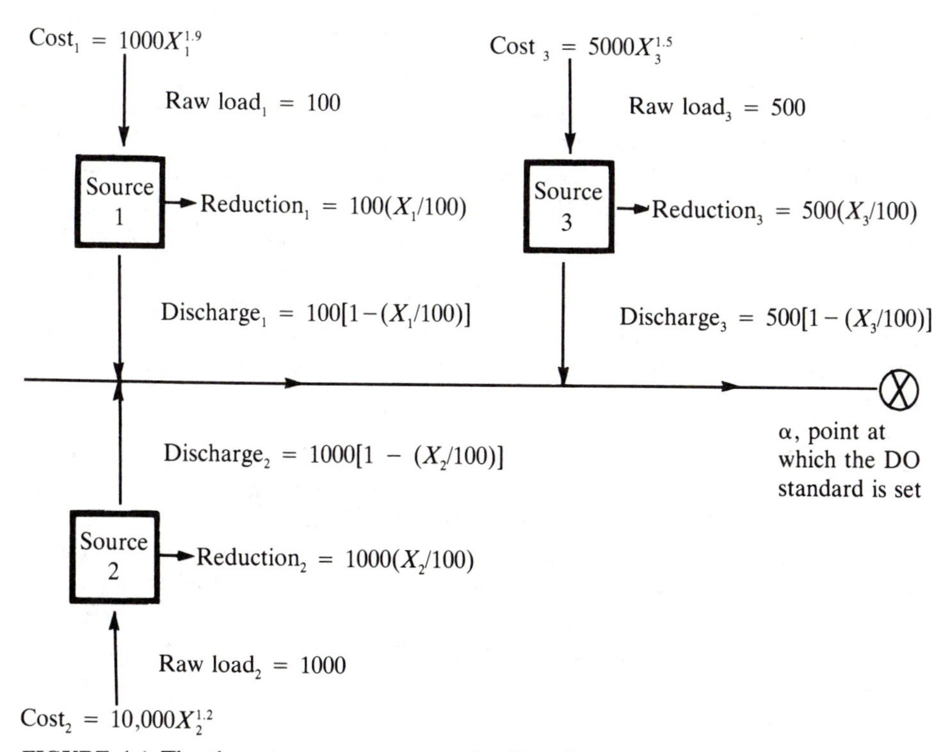

FIGURE 4.1 The three waste source example. *Note:* Loads, reductions, and discharges are in 1000 lb/day of BOD.

Figure 4.1 indicates the BOD reductions and discharges associated with any combination of treatment levels, X_1, X_2, and X_3. It shows, for example, that for source 1, 100 $(X_1/100)$ units are removed by end-of-pipe treatment and $\{100\ [1 - (X_1/100)]\}$ units are discharged to the stream. These results follow because X_1 is the percent of raw load at source 1 removed by treatment. Enumerating percentage removals, X_1, X_2, and X_3, is equivalent to specifying a set of effluent standards. For any X_i, the corresponding BOD discharge is simply the raw load at source i multiplied by $[1 - (X_i/100)]$. The corresponding effluent standard requires that the BOD discharge be no greater than this value.

Costs of Wastewater Treatment Figure 4.1 also indicates the costs of waste treatment. It shows, for example, that the cost to source 1 of removing X_1 percent of its raw BOD load is $1000\,X_1^{1.9}$. All costs in the figure represent the "present value" of costs to construct and operate treatment facilities. The present value idea is used because the costs to operate facilities occur at different future times. The concept is clarified by considering an ordinary savings account. At 6% interest, it would take only \$1.00 today to yield \$1.06 one year from now. In this case, \$1.06 one year from now has a present value of \$1.00.[3]

The total cost of effluent standards requiring removal percentages X_1, X_2, and X_3 is labeled $C_T(X_1, X_2, X_3)$. It is the sum of the costs in Figure 4.1:

$$C_T(X_1, X_2, X_3) = 1000X_1^{1.9} + 10{,}000X_2^{1.2} + 5000X_3^{1.5} \tag{4-1}$$

The individual terms in equation 4-1 depart somewhat from typical treatment costs. In general, real costs increase greatly as the percentage reduction approaches 100. This type of increase occurs to some extent for sources 1 and 3. However, the cost expression for source 2 has an exponent close to 1, and it does not show rapidly increasing costs. The costs in this example were chosen intentionally to dramatize the differences between economically efficient standards and those that appear equitable. Such distinctions occur because source 1 has a small quantity of waste that is very expensive to remove. Source 2, which discharges at the same point on the stream, can remove an equivalent quantity of waste for a much lower cost.

Effects of Discharges on Stream Quality Additional information is required in order to set effluent standards that attain the ambient standard of 6 mg/l DO. It is known that BOD discharges reduce stream DO, but a quantitative description of this effect is needed. The simplest analysis procedures involve "steady-state" conditions in which the stream flow and the waste discharges are all constant. In these circumstances, it is common to assume that a linear relationship exists between waste releases and stream dissolved oxygen at a particular location. For the hypothetical example, ΔDO, the change in DO at point α caused by the BOD discharges, is taken as

[3] When estimating the present value of costs for numerous future time periods, the computations can be tedious. Formulas and tables are used to simplify the calculations; see, for example, Grant, Ireson, and Leavenworth (1982).

$$\Delta DO = \phi_{1\alpha}\left[100\left(1 - \frac{X_1}{100}\right)\right] + \phi_{2\alpha}\left[1000\left(1 - \frac{X_2}{100}\right)\right] + \phi_{3\alpha}\left[500\left(1 - \frac{X_3}{100}\right)\right] \tag{4-2}$$

The expressions in brackets represent the BOD discharges (see Figure 4.1). The value of $\phi_{i\alpha}$, referred to as a "transfer coefficient," estimates the decrease in DO at α per unit increase in BOD discharge at source i. The units of $\phi_{i\alpha}$ are milligrams per liter of DO per 1000 lb/day of BOD. In this example, $\phi_{1\alpha} = \phi_{2\alpha} = 0.002$, and $\phi_{3\alpha} = 0.003$. The values of $\phi_{1\alpha}$ and $\phi_{2\alpha}$ are equal since the effluents from sources 1 and 2 are released at the same stream location. In this example, the value of $\phi_{3\alpha}$ is greater than the other two coefficients because source 3 is closer to point α.[4] The relationship between stream DO and distance from the waste source is such that after a certain distance the stream "regenerates" itself naturally. If no other wastes are added, natural stream reaeration may eventually cause the DO to rise to the concentration that exists upstream of sources 1 and 2.

Before examining alternative effluent standards, consider what the DO at α would be if *no* waste treatment were provided. This is determined using equation 4-2 with appropriate numerical values of $\phi_{i\alpha}$ and with $X_1 = X_2 = X_3 = 0$. Substituting these values in equation 4-2 gives the *change* in DO (in mg/l) as

$$\Delta DO = (0.002)(100) + (0.002)(1000) + (0.003)(500) = 3.7$$

With each source emitting its raw load, the stream dissolved oxygen at point α is 3.7 mg/l *below* the DO value that exists in the absence of any waste loads, 7 mg/l. The difference between the two numbers, 3.3 mg/l, is the resulting DO at point α. This value is significantly lower than the 6 mg/l standard.

If the DO decrease caused by all three effluents is limited to 1 mg/l, the 6 mg/l standard will be met. The DO decrease can be restricted in this way by choosing X_1, X_2, and X_3 such that

$$\phi_{1\alpha}\left[100\left(1 - \frac{X_1}{100}\right)\right] + \phi_{2\alpha}\left[1000\left(1 - \frac{X_2}{100}\right)\right] + \phi_{3\alpha}\left[500\left(1 - \frac{X_3}{100}\right)\right] = 1.0 \tag{4-3}$$

Equation 4-3 sets the ΔDO given by equation 4-2 equal to 1 mg/l. After substituting appropriate values for the transfer coefficients and simplifying, equation 4-3 can be written as

$$0.2X_1 + 2X_2 + 1.5X_3 = 270 \tag{4-4}$$

Any combination of X_1, X_2, and X_3 that satisfies equation 4-4 can be used to

[4]Thomann (1972) presents mass balance analyses for developing the transfer coefficients. The numerical values of $\phi_{i\alpha}$ depend entirely on the particular circumstances. As indicated by Thomann's analysis, it is possible to have a case in which $\phi_{3\alpha} < \phi_{1\alpha}$.

determine effluent standards that meet the 6 mg/l ambient standard. Because there are many such combinations, there are numerous possible sets of effluent standards.

Efficiency and Equity in Setting Effluent Standards

Equal Percent of Waste Removed One of the simplest approaches to setting standards requires that each source remove the same percentage of its raw load. Using this method, the necessary percentage reduction is found by setting $X_1 = X_2 = X_3 = X$ in equation 4-4 and solving for X. The solution is $X = 73\%$. If each source removes 73% of its raw load, the 6 mg/1 standard will be met. The total cost of this equal percentage removal plan is found using equation 4-1,

$$C_T(73, 73, 73) = 1000(73)^{1.9} + 10{,}000(73)^{1.2} + 5000(73)^{1.5} = \$8{,}310{,}000$$

Some people argue that an equal percentage removal policy is fair because each waste source is treated in the same way. This position is easily challenged. Dischargers have unequal costs of treatment and they are situated at different locations. Consequently, equity is not necessarily attained by imposing an equal percentage reduction requirement.

Inequities in the distribution of costs are demonstrated by considering sources 1 and 2. They are at the *same location,* and a unit of waste from each produces an identical effect on dissolved oxygen. In this sense, the sources appear as equals. Consider, however, what it costs to reduce a unit of BOD at each source when a 73% reduction is imposed. Table 4.1 presents the average per unit costs: total treatment costs divided by BOD loads removed. The average costs of reducing a unit of BOD discharge are shown to be much greater for source 1 than for source 2. Is it fair to force them to reduce their raw loads by the same percentage?

Inequities also result because an equal percentage removal policy does not account for differences in the locations of discharges. In this example, the effects of location are represented by the transfer coefficients: $\phi_{1\alpha} = \phi_{2\alpha} = 0.002$ and $\phi_{3\alpha} = 0.003$. A unit of BOD released by source 3 causes 1.5 times as much of

TABLE 4.1 Standards Based on Equal Percentage Removal Policy

Waste Source	Percent of Waste Removed	BOD Load Removed (10^3 lb/day)	Cost of Treatment (\$$10^3$)	Average Cost per Unit of BOD Load Removed (\$/10^3 lb/day)
1	73	73	3470	47.5
2	73	730	1720	2.4
3	73	365	3120	8.5
Total cost			8310	

a decrease in DO at α as a unit released from the other sources. In these circumstances, would it not be more equitable to have source 3 remove a higher fraction of its raw load?

The most common criticism of the equal percentage removal method is that it fails to utilize society's resources efficiently. Many alternative schemes can meet the 6 mg/l standard at much less than the $8.31 million cost of the equal percentage reduction arrangement. For example, a less expensive program can be designed by leaving source 3 at the 73% level and having source 1 treat nothing. To meet the ambient standard, source 2 must remove 73% of its own raw load *plus* 73% of the raw load at source 1. The percentage of waste removal by source 2 increases to

$$\left[\frac{(0.73)(1000) + (0.73)(100)}{1000}\right] \times 100 = 80.3\%$$

Using equation 4-1, the total cost of this scheme is

$$C_T(0, 80.3, 73) = 10{,}000(80.3)^{1.2} + 5000(73)^{1.5} = \$5{,}050{,}000$$

This represents a $3.26 million cost savings over the equal percentage removal policy. As shown below, even greater savings are possible.

Waste Treatment to Minimize Total Costs Effluent standards can also be based on the removal percentages that minimize total cost. Values of X_1, X_2, and X_3 that minimize cost must satisfy equation 4-4, the condition requiring a 6 mg/l dissolved oxygen content at point α. In mathematical terms, finding the cost minimizing effluent standards is equivalent to finding X_1, X_2, and X_3 to minimize equation 4-1 subject to the requirement represented by equation 4-4. This cost minimization problem is solved using standard procedures from calculus.[5] The results are given in Table 4.2.

[5]One solution procedure begins by solving equation 4-4 for X_3 and back substituting into equation 4-1 to obtain

$$\hat{C}_T(X_1, X_2) = 1000\, X_1^{1.9} + 10{,}000\, X_2^{1.2} + 5000(180 - 0.133\, X_1 - 1.33\, X_2)^{1.5}$$

The values of X_1 and X_2 that minimize $\hat{C}_T$ are found by solving $\partial \hat{C}_T/\partial X_1 = 0$ and $\partial \hat{C}_T/\partial X_2 = 0$, simultaneously. This yields $X_2 > 100\%$, which is not physically possible and can be interpreted to mean that source 2 must remove all its wastes if equation 4-1 is to be minimized. Next, the entire optimization problem is reexamined with $X_2 = 100\%$. Equation 4-4 is solved for X_3. Substituting the result into equation 4-1 gives

$$\tilde{C}_T(X_1) = 1000\, X_1^{1.9} + 10{,}000\, (100)^{1.2} + 5000(46.7 - 0.133\, X_1)^{1.5}$$

The value of X_1 that minimizes $\tilde{C}_T$ is determined by solving $d\tilde{C}_T/dX_1 = 0$. The result, $X_1 = 4.12\%$, together with $X_2 = 100\%$, is used in equation 4-4 to obtain $X_3 = 46.12\%$. The condition assuring that $X_1 = 4.12\%$ provides a minimum (and not a maximum) for $\tilde{C}_T(X_1)$ is that

$$\left.\frac{d^2\tilde{C}}{dX_1^2}\right|_{X_1 = 4.12} > 0$$

This condition is satisfied.

TABLE 4.2 Standards That Minimize Total Cost

Waste Source	Percent of Waste Removed	Percent of Waste Discharged	BOD Load Discharged (10^3 lb/day)	Decrease in DO at α (mg/1)	Cost of Treatment (\$$10^3$)
1	4.12	95.88	95.9	0.192	15
2	100.00	0	0	0	2510
3	46.12	53.88	269	0.808	1566
Total cost					4091

The cost minimizing effluent standards provide about a 50% cost reduction compared to standards in which all sources have the same percentage removal (compare Tables 4.1 and 4.2). Because of opportunities for substantial cost savings, it might be supposed that there is enthusiasm for the cost minimizing approach to setting effluent standards. This is *not* the case because costs are often distributed unevenly. For example, the cost minimizing standards require source 2 to remove virtually all of its wastes, whereas source 1 removes less than 5%. Since both discharges are at the same stream location, many people would consider this unfair.

Inequalities in the distribution of costs do not necessarily require the minimum cost approach to be abandoned. If inequities exist, they can potentially be eliminated by a system of taxes and subsidies. Suppose, for example, it is considered unfair to have source 2 remove its entire BOD load. Instead of disregarding the minimum cost policy, the government could tax sources 1 and 3 and subsidize source 2 so that the final cost distribution is equitable. In this way, a minimum of society's resources are devoted to waste treatment and the treatment costs are allocated fairly. Although this position seems reasonable, it has not been widely embraced. One reason is the difficulty in getting agreement on what constitutes a fair distribution of costs. Another is that it is often politically infeasible to implement the necessary taxes and subsidies.

Equal Percentage Removal within Zones Each of the above methods for setting effluent standards has shortcomings. The equal percentage removal approach is economically inefficient, and the minimum cost policy often yields an unfair distribution of waste removal costs. The concept of requiring equal removal percentages for sources in the same "zone" is an attempt to eliminate some of the inequities in the minimum cost scheme while being more economical than the equal percentage removal arrangement.

There are as many ways of designing a program with "equal percentage removal within zones" as there are ways of defining zones. If all sources are put into a single zone, the outcome is the equal percentage removal approach. If each source has its own zone, the minimum cost scheme is obtained. Intermediate policies are derived by defining zones that differ from these two extremes. For illustrative purposes, suppose the three waste sources in the example are placed in two zones: sources 1 and 2 are in one zone, and source 3 is in the other. This zoning recognizes that a unit of BOD from either source 1 or source 2 has the same effect on stream dissolved oxygen. It also recognizes that source 3 should not necessarily remove the same percentage of waste as the others, since a unit of BOD from source 3 causes a greater decrease in DO at point α.

A zoned equal percentage removal program that makes economical use of resources keeps the total cost to a minimum while requiring sources 1 and 2 to reduce their raw loads by the same fraction. Standards based on this approach are determined by finding X_1, X_2, and X_3 to minimize total cost, equation 4-1, while satisfying the conditions that $X_1 = X_2$ and that DO at point α equals 6 mg/l. This latter condition is represented by equation 4-4. The values of X_1, X_2, and X_3 are found using standard procedures from calculus and are shown in Table 4.3.[6]

The characteristics of the example are such that the *zoned* equal percentage removal requirements cost almost as much as the program with all sources removing 73% of their loads (compare Tables 4.1 and 4.3). The high total cost for the zoned scheme results because source 1 has very high average treatment costs and both sources 1 and 2 must make substantial BOD reductions to meet the ambient standard. In other circumstances, the use of zones might be much more economical than the equal percentage removal policy. Imagine a case in which some sources are so far from the point at which an ambient standard is imposed that their discharges have little influence on quality at that point. A

TABLE 4.3 Standards That Minimize Total Cost with $X_1 = X_2$

Waste Source	Percent of Waste Removed	Percent of Waste Discharged	BOD Load Discharged (10^3 lb/day)	Decrease in DO at α (mg/l)	Cost of Treatment ($\$10^3$)
1	61.3	38.7	38.7	0.077	2490
2	61.3	38.7	387.0	0.774	1400
3	89.9	10.1	50.5	0.152	4260
Total cost					8150

[6]The solution procedure solves for X_3 using equation 4-4. This result, together with the constraint that $X_1 = X_2$ is substituted into equation 4-1 to yield

$$\bar{C}(X_1) = 1000\, X_1^{1.9} + 10{,}000\, X_1^{1.2} + 5000(180 - 1.47\, X_1)^{1.5}$$

The cost minimizing value of X_1 is found by solving $d\bar{C}/dX_1 = 0$. This value, $X_1 = 61.3\%$, also satisfies the condition required for X_1 to be a minimum:

$$\left.\frac{d^2\bar{C}}{dX_1^2}\right|_{X_1=61.3} > 0$$

Using $X_1 = X_2 = 61.3\%$ in equation 4-4 yields $X_3 = 89.9\%$.

zoned equal percentage removal scheme would mandate low levels of treatment for such sources. Consequently, the total cost of meeting the ambient standard could be much less than when all sources reduce their raw loads by the same proportion.

The zoned equal percentage removal approach may not eliminate all unevenness in the distribution of waste reduction costs. In the hypothetical example, apparent inequities remain even after putting sources 1 and 2 into the same zone. The average cost per unit of BOD removed for source 1 is many times higher than the corresponding cost for source 2.

Zones need not be delineated on a geographical basis. To provide a more equitable outcome, zones may be defined to include waste dischargers of the same type. Suppose, for example, sources 1 and 3 are municipalities and source 2 is a petroleum refinery. It might be argued that sources 1 and 3 should remove the same fraction of their raw loads because both are municipalities and their treatment costs should thus be similar. This approach to zoning does not eliminate all apparent inequities. Since source 3 has a greater impact on DO at α than source 1, requiring $X_1 = X_3$ might not be considered fair.

The three waste source example demonstrates that no one set of effluent standards will be both economically efficient and provide a fair distribution of costs. Compromises between equity and economic efficiency must be made.

Setting Effluent and Ambient Standards for the Delaware Estuary

Although the example above is hypothetical, it has several features in common with an effort to determine effluent standards on the Delaware Estuary during the late 1960s. This effort was led by the Delaware Estuary Comprehensive Study (DECS), a part of what was then the Federal Water Pollution Control Administration. The discussion of effluent standards is extended by examining selected aspects of the DECS analysis. Unlike the three waste source example, the DECS investigation did not begin with a given ambient standard. In fact, an aim of the DECS analysis was to find the costs and benefits of meeting each of several ambient standards. Costs to meet each ambient standard were determined using different policies for setting effluent standards. The benefits associated with each ambient standard were considered, but most benefits could not be measured in monetary terms. The final selection of an ambient standard reflected the various costs and gains involved, but a rigorous benefit–cost analysis was not used.[7]

As in the three waste source example, the Delaware Estuary Comprehensive Study focused on BOD and DO. Forty-four sources of BOD were considered. Collectively, the sources accounted for about 95% of the biochemical oxygen demand discharged to the estuary. The DECS obtained estimates of how much

[7]Ackerman et al. (1974) discuss the DECS analyses and how they influenced ambient and effluent standards on the Delaware Estuary. Thomann (1972) provides detailed information on the mathematical and engineering aspects of the DECS investigation.

it would cost to remove different percentages of the BOD discharge at each source. The costs were represented using straight-line segments as in Figure 4.2. The DECS was thus able to express the total cost of any BOD reduction program as a linear equation.

In addition to cost information, the DECS also determined transfer coefficients similar to the ones used above. To do this, the 86-mile Delaware Estuary was divided into 30 sections, each 10,000 or 20,000 ft in length. For each section, two equations representing the law of conservation of mass were written, one for dissolved oxygen and one for biochemical oxygen demand. Steady-state conditions were assumed. The resulting 60 equations were solved simultaneously to yield the transfer coefficients. The latter provided estimates of how much a decrease in BOD input to any one section of the estuary would increase the DO in each of the 30 sections. Hundreds of transfer coefficients were needed since the ambient quality standards were set for 30 sections, not just a single point. In addition, the influence of BOD releases on DO both downstream *and* upstream of the discharge points was taken into account. Estuaries are coastal waters in which both tides and fresh water flows are present. The tidal influence causes wastes to be transported upstream as well as downstream of the discharge locations.

The Delaware Estuary Comprehensive Study examined five sets of ambient standards. Each was referred to by the DECS as an "objective set." Table 4.4 displays portions of the objective sets for dissolved oxygen in several estuary sections. Objective set 5 represents the 1964 conditions in the estuary. A complete description of the objective sets includes all 30 estuary sections, and it also includes limits on other variables such as chlorides, turbidity, and pH. Although many water quality indicators were considered by the DECS, dissolved oxygen received the most attention.

Alternative effluent standards that would attain the water quality levels represented by each objective set were determined. The DECS investigated three

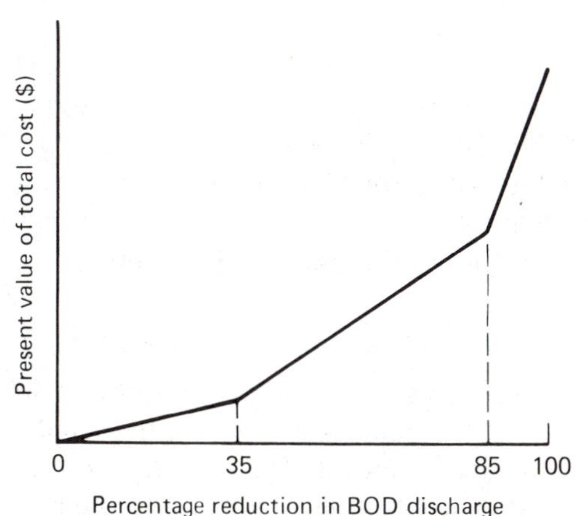

FIGURE 4.2 Costs of BOD reduction as straight-line segments.

TABLE 4.4 Alternative Ambient DO Standards Used in the Delaware Estuary Analysis[a]

	Selected Locations (Section Numbers)				
Objective Set Number	Trenton (1)	Philadelphia (14)	Chester (18)	Wilmington (21)	Liston Point (30)
1	6.5	4.5	5.5	6.5	7.5
2	5.5	4.0	4.0	5.0	6.5
3	5.5	3.0	3.0	3.0	6.5
4	4.0	2.5	2.5	2.5	5.5
5	7.0	1.0	1.0	4.0	7.1

[a]The DO standards, which are for summer average conditions, are given in the table in terms of milligrams per liter. These are taken from a more complete description by Kneese and Bower (1968, p. 227).

strategies for establishing effluent standards: equal percentage removal, minimum cost, and equal percentage removal within zones. Essentially, the analysis used in the three waste source example was conducted for each of the five objective sets.

An effluent limitation for waste source *i* was represented by X_i, the percentage removal of BOD load by that source. For a set of effluent standards to meet the DO requirements of a particular objective set, the values of $X_1, X_2, \ldots, X_{44}$ had to satisfy various "water quality constraints." For any section "β" in the estuary, the water quality constraint took the form

$$\phi_{1\beta} W_1 \left(\frac{X_1}{100}\right) + \phi_{2\beta} W_2 \left(\frac{X_2}{100}\right) + \cdots + \phi_{44\beta}\left(\frac{X_{44}}{100}\right) \geqslant K_\beta \qquad \textbf{(4-5)}$$

where

W_i = total current BOD load discharged by source i

$\phi_{i\beta}$ = increase in DO in section β resulting from a 1-unit decrease in BOD load discharged by source i

K_β = DO increase in section β required to meet the dissolved oxygen standard

The left-hand side of expression 4-5 is analogous to equation 4-2 above. For any source i, W_i multiplied by $(X_i/100)$ is the cutback in the existing BOD discharge associated with an X_i percentage load removal. The discharge reduction, $W_i(X_i/100)$, multiplied by the transfer coefficient $\phi_{i\beta}$, indicates the consequent increase in DO in section β. The left hand side of expression 4-5 aggregates the DO increase in section β caused by each of the load reductions. The inequality assures that the total improvement in dissolved oxygen in section β caused by $X_1, X_2, \ldots, X_{44}$ is at least as great as the improvement required to satisfy the DO requirements of a particular objective set.

The equal percentage removal program for setting effluent standards was determined by trial and error. First a percentage was picked. Then water quality constraints of the type shown in expression 4-5 were checked to see if the DO standards in each of the 30 sections were attained. If the standards were violated in one or more sections, the percentage removal was increased by a small amount, and the water quality constraints were checked again. If the standards were met throughout, then the percentage was decreased by a small amount and the constraints were rechecked. This procedure yielded the single percentage removal that *just met* the dissolved oxygen requirements.

To find the cost minimizing values of $X_1, X_2, \ldots, X_{44}$, it was necessary to sum the straight-line segments representing BOD reduction costs for each source (see Figure 4.2). The resulting total cost equation was a sum of terms that were *linear* with respect to removal percentages. In other words, each percentage removal was raised to the first power only. The water quality constraints consisted of 30 linear inequalities, one for each section; expression 4-5 illustrates a typical constraint. Because all the equations and inequalities consisted of terms that were linear with respect to percentage removals, the cost minimizing values of $X_1, X_2, \ldots, X_{44}$ could be calculated using standard mathematical techniques.[8] The cost minimization problem was easily solved by the DECS using a digital computer.

The third approach to setting effluent standards, the equal percentage removal within zones policy, involved a more sophisticated cost minimization exercise. To find the appropriate values of $X_1, X_2, \ldots, X_{44}$, the cost minimization problem above was expanded to require all of the sources in any particular zone to remove the same percent of their discharges. There are an enormous number of ways to group 44 sources into zones, and there is little theoretical basis for arguing that one grouping makes more sense than another. Several zoning schemes were examined by the DECS. Although the requirements for equal percentage removal within zones increased the mathematical complexity of the cost minimization problem, the DECS staff devised a special procedure to solve it.

A portion of the results from the Delaware Estuary Comprehensive Study is displayed in Table 4.5. Costs shown include both construction costs and operation and maintenance costs and reflect waste loads that were anticipated for the 1975–1980 period. The minimum cost programs are generally much less expensive than the equal percentage removal schemes. The costs of programs involving equal percentage removal within zones depend on how the zones are

[8]In addition to the water quality constraints, it was necessary to restrict each X_i to be between 0 and 100%. The resulting cost minimization problem took the form of a "linear program." Such problems are solvable using well-established procedures. A complete formulation of the linear programming problem to find the minimum cost set of $X_1, X_2, \ldots, X_{44}$ is given by Thomann (1972).

defined. The costs in the last column of Table 4.5 are for the zones finally used by the DECS.

The outcome of the standard setting process for the Delaware Estuary was the adoption of a new set of ambient standards representing a compromise between objective sets 2 and 3. The regulatory approach to implement this water quality standard was designed by the Delaware River Basin Commission (DRBC), an interstate agency with broad powers and responsibilities. Their scheme involved four zones, with each waste source within any one zone removing the same percent of its raw BOD load. These percentages ranged from 86.0 to 89.25.

Aspects of the final outcome illustrate some of the difficulties in implementing effluent standards. Consider the problem of setting limits for the BOD discharges from two petroleum refineries in the same zone. Suppose that at the time of implementing the equal percentage removal within zones policy, one refinery had already implemented costly waste treatment while a neighboring refinery was discharging all of its wastes without treatment. It would be unfair to require each refinery to reduce its existing discharge by the same percent. To deal with this, the DRBC determined "hypothetical raw loads" on a case-by-case basis. The hypothetical loads were used by DRBC to determine numerical limits ("quotas") on the BOD that could be released by each source. Dischargers often argued that the quotas initially set by DRBC were too low. The final quotas resulted after a series of bargaining sessions involving the dischargers and DRBC. Baumol and Oates (1979, p. 236) observe

> *The anomalies in the final set of permits are quite striking: for example, the final pollution quotas for petroleum refineries on the estuary ranged from 692 pounds to 14,400 pounds of BOD per day!*

TABLE 4.5 Total Costs of Achieving Objective Sets for the Delaware Estuary ($\$10^6$)[a]

Objective Set	Equal Percentage Removal	Minimum Cost	Equal Percentage Removal within Zones
1[b]	460	460	460
2	315	215	250
3	155	85	120
4	130	65	80

[a]Based on information in Kneese and Bower (1968, p. 230). All costs are "present values" reported in 1968 dollars.

[b]The cost figures for objective set 1 are the same for each of the three policies. The high level of DO associated with this objective set required very high levels of treatment and the use of mechanical devices to reoxygenate the estuary artificially.

The wide variations in the final quotas reflect the difficulties in developing effluent standards that are equitable, especially where judgments must be made to identify hypothetical raw loads.

Another problem in implementing the effluent requirements on the Delaware Estuary arose over challenges to the validity of the transfer coefficients. The DECS investigation assumed that these coefficients provided a suitable basis for estimating how much dissolved oxygen would improve if particular BOD loads were reduced. Several dischargers argued that the coefficients greatly oversimplified the complex functioning of the estuary. They felt that precise quantitative predictions of how dissolved oxygen would change could not be made with available knowledge of estuarine behavior. These dischargers objected to having costly BOD release quotas imposed on the basis of what they viewed as unreliable and overly simple procedures for predicting changes in water quality.

The dispute over the validity of the transfer coefficients illustrates a general problem encountered in setting effluent standards to meet an ambient standard. This method of deriving effluent standards requires an explicit judgment regarding how much ambient quality would improve if a proposed residuals reduction were implemented. Because relationships between discharges and ambient environmental quality are incompletely understood, these judgments can be disputed. Challenges may involve costly legal actions and long delays in implementing residuals reduction measures.

TECHNOLOGY-BASED EFFLUENT AND EMISSION STANDARDS

There is a way of setting effluent and emission standards that relieves governments from judging how much improvement in environmental quality would result if a *particular* source of residuals is reduced by a specific amount. This approach involves using technology as a basis for determining cutbacks required by particular sources. The resulting "technology-based" standards often take account of the costs to achieve different levels of reduction. Sometimes, however, governments impose requirements calling for the "best" available techniques for reducing discharges. These are usually the most expensive.

Emission Standards for Automobiles

The regulation of emissions from automobiles in the United States in the 1970s illustrates a case in which standards were promulgated without relating discharges to ambient environmental quality, except in a general way. The emission standards for autos were based on congressional judgments of what was technologically feasible and how much capacity the auto industry had for absorbing costs. The approach adopted by Congress assumed that the benefits of high levels of auto emission control were worth the substantial costs. Many people challenged this assumption. Congress' policies were also criticized for leading to frequent confrontations between the auto industry and the regulating agency, the U.S. Environmental Protection Agency (EPA).

To illustrate some of the problems in implementing Congress' policy, consider the emission decreases mandated by the Clean Air Act of 1970. The act called for a 90% reduction in hydrocarbon (HC) and carbon monoxide (CO) emissions (based on 1970 levels) for 1975 model year cars, and a 90% reduction in oxides of nitrogen (NO_x) emissions (based on 1971 levels) for 1976 model year cars. These reductions were "technology forcing"; they went beyond what was technologically feasible at the time. In passing the act, Congress attempted to force auto companies to quicken the pace of their research on emission controls. The 1970 act granted the EPA administrator the ability to allow a 1-year extension for meeting the 1975–1976 standards. The history of how this extension was granted is filled with "eleventh hour" agreements that just barely averted the government's threatened closure of a major auto company's plants, and seemingly endless arguments between government and industry experts regarding what levels of emission control were technically feasible.[9]

The technology-based standards to control auto emissions did not satisfy the goals envisioned by Congress in 1970. Ninety percent reductions in CO, HC, and NO_x, which the act required by 1975–76, were not attained until the early 1980s. Moreover, in regulating only emissions from brand new cars, Congress' policy did not assure that emission standards would be met once cars were in use.

Interestingly, it has been argued that the technology-based auto emission standards provided the U.S. auto industry with some incentives *against* making a major advance in emission control technology.[10] Why would a manufacturer try to decrease emissions further if the new reductions could eventually be used as a basis for an even more stringent standard? Such an outcome is conceivable since the auto emission standards were largely based on technological feasibility, not on practical cost considerations. It has also been said that the auto industry did not initially pursue its emission control research with intensity because there were many avenues available to press for a relaxation or postponement of the standards. The auto industry could try to influence the regulations by negotiating with federal administrators, challenging control requirements in the courts, and lobbying the Congress.

Some have argued that the technology forcing characteristics of the Clean Air Act of 1970 distorted the research activities of U.S. auto manufacturers. In the early 1970s, there were several high-risk research opportunities for reducing auto emissions. The U.S. manufacturers opted to develop the catalytic-converter technology, since it had a high probability of meeting the federal emission requirements. Mills and White (1978) have indicated that other riskier tech-

[9]For an "insider's" view of these events, see Quarles (1976), who was then among the EPA officials acting on the auto industry's application for an extension.

[10]The discussion of the influence of federal auto emission regulations on the research efforts of auto manufacturers relies on Mills and White (1978).

nologies, such as the stratified-charge engine, may have been superior. Foreign auto manufacturers, less dependent on meeting the standards by a particular date, explored alternative technologies. They developed some of the earliest and least costly vehicles capable of meeting applicable standards. Another view, offered by Garwin (1978), is that foreign manufacturers felt less capable than their American counterparts of lobbying against the federal standards. This provided foreign manufacturers with substantial motivation to meet the standards by the 1975–1976 deadlines specified in the Clean Air Act.

Effluent Standards to Control Wastewaters

Prior to the Federal Water Pollution Control Act Amendments (FWPCAA) of 1972, the federal water quality management strategy used effluent standards based on ambient standards set by the states. The 1972 amendments provided a new component in the federal strategy: a program requiring all significant sources to obtain wastewater discharge permits that satisfied technology-based requirements.

To examine these technology-based effluent restrictions, consider the part of the 1972 amendments mandating the "best practicable control technology currently available" (BPT). This technology was defined in "effluent limitations guidelines" issued by EPA. When a discharger applied for a permit in the 1970s, effluent restrictions based on BPT were determined. Before a permit was issued, however, a check was made to assure that the use of BPT would satisfy applicable ambient water quality standards. When it was possible, this check was made using quantitative procedures, such as the equation used to predict dissolved oxygen in the three waste source example. If the BPT requirements were not sufficient to meet the ambient standards, permit conditions had to be made even more stringent.

Table 4.6 shows some of the BPT requirements for municipal wastewater treatment facilities. These limits reflect the judgment of EPA that the BPT (in

TABLE 4.6 Effluent Limitations for Publicly Owned Treatment Works

Water Quality Measure	Quality of Discharge	Efficiency of Treatment Plant (%)
BOD (mg/l)	30[a]	85
Suspended Solids (mg/l)	30[a]	85
pH	6–9	—
Fecal coliform bacteria (per 100 ml)	200	—

Note. Modified from *Environmental Impact Assessment* by L. W. Canter. Copyright © 1977 by McGraw–Hill. Used with the permission of McGraw–Hill.

[a]These are 30-day average values; the limit for any 7-day period is 45 mg/l.

the 1970s) was capable of removing 85% of the BOD and suspended solids entering a municipal treatment plant.

Effluent limitations guidelines defining BPT were issued for many industries. In each case, the guidelines categorized types of firms and production process. As an example, consider the shrimp processing industry. Plants for processing shrimp were divided into categories, such as nonremote Alaskan shrimp, remote Alaskan shrimp, and breaded shrimp in contiguous states. For each category, restrictions were set for selected water-borne residuals. In the case of nonremote Alaskan shrimp processing plants, for example, discharge permits issued in the 1970s reflected the conception of BPT shown in Table 4.7. The full description of BPT requirements accounted for variations in the age and size of plants, manufacturing processes used, products produced, and treatment technology available. The agency issuing a permit could modify the BPT restrictions in the published guidelines, but only after the permit applicant had demonstrated that the guidelines did not adequately reflect its particular plant's characteristics.

Technology-based effluent standards required by the FWPCAA of 1972 were problematic in one important respect. Many of the industries disagreed with EPA's judgments regarding what constituted BPT. In the many legal challenges that ensued, industries argued that EPA had an incorrect view regarding which control technologies were "currently available." Because there were many court actions, the effective implementation of the permit program based on BPT requirements was delayed for a few years.

Although the work load in issuing permits based on BPT was substantial, the technology-based strategy had one major administrative advantage. Permits could be written without necessarily predicting how the ambient environment would respond to the discharge reductions called for by the BPT requirements. If the scientific knowledge to make such predictions was unavailable, the entire permit could be written using only the BPT guidelines.

A major weakness in using technology-based effluent restrictions is that economic efficiency is not attained. This occurs because technology-based standards do not account for differences in discharge *location* or in *costs* of residuals reduction. The effluent requirements apply uniformly to all facilities of a given type. For example, the BPT limitations in Table 4.6 require small cities with

TABLE 4.7 Effluent Guidelines Defining BPT for Nonremote Alaskan Shrimp Processing Plants[a]

Water Quality Measure	Quality of Discharge
Total suspended solids	320 lb/1000 lb
Oil and grease	51 lb/1000 lb
pH	6–9

[a]Based on information in *Environmental Impact Assessment* by L. W. Canter. Copyright © 1977 by McGraw–Hill, New York.

high average treatment costs to remove the same 85% of their raw BOD and suspended solids as large municipalities with low average treatment costs. There are substantial opportunities for reducing the *total* cost to meet a particular ambient standard by tailoring effluent requirements to reflect both discharge location and treatment costs. These opportunities are demonstrated in Table 4.5 for the Delaware Estuary.

EFFLUENT CHARGES TO ATTAIN AMBIENT STANDARDS

Many economists have advocated effluent charges as a more effective means of managing residuals than effluent standards.[11] They argue that residuals are discharged without regard to their negative impacts because the price charged for using the environment's waste receptor services is zero. The excessive use of these services can be curtailed by collecting a charge (or "fee") for each unit of waste emitted. The charge can be set to meet any particular environmental quality goal. In contrast to standards, effluent charges give individual waste dischargers the freedom to decide which specific residuals reductions make the most sense. In addition, charges provide an incentive for research on new methods to reduce residuals. Dischargers can decrease their total charge if they more effectively reduce their waste loads.

The most theoretically interesting use of effluent charges involves neither ambient standards nor effluent standards. In this case, each waste source is required to pay a charge based on the social cost (or "damage") caused by a unit of its discharge. As demonstrated by Kneese and Bower (1968), a source facing such charges will reduce its raw load to the point that yields a maximum difference between the total benefits and costs of residuals reduction. To illustrate this, consider the Cedro River example of Chapter 3. If the Margarita Salt Company had to pay effluent charges based on the damages it caused to the Long Shot Brewery, the theory demonstrated by Kneese and Bower indicates that chloride discharges would be reduced by 30%. Margarita Salt Company would make this reduction to minimize its total cost, which includes the chloride reduction cost plus the effluent charge. Because of the impracticality of measuring the damages caused by residuals, this approach to setting charges is primarily of theoretical interest. It is not considered further. The use of charges together with ambient standards has much greater potential for application.

Setting Effluent Charges to Meet an Ambient Standard

The three waste source example demonstrates how effluent charges can be used to attain an ambient standard. The example, summarized here for convenience, finds the BOD removal percentages at each of three sources needed to meet a

[11]The term *effluent charge* is used here because the example below concerns water-borne residuals. The concepts introduced apply equally well to the management of air-borne residuals. In that context, the term *emission charge* is used.

6 mg/l DO standard at a point α. With this standard, 1 mg/l of dissolved oxygen in the stream provides waste receptor services. The removal percentages that meet the ambient standard at the lowest total treatment cost are $X_1 = 4.12\%$, $X_2 = 100\%$, and $X_3 = 46.12\%$, where X_i represents the reduction in BOD discharge at waste source i. As indicated in Table 4.2, this scheme costs \$4.091 million. Alternative programs are shown in Tables 4.1 and 4.3. Using effluent standards, a particular percentage reduction is implemented by direct regulation. If a discharger does not comply with waste reduction requirements, then various penalties involving judicial and administrative procedures are imposed.

As an alternative to effluent standards, a government agency could meet the 6 mg/l DO target by charging each source a single, constant price per unit of BOD released. To see how an appropriate price could be set, assume the agency knows the cost of BOD reduction for each source, and it knows how each discharge affects the DO at point α. With this information, the agency can use trial and error to calculate the effluent charge, \$$p$ per 1000 lb of BOD discharged, that leads to attainment of the 6 mg/l DO standard.

To find the value of p, the agency assumes each discharger will decrease its raw load so that the sum of BOD reduction costs and effluent charges is at a minimum. In formulating an equation to represent these costs and charges, it is convenient to use the "capital recovery factor" (C), defined as

$$C = \frac{r(1 + r)^n}{(1 + r)^n - 1} \tag{4-6}$$

where r is the interest rate and n is the number of years over which annual costs are considered in the discharger's analysis. This factor converts a cost of \$$K$ incurred at the present time to an equivalent uniform series of annual \$$CK$, incurred over n years.[12] The definition of C is such that the present value of these n years of costs is equal to \$$K$. Suppose source 1 uses $r = 0.10$ and $n = 30$ years in its cost analyses. Substituting these values into equation 4-6 gives $C = 0.106$. The present value of BOD reduction cost for source 1 is \$1000 $X_1^{1.9}$. This is equivalent to a 30-year series of annual treatment costs equal to:[13]

$$0.106(1000\,X_1^{1.9}) \tag{4-7}$$

The charge p that leads to attainment of the 6 mg/l dissolved oxygen standard is found using an equation for the treatment costs and effluent charges faced by each discharger. Once more, consider source 1. The annual cost of BOD re-

[12]For an explanation of the capital recovery factor, see Grant, Ireson, and Leavenworth (1982, pp. 36–37). In computing C, r is the percent interest divided by 100.

[13]A more sophisticated analysis of a firm's cash outlays for treatment would include opportunities for tax writeoffs. Johnson's (1967) study of effluent charges on the Delaware Estuary shows how the computations here must be adjusted to account for tax considerations.

duction is indicated by expression 4-7. The annual cost of the effluent charge is

$$365\,p\left[100\left(1-\frac{X_1}{100}\right)\right] \tag{4-8}$$

The term in brackets represents the daily release by source 1 in 1000 lb/day of BOD (see Figure 4.1). For a single day, the effluent charge is equal to the daily BOD discharge times p ($/1000 lb of BOD). Multiplying this product by 365 days/year gives the annual charge in expression 4-8.

During any 1 year, source 1 faces a total cost, $h_1(X_1)$, given by

$$h_1(X_1) = 106\,X_1^{1.9} + 36{,}500\,p\left(1-\frac{X_1}{100}\right) \tag{4-9}$$

Equation 4-9 is a simplified form of the sum of expressions 4-7 and 4-8, representing treatment cost and effluent charge, respectively. Using calculus,[14] the value of X_1 that minimizes the total annual cost is found to be

$$X_1 = (1.81\,p)^{1.11} \tag{4-10}$$

Repeating the cost minimization computations for sources 2 and 3, the agency determines the percentages of BOD reduction for these sources as

$$X_2 = (2.86\,p)^5 \tag{4-11}$$

and

$$X_3 = (2.29\,p)^2 \tag{4-12}$$

To find the charge needed to meet the 6 mg/l DO standard, a trial value of p is selected and the corresponding values of X_1, X_2, and X_3 are computed using equations 4-10 to 4-12. A check is then made to see if these removal percentages meet the 6 mg/l DO standard. Recall from the previous discussion of the three waste source example that the DO decrease at point α is computed using equation 4-2. It is repeated here for convenience.

$$\Delta DO = \phi_{1\alpha}\left[100\left(1-\frac{X_1}{100}\right)\right] + \phi_{2\alpha}\left[1000\left(1-\frac{X_2}{100}\right)\right] + \phi_{3\alpha}\left[500\left(1-\frac{X_3}{100}\right)\right]$$

[14]The value of X_1 that minimize $h_1(X_1)$ satisfies

$$\frac{dh_1(X_1)}{dX_1} = 1.9(106)\,X_1^{0.9} - 365\,p = 0$$

Solving for X_1 yields the value in equation 4-10. The second derivative of $h_1(X_1)$ is positive for positive values of X_1, and this assures that X_1 found using equation 4-10 is a minimum.

If the DO decrease is greater (less) than 1 mg/l, p is increased (decreased) and the entire calculation is repeated. For any waste source i, if the value of p results in X_i greater than 100%, X_i is simply set at 100%. The procedure is completed when a p is found that gives $\Delta DO = 1$ mg/l.

Several steps in the trial-and-error process are shown in Table 4.8. The first trial value, $p =$ \$1/1000 lb, is too low. It leads to a DO decrease of 1.62 mg/l, which exceeds the 1.0 mg/l limit. The effluent charge is increased in increments of 1.0 until it yields X_1, X_2, and X_3 that provide more than enough BOD cutback. This occurs at $p =$ \$3/1000 lb, a charge that produces a DO decrease of 0.98 mg/l. The charge that attains the 1 mg/l limit exactly is $p =$ \$2.96/1000 lb.

Now consider whether the final value of p yields removal percentages that meet the 6 mg/l DO standard at a minimum total cost. This is examined using equation 4-1 to compute the treatment costs that result when $p =$ \$2.96/1000 lb. These costs, as well as those for the minimum cost solution in Table 4.2, are given in Table 4.9. The total cost using the charge program is greater than the overall minimum cost. This occurs because a single, uniform effluent charge cannot account for differences in the locations of the sources. Effects of source location are reflected by relationships among the transfer coefficients: $\phi_{3\alpha} = 1.5\ \phi_{1\alpha} = 1.5\ \phi_{2\alpha}$.

A program with nonuniform charges can yield the cost minimizing values of X_1, X_2, and X_3 by accounting for the way different sources influence DO. This involves setting the charge for source 3 at 1.5 times the charge for sources 1 and 2. When this is done, and the computations for finding the charges are repeated, the removal percentages meeting the DO standard are precisely the cost minimizing values. Some of the trial-and-error calculations are shown in Table 4.10.

Practical Problems in Determining Effluent Charges

The above example demonstrates how, in theory, effluent charges can be set to attain an ambient standard. It also shows that a system of nonuniform charges can be designed to meet an ambient standard at minimum cost.

In a real situation, the correct charges could not be determined using the above procedure. One reason is that a government agency would have only a rough idea of the costs of residuals reduction at each source. In the best of

TABLE 4.8 Using a Single Effluent Charge to Attain a DO Standard

Trial Value of p (\$/1000 lb)	$X_1 = (1.81\ p)^{1.11}$ (%)	$X_2 = (2.87\ p)^5$ (%)	$X_3 = (2.29\ p)^2$ (%)	Decrease in DO at α (mg/l)
1.0	1.93	100	5.2	1.62
2.0	4.17	100	21.0	1.38
3.0	6.54	100	47.2	0.98
2.96	6.43	100	45.8	1.00

circumstances, the agency would have an even more approximate estimate of how changes in discharges influence ambient environmental quality. For many water quality indicators, information similar to the transfer coefficient, $\phi_{i\alpha}$, would not be available.

Another practical complication concerns the *assumption* that each discharger sets its percentage removal so that the waste reduction cost plus the effluent charge is at a minimum. This assumption is the basis for the trial-and-error process used in computing charges. Other factors, such as the dischargers' inability to raise money for facility construction, may influence the selection of a percentage removal. In such cases, the decisions made by the waste source will not be predicted correctly using the above theory.

Because of the above-noted difficulties, only luck would lead a real agency to set effluent charges so that ambient standards were met precisely. If the charge were set too low, the standard would be violated. If, in the above example, a single uniform charge of \$1/1000 lb were used (instead of the required \$2.96/1000 lb), then $\Delta DO = 1.62$ mg/l instead of the maximum 1.0 mg/1 decrease associated with meeting the DO standard (see Table 4.8). If the charge were set too high, the stream DO would be greater than required. More resources would be devoted to BOD reduction then are called for by the standard.

TABLE 4.9 Treatment Costs Using Effluent Charges and Minimum Cost Approach

	Removal Percentages That Minimize Cost		Removal Percentages Using p = \$2.96/1000 lb	
Waste Source	Percentage of Waste Removed	Cost (\$$10^3$)	Percentage of Waste Removed	Cost (\$$10^3$)
1	4.12	15	6.43	34
2	100	2510	100	2510
3	46.12	1566	45.82	1551
Total cost		4091		4095

TABLE 4.10 Nonuniform Effluent Charges to Attain Ambient Standard at Minimum Cost

Trial Value of p_1 (and p_2) (\$/1000 lb)	$p_3 = 1.5p_1$ (\$/1000 lb)	$X_1 = (1.81\ p_1)^{1.11}$ (%)	$X_2 = (2.87\ p_2)^5$ (%)	$X_3 = (2.29\ p_3)^2$ (%)	Decrease in DO at α (mg/l)
1.0	1.5	1.93	100	11.8	1.52
2.0	3.0	4.17	100	47.2	0.98
1.977	2.966	4.12	100	46.13	1.00

Some economists believe that problems in setting the theoretically correct effluent charges can be overcome using an iterative scheme. For instance, suppose the agency's cost estimates are in error. The agency sets the charge at \$1/1000 lb, thinking that this would be just enough to meet the 6 mg/l DO standard. After the waste sources respond to the effluent charges, the agency observes that the DO decrease at α is 1.62 mg/l instead of the expected 1.0 mg/l. In these circumstances, the agency could revise the charge upward, wait for the dischargers to respond to the increase, and then check to see if the 6 mg/l standard is met. Continuing with this iterative approach, the charge would eventually increase to a level which just meets the standard.

An iterative process for setting charges may be difficult to implement because it confounds the decision making of firms and municipalities. They may respond to charges by making major investments in facilities and equipment. Decisions based on one set of charges may be very difficult to modify in response to new, higher charges. Because long-term resource commitments are involved, investments made in reaction to the initial, lower charges might have to be built into subsequent decisions involving facilities and equipment. Dischargers cannot be expected to react favorably to a regulatory program using effluent charges that change frequently.

Comparison of Effluent Standards with Charges

Effluent charges exhibit both advantages and disadvantages when compared to effluent standards. Table 4.11 summarizes points frequently raised in considering them as alternative means of attaining an ambient standard. The first two items show a clear advantage for effluent standards. One favorable point is that people easily understand what effluent standards are and how they work. Effluent standards have been used for decades. They are a well-established part of the environmental management programs of many countries. In contrast, effluent charges are unfamiliar to most people, and it takes some effort to figure out how and why they work.

A second advantage of effluent standards is that, in principle at least, there is no uncertainty regarding the magnitude of *expected* discharges. If the residuals released are not within the limits established by the standards, governments can usually take administrative or judicial actions to encourage compliance. In contrast, there can be great uncertainty regarding how much waste will be released when effluent charges are used. Once charges are set, a government agency must wait and see how the waste sources respond. Because of the agency's uncertainties regarding how residuals are transported in the environment and what it costs different sources to control residuals, the initial charges might be set so low that ambient standards are violated.

The next three items in Table 4.11 show characteristics that make charges attractive. Unlike standards, charges do not provide incentives to delay implementing residuals reduction measures. As noted in the previous discussion of

TABLE 4.11 Standards versus Charges: A Comparative Analysis

	Effluent Standards	Effluent Charges
Comprehension	Traditionally used and easily understood	Unfamiliar and hard to understand
Uncertainty in discharges	Residuals discharges known (in principle)	Residuals discharges uncertain; ambient standard may be violated
Delayed compliance	Financial incentives to delay compliance	Eliminates advantages of noncompliance
Incentives for research and development	Distorts R&D programs; new technology may become new standard	Positive incentive to invent cost-cutting technologies
Freedom in decision making	Restricts freedom, especially if required control technology is specified	Individuals free to decide on levels and methods of residuals reduction
Legal rights to pollute	Grants rights, but they seem unobjectionable	Grants rights to pollute if you pay, which are objectionable to some
Cost burden on dischargers	Dischargers pay only for residuals reductions	Dischargers pay for residuals reductions plus charges
Revenue generation	No new revenues to government	Provides revenues for possible use in improving environmental quality or compensating damaged parties

federal auto emission requirements, firms can exert considerable pressure to have standards relaxed or compliance deadlines extended. Dischargers have economic incentives to delay meeting the standards. Effluent charges eliminate these incentives. If a firm or municipality fails to reduce its residuals, it simply pays higher charges.

Another attribute of effluent charges is that they can motivate firms to increase research and development (R&D) on residuals reduction techniques. This advantage is tied closely to another item in Table 4.11: charge programs give individuals freedom to decide on levels and methods of residuals reduction. Under an effluent charge plan, firms have strong incentives to engage in research on new methods for residuals reduction. As firms discover and implement more efficient techniques for decreasing residuals, they lower their total residuals management costs. By comparison, effluent standards constrain the decision-making ability of firms and can distort the types of research a firm pursues. This is especially apparent when technology-based effluent standards are used, since a firm's discovery of new residuals reduction methods could be used to design even more stringent effluent requirements.

For the remaining items in Table 4.11, it is not clear whether the advantage is for charges or for standards. Consider the legal rights issue. Some people object to effluent charges because they grant a legal right to "pollute if you pay." This argument is not convincing, since effluent standards also impart a right to discharge residuals.

The final points of comparison concern financial matters, such as the cash outlays required using charges and standards. For any particular percentage reduction, sources incur greater costs with charges than with standards. The actual cost of residuals reduction is the same using either approach. However, with a charge scheme, costs include residuals reduction costs *plus* effluent charges. Thus it is not surprising that industries often oppose the use of charges.

Another financial issue concerns the fairness of imposing effluent charges. If it is assumed that no one has a right to pollute, then it seems just to impose costs on discharges that decrease environmental quality. Since the environment's waste receptor services are being used, why shouldn't payments be required? Matters concerning equity are rarely this simple. Would the added cost be fair if the charges were passed along to consumers in the form of higher prices? What if the price increases were for products which demanded a notable fraction of the income of low-income groups? This last point makes questions of fairness more difficult to answer.

Fairness is also an issue in deciding what should be done with the government revenues received as effluent charges. The revenues could be used to compensate people damaged by the discharges. Alternatively, the government might use the money to construct treatment facilities and thereby diminish the negative effects caused by residuals. Still other uses of the collected funds can be imagined. There are, however, two important considerations. One is that the revenues should not be redistributed in a way that offsets the incentives for residuals reduction provided by the effluent charges. Another factor is that the amount of revenue generated through a charge system may change frequently. As new residuals reduction techniques are discovered and implemented, both the discharges *and* the revenues will decrease. This could lead to disruptions in activities supported by these revenues.

CHARGES AS A SUPPLEMENT TO STANDARDS

Despite their advantages, effluent charges are *not* widely used as an *alternative* to effluent standards. However, charges are more than academic curiosities. Programs that use charges and standards in tandem have received considerable attention. The discussion below introduces two examples in which effluent charges increased the effectiveness of programs that originally relied only on standards.

Charges to Assure Rapid Compliance with Standards

In the "Connecticut Plan," emission charges force firms to comply rapidly with emission standards. (The term *emission* is used instead of "effluent" since air-borne residuals are involved in this case.) Before the plan was implemented in the 1970s, the strategy for controlling air-borne residuals in Connecticut relied on emission standards. If a firm failed to meet a standard, the state agency in charge of air quality had several possible responses. For example, it could negotiate changes in the requirements or initiate legal action to curtail the emissions. Long time periods ensued while disagreements were resolved. There were substantial financial advantages to delay in meeting emission requirements. Frequently, a firm could avoid installing required emission controls for a year or more and save the cost of meeting the standard during that period.

Implementation of an emission charges program increased the rate at which firms complied with emission standards. If a firm failed to comply by a specified date, the responsible agency imposed a charge equal to the amount of money the firm was saving by not meeting the standard. The charge was calculated by accounting for: savings on delayed capital investments, avoided operation and maintenance costs, the effects of taxes and inflation, and the revenues the firm could earn by investing the money saved by noncompliance.[15]

The Connecticut Plan substituted an administrative enforcement procedure for the usually slow and cumbersome reliance on legal action. The plan received such wide acclaim that a modified version of it was written into the U.S. Clean Air Act Amendments of 1977. The amendments authorized the EPA to impose a civil "noncompliance penalty" on major stationary sources of air pollution that failed to satisfy applicable restrictions on their air-borne residuals discharges. The basis for calculating the noncompliance penalty was similar to that used in the Connecticut Plan.

Charges as Sources of Revenues

Effluent charges have been used for many years in the water quality management programs of France, West Germany, Hungary, and the Netherlands. As reported by Johnson and Brown (1976), none of these countries set charges according to the theoretical principles economists employ to rationalize the use of charges. Moreover, in each case, a charge was not the only factor motivating dischargers to reduce their waste loads. Typically, charges were used as part of a management scheme including ambient and effluent standards and government subsidies to offset partially the costs of residuals reduction.

The effluent charge system in France serves primarily to generate revenues that support the six French river basin agencies created in the 1960s.[16] These

[15]Procedures for calculating the emission charges are discussed by Anderson et al. (1977). The charge applicable to any firm can be calculated quickly, using standard procedures, by the agency's inspectors and engineers.

[16]This discussion of the effluent charge program in France draws heavily from Barré and Bower (1981).

agencies do not construct facilities or issue regulations. Their functions are limited to planning and research, and to providing loans and grants for water and wastewater management projects. The agencies also subsidize the costs of wastewater treatment by firms and municipalities.

The effluent charges that support a basin agency are set through a complex negotiation process involving the agency staff, the agency's "basin committee" and a host of government officials. The basin committee includes representatives from various French ministries and from municipalities and other water users. The effluent charges are *not* set using economic theory. They are based mainly on estimates of the funds needed by the basin agency to meet its proposed "multiyear action program." The charges can be raised or lowered depending on the scope of the action program.

During the 1970s, the effluent charges used by the basin agencies were much lower than required to motivate dischargers to treat their wastewaters. Why then did so many firms and municipalities implement high levels of treatment during that period? The answer is found by considering a complementary part of French water quality management strategy, effluent standards. Since 1917, the prefects of French "departments" have issued permits controlling the discharge of wastewaters.[17] Penalties for permit violations include fines and court-imposed sanctions on dischargers. Before the basin agencies were formed in the 1960s, the permit system was judged ineffective. Even though the basin agencies do not issue permits, their programs of charges and subsidies have played a key role in motivating dischargers to treat their wastes in compliance with permit conditions. The use of a two-part strategy, economic incentives administered by basin agencies and permit requirements administered by prefects, has led to substantial water quality improvements.

MARKETABLE RIGHTS TO POLLUTE

An alternative way of charging for use of the environment's waste receptor services involves creating legal "rights to pollute" and allowing the rights to be bought and sold like ordinary commodities. Suppose, for example, that a government agency estimates the maximum amount of a residual that can be discharged to a given zone without violating an ambient standard. Assume also that legal rights to discharge only this maximum amount are established and the government allocates those rights by one means or another. The rights could be granted free to existing dischargers. Alternatively, they might be sold to the highest bidders, and firms that had plans to locate in the zone could take part in the bidding. If they could raise the money, conservation groups could also purchase rights and thereby prevent them from being used. This would lead to

[17]France is divided into 91 departments, which average about 5000 km^2 in area. An exception is the Seine department (City of Paris), which is approximately 150 km^2. A prefect representing the national government heads each department.

environmental quality levels that were higher than the ambient standard. Regardless of how the rights are allocated initially, the total number of rights is limited such that the ambient standard cannot be violated.[18]

A program based on marketable rights to pollute can be organized so that the residuals reduction costs incurred in meeting a particular ambient standard are at a minimum. For this to occur, however, a market for rights to pollute must develop and the transaction costs for trading in the market must be zero. In addition, these rights must be given an unambiguous legal definition so buyers can be sure of what they are purchasing.

A Trading Equilibrium in Rights to Discharge BOD

To further examine how marketable pollution rights might work, consider the three waste source example again. Recall that the raw loads from the three sources combined causes a violation of the 6 mg/l dissolved oxygen standard. To set up pollution rights, it is first necessary to compute the quantity of BOD that can be released without violating the 6 mg/l DO standard. This quantity depends on the locations of the effluents. To compute the maximum permissable BOD discharge, assume for a moment that source 3 does not exist. Under these circumstances, the maximum allowable release multiplied by the transfer coefficient, $\phi_{1\alpha} = \phi_{2\alpha} = 0.002$, equals the allowable 1 mg/l decrease in DO at point α. This relationship indicates a maximum allowable BOD discharge at the location of source 1 or 2 of 500 units.

The responsible government agency could issue 500 certificates entitled "rights to discharge 1000 lb/day of BOD at the location of source 1 or 2." These certificates, or "Rights," could *not* be used by source 3 without first adjusting for the way its BOD release influences DO at ponit α. This adjustment is readily made. A ratio of transfer coefficients,

$$\frac{\phi_{1\alpha}}{\phi_{3\alpha}} = \frac{0.002}{0.003} = \frac{2}{3}$$

reflects the relatively greater influence that BOD from source 3 has on dissolved oxygen at α. Thus, for source 3, a Right can only be used to release two-thirds of a unit of BOD (667 lb/day). To demonstrate the validity of this adjustment, suppose that source 3 has all 500 Rights. In this case, there are no discharges from sources 1 and 2, and source 3 releases two-thirds of 500 units of BOD. This quantity, multiplied by $\phi_{3\alpha} = 0.003$, yields the allowable DO decrease of 1 mg/l.

Next consider the initial allocation of the 500 Rights. Under conditions in which the transaction costs of buying and selling rights are zero, Krupnik, Oates,

[18]This discussion relies on a more extensive description of marketable pollution rights by Dales (1968). Recent research on this topic is summarized by Joeres and David (1983).

and Van De Verg (1983) have shown that the initial allocation of Rights does not affect economic efficiency. They demonstrate that dischargers will make mutually advantageous purchases and sales of Rights. The resulting "trading equilibrium" will be one in which the ambient standard is attained at a minimum residuals reduction cost. Krupnik, Oates, and Van De Verg describe a trading equilibrium as a point at which dischargers cannot realize further economic gains by the mutually beneficial trading of rights.[19]

For illustrative purposes, suppose the government initially allocates the 500 Rights by auctioning them off one at a time to the highest bidder. Assume that only the three dischargers participate in the auction and that each knows its *own* costs of BOD reduction. Furthermore, assume that the highest sum a source would offer for any Right is the amount its treatment cost would be reduced by owning the Right. The sources do not speculate in Rights. They only purchase what they currently need. Under these conditions, the agency's auction will result in removal percentages that minimize the *total* treatment costs for the three waste sources.

For each source, Table 4.12 shows the maximum and minimum treatment cost savings that would result from purchasing one additional Right. In this example, the cost of removing an additional unit of BOD increases as the removal percentage increases. The first Right purchased is the most valuable one since it leads to the greatest decrease in treatment costs. The least valuable Right is the one purchased when the discharger is only one short of having all the Rights it can use to reduce treatment costs.

The computations of incremental savings in treatment cost are demonstrated for source 1 when it owns no Rights. If source 1 purchased one Right, its savings

TABLE 4.12 Range of Incremental Savings in BOD Reduction Costs

Waste Source	Hypothesized Number of Rights Held	Cost Savings from Holding One More Right
1	0	119,000
	99	1,000
2	0	3,000
	499	2,600
3	0	10,000
	499	6,000

[19]As shown by Krupnik, Oates, and Van De Verg (1983), a trading equilibrium corresponding to a minimum total cost of residuals reduction will result regardless of the initial allocation of Rights if (1) the total number of Rights is limited such that the ambient standard is attained; (2) for each source, the incremental cost of residuals cutback increases with increasing removal percentages; and (3) the transaction costs associated with buying and selling Rights are zero.

would equal the cost of removing 100% of its raw BOD load minus the cost of removing 99%. Using the cost equations in Figure 4.1, this difference is

$$1000(100)^{1.9} - 1000(99)^{1.9} = \$119{,}000$$

A slightly more complex computation is required for source 3 since a Right covers only two-thirds of a unit of discharge. For example, if source 3 owns 499 Rights, it is entitled to discharge two-thirds of 499 units of BOD. Its required BOD reduction is 167.3 units, the difference between its raw load (500) and its allowable discharge. This corresponds to $X_3 = 33.47\%$. If source 3 purchased one more Right, its required removal would decrease to 33.33%, and its cost savings would be

$$5000(33.47)^{1.5} - 5000(33.33)^{1.5} = \$6000$$

Table 4.12 shows that the incremental savings in BOD reduction costs for source 2 are always lower than the corresponding savings for source 3. Therefore, source 3 could outbid source 2 for each of the 500 Rights being auctioned. Consequently, source 2 purchases no Rights and removes 100% of its raw load.

The competition for Rights between sources 1 and 3 involves only the last 100 Rights auctioned. This occurs because the raw load at source 1 is only 100 units. Suppose source 3 has 400 Rights in hand as it begins to compete with source 1. To determine how much each source would bid for Rights, it is necessary to compute the incremental savings in treatment costs if one more Right is purchased. For each of the first 96 Rights, the incremental savings are greater for source 1 than for source 3. Consequently, these Rights are purchased by source 1. If source 1 obtained the ninety-seventh Right, its treatment costs would be reduced by

$$1000(4)^{1.9} - 1000(3)^{1.9} = \$6000$$

If source 3 obtained this Right, it would have to reduce 46.67% instead of 46.53% of its raw load. The corresponding cost saving is

$$5000(46.67)^{1.5} - 5000(46.53)^{1.5} = \$7000$$

Because its incremental savings are higher, source 3 purchases the ninety-seventh Right. This outcome can also be shown to hold for each of the remaining 3 Rights. The final trading equilibrium has source 1 with 96 Rights, source 2 with no Rights, and source 3 with 404 Rights. This outcome is essentially equivalent to the minimum cost removal percentages in Table 4.2.

Attainment of the 6 mg/l DO standard at minimum cost is only one of the advantages of employing marketable pollution rights. Use of marketable rights is also attractive because, in theory, it *automatically* adjusts for either growth in existing sources or introduction of completely new sources. The ambient standard is attained regardless of future conditions because the number of available Rights is fixed. The marketable rights approach also gives firms an incentive to invent new residuals reduction technologies. If a firm reduces its residuals, the Rights it no longer needs can be sold at a profit.

Implementing a Marketable Pollution Rights Approach

A crucial step in implementing a system of marketable pollution rights is for the government to make a clear and legally enforceable statement about what the bearer of a right is entitled to do. In the three waste source example, the holder of a Right was entitled to discharge 1 unit of BOD at the location of source 1 or 2 or two-thirds of a unit at the location of source 3. Rights could not be defined this simply in practice. Because the use of marketable rights to discharge is a fairly novel idea, the process of defining and issuing such rights will require innovations in the legal system. In addition, for rights to be meaningful, the government must assure that those without rights do not discharge. It may be difficult to provide these assurances when there are many waste sources involved.

Another implementation problem concerns the number of rights to issue. In the example, information on transfer coefficients provided the basis for the decision to issue 500 Rights. In practice, ambient standards are established at numerous places, not just a single location such as point α in the example. Thus, to decide on the number of rights requires that transfer coefficients be known for all combinations of waste sources and locations at which the standard is set. This information is often unavailable in real situations. Even when there is a basis for predicting how discharges affect ambient environmental quality, the predictions are highly uncertain. Using this information to determine the total number of rights leads to uncertainties regarding whether the ambient standard will be met when the rights are exercised.

Once rights are legally defined and initially allocated, there may be a continuing role for government in facilitating market transactions. As in any other market, the buyers and sellers of pollution rights must spend time and money in locating each other and in learning about potential offers and selling prices. For a product as unusual as pollution rights, the cost of obtaining this information may be high. One possible role for government is as a "broker" that brings buyers and sellers together. The cost of exchanging a right could include a fee to offset the expense of maintaining an agency to administer the program.

Dales (1968) has suggested ways for the government to "rig the market" to maintain stable prices for rights to pollute. Price stability is important since the price of pollution rights influences the decisions of firms and municipalities to invest in residuals reduction measures. If prices fluctuate widely, investment decisions cannot be made rationally and confidently. Government could help stabilize prices by not frequently changing the maximum number of rights issued.

A sudden increase in the number of rights would weaken the confidence that holders of rights had in the government's program. In addition, it would lead to violations of the ambient standard. The government might also buy rights if sellers have difficulties in finding buyers or if too many rights are put up for sale at any one time. These actions would all help to stabilize the prices of rights.

Controlled Trading in Air Pollution Emissions

During the 1970s, a limited policy involving marketable rights to pollute emerged in the federal program for managing air quality in the United States. This policy did not originate with a desire to implement a marketable rights approach. Instead, it represented a means of accommodating new growth in "nonattainment areas," regions that violated one or more of the national ambient air quality standards (NAAQS) established under the Clean Air Act of 1970.

The Clean Air Act Amendments of 1977 responded to this growth dilemma with special requirements for significant new sources of air pollution that wanted to locate in nonattainment areas. The amendments included definitions clarifying which new sources would be considered "significant." Two conditions had to be met before a permit could be issued for a significant new source in a nonattainment area. First, the discharger had to satisfy very stringent technology-based emission standards. Second, the discharger had to demonstrate that the proposed emissions of pollutants for which the NAAQS were violated in the region would be offset by reductions in emissions from other sources. The *net effect* of the new emissions and the emission reductions (or "offsets") had to represent "reasonable progress" toward meeting the NAAQS.

The use of emission offsets and their relationship to marketable pollution rights are clarified by an example. Following the passage of the 1977 Clean Air Act Amendments, the Wickland Oil Company applied to the Bay Area Air Quality Management District (BAAQMD) for a permit to construct a new 40,000 barrel/day terminal at a site on the San Francisco Bay in Contra Costa County.[20] Fuel oil, received via tanker ship or pipeline, was to be stored in floating roof tanks on the site for later shipment by tank trucks or pipeline. At the time the permit application was processed, Contra Costa County was a nonattainment area for three air pollutants: sulfur dioxide (SO_2), carbon monoxide (CO), and hydrocarbons (HC). Moreover, the proposed terminal was a significant new source of air pollution and thereby subject to requirements calling for emission offsets.

The base conditions used in calculating the necessary emission offsets were the technology-based emission requirements for new sources in nonattainment areas. The first row in Table 4.13 shows the discharges that would result *after* Wickland Oil implemented various measures required to meet these emission

[20]The discussion of the Wickland Oil Company example is based on an unpublished report by Lance M. Goto in 1980 while he was a civil engineering student at Stanford University, Stanford, California.

TABLE 4.13 Wickland Oil Proposal – Estimated Emissions and Proposed Offsets (Tons per Year)[a]

	SO_2	HC	CO
Total estimated emissions before offsets	24.7	83.2	1.31
City of Paris offset—new equipment		−151.4	
Virginia Chemicals Inc. offset—plant shutdown	−7.4		
Offsets from ships and vehicles burning low-sulfur fuel	−22.2		
Total estimated emissions after offsets	−4.9	−68.2	1.31

[a]From "Notice Inviting Written Public Comment on the Authority to Construct the Wickland Terminal," Bay Area Air Quality Management District, San Francisco, California, March 15, 1979.

requirements. The expected emissions of CO were low enough to be disregarded. However, the emissions for SO_2 and HC shown in Table 4.13 were so high that BAAQMD required offsets.

Before the Wickland Oil Company could receive a permit from the BAAQMD it had either (1) to reduce its emissions below the levels used by the BAAQMD to designate a new source as "significant" or (2) to obtain enough emission offsets to demonstrate that the net effect of its actions would be progress in attaining the NAAQS for sulfur dioxide and hydrocarbons. The first option was prohibitively expensive. Wickland Oil therefore set out to acquire the necessary offsets. Means of obtaining offsets included

1 Reducing or eliminating emissions from an existing Wickland Oil facility in the vicinity of the proposed terminal. These were termed *internal offsets*.
2 Utilizing "banked offsets," emission offsets obtained by Wickland Oil in the past for bringing emissions at nearby facilities below levels required by the BAAQMD.
3 Reducing or eliminating emissions from existing facilities of *other firms*. These "external offsets" could be obtained by paying other firms to decrease emissions or by buying their facilities and closing them down. The BAAQMD did not accept external offsets at full value; for the Wickland Oil proposal, 1.2 tons/year of external emission reduction was required to offset 1 ton/year at the proposed terminal.
4 Purchasing offsets that had been banked by other firms.

The Wickland Oil Company arranged for a 181.7 tons/year reduction in hydrocarbon emissions at the City of Paris dry cleaners in San Francisco. Wickland Oil agreed to purchase new dry cleaning and solvent reclamation equipment that would bring the City of Paris emissions substantially below those required by the BAAQMD. The oil company was credited with 151.4 tons/year of hydrocarbon offsets (181.7 tons/year divided by the 1.2 factor applied to external offsets). The sulfur dioxide offset requirements were met in two ways. First,

Wickland Oil purchased and closed down a plant owned by Virginia Chemicals Inc. that was on the site to be developed by Wickland Oil. After applying the 1.2 factor, this accounted for an offset of 7.4 tons/year. Second, Wickland Oil arranged for a number of ocean going oil tankers and surface motor vehicles to switch to the use of low-sulfur fuels. This provided an additional 22.2 tons/year of offsets for SO_2. As indicated in Table 4.13, the net result of these offsets is an estimated *reduction* of 4.9 tons/year of sulfur dioxide and 68.2 tons/year of hydrocarbons.

The Wickland Oil Company was granted a permit by the Air Pollution Control Officer (APCO) of the BAAQMD in 1979. In May 1980 the APCO's decision was overruled by the BAAQMD Hearing Board based on objections raised by citizens. Subsequent negotiations between the Wickland Oil Company and the objecting citizens led, eventually, to a mutually acceptable solution. Wickland Oil Company agreed to eliminate two large storage tanks from its original proposal and thereby reduce the expected emissions of hydrocarbons by an additional 10 tons/year. This new proposal was acceptable to the Hearing Board, and a permit to construct the oil terminal was issued.

The Wickland Oil Company example demonstrates that transaction costs can be very high using a system of tradeable emission offsets. According to a report by the U.S. General Accounting Office (1982), Wickland Oil contacted over 150 firms in its efforts to secure the needed offsets for its hydrocarbon emissions. Most of the firms were dry cleaning companies that did not use a cleaning technology that would allow for inexpensive further reductions in hydrocarbon emissions. Eventually, the opportunity to reduce hydrocarbon emissions economically at the City of Paris dry cleaners was discovered and an agreement was negotiated.

Transaction costs have also been high for firms in other parts of the United States that sought emission offsets to meet requirements of the 1977 Clean Air Act Amendments. As elaborated by Liroff (1980), the concept of an "offset bank" emerged in the late 1970s as a mechanism for reducing transaction costs and helping in the formation of active markets in offsets. These banks were created by regional air quality agencies such as the BAAQMD. They provided a convenient first stop for buyers seeking sellers of offsets. Offset banks also played an important role in clarifying the legal status of offsets and the offset policies of regional air quality agencies. Without such clarification, firms with potential offsets would hesitate to make the investments in residuals reduction facilities needed to create offsets.

The markets in emission offsets that developed in response to the 1977 Clean Air Act Amendments were not the same as the one envisioned in the three waste source example. The offset markets were constrained in that technology-based emission standards formed the basis for defining offsets. This outcome is not surprising. Residuals management programs rarely consist of a single form of economic incentive or direct regulation. These programs are often made up of combinations of ambient and effluent standards and market-based strategies.

KEY CONCEPTS AND TERMS

EFFLUENT STANDARDS BASED ON AMBIENT STANDARDS
- Economic efficiency
- Equity in distributing costs
- Biochemical oxygen demand
- Dissolved oxygen
- Transfer coefficients
- Minimize total treatment costs
- Equal percentage removal policy
- Zoned equal percentage removal

TECHNOLOGY-BASED EFFLUENT AND EMISSION STANDARDS
- Effluent limitations guidelines
- Best practicable control technology
- Distortions in pollution control research

EFFLUENT CHARGES TO ATTAIN AMBIENT STANDARDS
- Discharger's freedom in decision making
- Present value of costs
- Minimize charges plus treatment costs
- Iterative method for estimating charges
- Effects of underestimating charges
- Uniform versus nonuniform charges

CHARGES AS SUPPLEMENTS TO STANDARDS
- Noncompliance penalties
- Revenues from charges
- Creating subsidies using charges

MARKETABLE RIGHTS TO POLLUTE
- Maximum allowable discharge
- Legal status of rights
- Trading equilibrium
- Incremental cost of residuals reduction
- Stability in prices of rights
- Transaction costs
- Emission offsets
- Offset banks

DISCUSSION QUESTIONS

4–1 Consider the following information from the three waste source example:

	Minimum Cost (% reduction)	Equal Percentage Removal (% reduction)	Zoned Equal Percentage Removal (% reduction)
Waste Source			
1	4	73	61
2	100	73	61
3	46	73	90
Total cost ($\$10^6$)	4.091	8.31	8.15

Which of these three programs would you advocate and why? Your response should consider both equity and economic efficiency.

4–2 The federal strategy for managing water quality in the United States during the 1970s relied heavily on technology-based effluent standards. Effluent limitations guidelines were used to decide on conditions included in effluent permits. What are the principal advantages and disadvantages of this approach in comparison with effluent standards designed to meet an ambient standard?

4–3 Discuss the ways that effluent charges could be used to manage water quality on the Delaware Estuary. Consider charges as an alternative to effluent standards in meeting an ambient standard. Also discuss the possibilities for using charges as a supplement to effluent standards.

4–4 Consider three alternative strategies for air quality management: emission standards, emission charges, and marketable rights to discharge air pollutants. Which of these provides the greatest incentives for private firms to develop new residuals control technologies? Explain your answer.

4–5 Several countries in Europe have used effluent charges to generate revenues which are used to construct wastewater treatment facilities. Typically, a basin-wide agency with broad powers is formed to administer these charge programs. Why do you think this approach has not been followed in the United States?

4–6 Suppose the initial allocation of Rights in the three waste source example is as follows: source 1 has 100 Rights, source 2 has 400 Rights, and source 3 has no Rights. Use incremental savings in treatment costs for the three sources to demonstrate that this is not a trading equilibrium. Describe a mutually advantageous purchase and sale of Rights that would cause a change from the initial allocation.

4–7 Consider a newly formed, small nation that is planning for rapid industrial growth. What would you imagine to be the principal advantages and disadvantages in using marketable air pollution rights as a way of maintaining air quality at reasonable levels?

REFERENCES

Ackerman, B. A., S. R. Ackerman, J. W. Sawyer, Jr., and D. W. Henderson, 1974, *The Uncertain Search for Environmental Quality*. Free Press, New York.

Anderson, F. R., A. V. Kneese, P. D. Reed, R. B. Stevenson, and S. Taylor, 1977, *Environmental Improvement Through Economic Incentives*. Johns Hopkins University Press for Resources for the Future, Baltimore, Md.

Barré, R., and B. T. Bower, 1981, Water Management in France, with Special Emphasis on Water Quality Management and Effluent Charges, *in* B. T. Bower, R. Barré, J. Kühner, and C. S. Russell (eds.), *Incentives in Water Quality Management, France and the Ruhr Area,* Research Paper R-24, pp. 31–209. Resources for the Future, Washington, D.C.

Baumol, W. J., and W. E. Oates, 1979, *Economics, Environmental Policy and the Quality of Life*. Prentice–Hall, Englewood Cliffs, N.J.

Dales, J. H., 1968, *Pollution, Property and Prices*. University of Toronto Press, Toronto.

Garwin, R. L., 1978, Comments on Government Policies Toward Automotive Emissions Control, *in* A. F. Friedlaender (ed.), *Approaches to Controlling Air Pollution*. MIT Press, Cambridge, Mass.

Grant, E. L., W. G. Ireson, and R. S. Leavenworth, 1982, *Principles of Engineering Economy,* 7th ed. Wiley, New York.

Joeres, E. F. and M. H. David (eds.), 1983, *Buying a Better Environment,* University of Wisconsin Press, Madison.

Johnson, E., 1967, A Study in the Economics of Water Quality Management. *Water Resources Research* **3** (2), 291–305.

Johnson, R. W., and G. W. Brown, Jr., 1976, *Cleaning up Europe's Waters: Economics, Management and Policies*. Praeger, New York.

Kneese, A. V., and B. T. Bower, 1968, *Managing Water Quality: Economics Technology and Institutions*. John Hopkins University Press for Resources for the Future, Baltimore, Md.

Krupnik, A. J., W. E. Oates, and E. Van De Verg, 1983, On Marketable Air-Pollution Permits: The Case for a System of Pollution Offsets. *Journal of Environmental Economics and Management* **10** (3), pp. 233–247.

Liroff, R. A., 1980, *Air Pollution Offsets: Trading, Selling and Banking*. The Conservation Foundation, Washington, D.C.

Mills, E. S., and L. J. White, 1978, Government Policies toward Automotive Emissions Control, A. F. Friedlaender (ed.), *Approaches to Controlling Air Pollution*. MIT Press, Cambridge, Mass.

Quarles, J., 1976, *Cleaning up America: An Insider's View of the Environmental Protection Agency*. Houghton–Mifflin, Boston.

Thomann, R. V., 1972, *Systems Analysis and Water Quality Management*. Environmental Science Services Division, New York (reissued by McGraw–Hill, New York).

U.S. General Accounting Office, 1982, A Market Approach to Air Pollution Control Could Reduce Compliance Costs without Jeopardizing Clean Air Goals, Summary, No. PAD-82-15A, Washington, D.C. (March 23).

Velasquez, M. G., 1982, *Business Ethics, Concepts and Cases*. Prentice–Hall, Englewood Cliffs, N.J.

Velz, C. J., 1976, Stream Analysis—Forecasting Waste Assimilation Capacity, *in* H. W. Gehm and J. I. Bregman (eds.), *Handbook of Water Resources and Pollution Control*, pp. 216–261. Van Nostrand Reinhold, New York.

CHAPTER 5

AIR AND WATER QUALITY MANAGEMENT IN THE UNITED STATES: AN HISTORICAL PERSPECTIVE

Many of the concepts introduced in previous chapters are illustrated by programs to manage air and water quality in the United States. This chapter describes how these programs evolved. In providing an historical perspective, the chapter gives a basis for examining contemporary management strategies. Details on government pollution control programs are not furnished. Federal environmental regulations are revised so frequently that a presentation emphasizing the latest fine points would be obsolete in a short time[1].

WATER QUALITY MANAGEMENT

Early Dominance of State and Local Programs

Prior to 1948, the principal responsibilities for controlling water pollution were assumed by the states and by various local and regional agencies. The first

[1]Up-to-date discussions of U.S. water and air quality management programs are given in recent issues of the *Journal Water Pollution Control Federation* and the *Journal of the Air Pollution Control Association*, respectively. The annual reports of the Council on Environmental Quality provide yearly summaries of residuals management activities in the United States.

institutions to deal with water pollution problems were created soon after the "sanitary awakening" of the 1850s. At that time, scientists clarified the role of contaminated water in transmitting diseases. The Massachusetts Board of Health, formed in 1869, was the first state agency with responsibilities for water pollution control. During the latter part of the nineteenth century, state Boards of Health often administered water pollution control programs. A major concern of the early state programs was the control of water-borne infectious diseases like typhoid fever and cholera.

Although an effective technology for treating municipal wastewater existed by the late 1870s, only 4% of the nation's population had its wastes treated as of 1910.[2] By 1939, about half of the nation's urban population still discharged its waste untreated. Hey and Waggy (1979) argue that two factors impeded the use of wastewater treatment technology: (1) the ability to protect drinking-water supplies economically using chlorine for disinfection, a practice initiated in 1911, and (2) the widely shared attitude that it was appropriate to use natural waterways as receptors for wastes.

The federal role in water pollution control began with the Public Health Service Act of 1912. This act established the Streams Investigation Station at Cincinnati to carry out water pollution research. The Oil Pollution Act was passed in 1924 to prevent oily discharges on coastal waters. During the 1930s and 1940s, there was a continuing debate over whether the federal government should take a greater role in controlling water pollution. This debate led to the limited expansion of federal powers expressed in the Water Pollution Control Act of 1948. Table 5.1 highlights the key features of this act and other important federal water quality laws.

The 1948 act, administered by the U.S. Public Health Service, provided funds and technical services to states to strengthen their water quality control programs. Although the 1948 act included a provision for legal action by the federal government against polluters, this provision could not be used without the "consent of the state in which pollution was alleged to originate." The 1948 act formally recognized the primacy of the states in water quality management. Its preamble declared a national policy "to recognize, preserve and protect the primary responsibilities and rights of the states in preventing and controlling water pollution." Despite this declaration, the three decades that followed saw many of the traditional "rights and responsibilities" of the states overtaken by an increasingly dominant set of federal interventions to restrain water pollution.[3]

Early Federal Strategy: The Federal Water Pollution Control Act of 1956

The Federal Water Pollution Control Act (FWPCA) of 1956 was the cornerstone of the early federal efforts to reduce pollution. Key elements of the act included

[2]In this paragraph, statistics on population served by wastewater treatment plants are from Hey and Waggy (1979, p. 128).

[3]Information in this paragraph is from Cleary's (1967, pp. 251–252) summary of early federal water pollution control laws.

TABLE 5.1 Key Federal Laws Controlling Water Pollution

Year	Title	Selected *New* Elements of Federal Strategy[a]
1948	Water Pollution Control Act	Funds for state water pollution control agencies Technical assistance to states Limited provisions for legal action against polluters
1956	Federal Water Pollution Control Act (FWPCA)	Funds for water pollution research and training Construction grants to municipalities Three-stage enforcement process
1965	Water Quality Act	States set water quality standards States prepare implementation plans
1972	FWPCA Amendments	Zero discharge of pollutants goal BPT and BAT effluent limitations NPDES permits Enforcement based on permit violations
1977	Clean Water Act	BAT requirements for toxic substances BCT requirements for conventional pollutants
1981	Municipal Waste Treatment Construction Grants Amendments	Reduced federal share in construction grants program

[a]The table entries include only the *new* policies and programs established by each of the laws. Often these provisions were carried forward in modified form as elements of subsequent legislation.

a new program of subsidies for municipal treatment plant construction and an expanded basis for federal legal action against polluters. Increased funding for state water pollution control efforts and new support for research and training activities were also provided. Each of these programs was continued in the many amendments to the Federal Water Pollution Control Act in the 1960s and 1970s.

The expansion in federal actions against polluters took the form of a three-stage "enforcement process." It allowed the surgeon general of the Public Health Service to act against firms and municipalities whose wastewaters were interfering with the health or welfare of water users in a downstream state. The first stage involved a conference of federal and state water pollution control officials. During a conference, the Public Health Service presented evidence documenting the interstate water pollution problem. Agreements were then reached with the relevant state agencies regarding appropriate remedial measures and a timetable for their implementation. If state activities failed to yield the agreed-upon measures, the surgeon general could then implement the second stage of the enforcement process by holding a formal hearing on the matter. The third stage, entered only if the hearing failed to yield a satisfactory outcome, involved court actions by the federal government. The second and third stages of the enforcement process were not used frequently.[4] Often, when the deadlines set at a conference were not met, new conference sessions were convened and the deadlines were renegotiated. Although dischargers were encouraged to clean up, the resulting cutbacks in effluents were much less than expected.

The 1956 act's program of "construction grants," subsidies to help pay for municipal treatment facilities, was a response to pressures from municipalities. The cities felt that federal funds should be used to help pay for the treatment plants required by federal enforcement activities. Grants made under the 1956 act could subsidize as much as one third of the construction cost for a municipal plant, but the maximum grant for a single project was limited to $600,000. A total of $50 million/year was authorized for the program.

The effectiveness of the construction grants program was challenged frequently. Critics argued that the total budget authorized for grants was woefully inadequate and the maximum funds available for any one project were too small to influence the decisions of large cities. The numerous modifications to the FWPCA of 1956 throughout the 1960s and 1970s responded to these concerns by increasing substantially both the total funding and the maximum subsidy for any one project.

Another criticism of the construction grants program was that municipalities sometimes used the lack of available federal funds as an excuse for not constructing facilities quickly. Because so many cities needed to constructs treatment plants, there was often a long wait to receive the limited grant monies. In addition, delays were also caused by complex administrative procedures for processing grant applications and obligating funds. Treatment plant construction was sometimes delayed a few years because of the grant application and review process.[5]

The following additional points have been raised in assessments of the construction grants program during the 1960s and 1970s:

1 The program is biased toward end-of-pipe treatment since only treatment plant construction is subsidized. Other options, such as changing the timing or location of discharges, may yield the same ambient water quality at a much lower cost.
2 Many federally subsidized treatment plants are operated inefficiently, because cities fail to maintain equipment and hire adequately trained plant operators.[6]

[4]Freeman, Haveman, and Kneese (1973, p. 116) report that between 1956 and 1965 only one federal enforcement case reached the third stage.

[5]Evidence for this is offered by Whipple (1977, pp. 4–5), who cites an analysis by the National Utility Contractors Association (1975).

[6]Freeman, Haveman, and Kneese (1973, p. 119) give statistics documenting this inefficiency in plant operations.

3 The grants provide an indirect subsidy to firms discharging to municipal sewers, and this weakens the incentives for firms to seek minimum cost residuals reduction schemes. (This indirect subsidy was substantially reduced by 1972 amendments to the FWPCA which required firms to pay municipalities charges reflecting the additional costs of waste treatment.)
4 Since municipal treatment plants are traditionally designed with capacities much higher than their initial wastewater loadings, the construction grants program is subsidizing future growth that could lead to further deterioration in environmental quality.[7]

A Shift in Strategy to Ambient Standards

The Water Quality Act of 1965 carried forward many provisions of the earlier federal legislation, generally with an increase in levels of funding. The 1965 act also introduced important *new* requirements for states to establish ambient water quality standards and detailed plans indicating how the standards would be met. The act also shifted responsibilities for administering the federal water quality program from the Public Health Service to a separate agency, the Federal Water Pollution Control Administration, within the Department of Health, Education, and Welfare (HEW). This was not a permanent change. In 1970, a presidential reorganization order placed the water pollution control activities and several other federal environmental programs in a newly created Environmental Protection Agency (EPA).

Ambient water quality standards existed for many watercourses long before the Water Quality Act of 1965. The standard setting process generally started by determining the water uses to be "protected" on a given stretch of river. Professional judgment was employed to establish *criteria,* bases for assessing the suitability of water for different activities. Knowing the uses to be accommodated on a particular river section, and the criteria associated with protecting those uses, it was possible to set ambient standards.

The Water Quality Act of 1965 required states to establish standards for *all* interstate waters by June 1967. If this was not accomplished, the Secretary of HEW would promulgate the standards. Many states had trouble meeting the June 1967 requirement, because the time allotted for standard setting was short, and there were not much data describing existing water quality. In addition to establishing ambient standards, states were required to prepare "implementation plans." For each waste source, these plans specified the steps required to reduce wastewater discharges so that the ambient standards would be met in a timely manner. The earlier federal enforcement process was modified by the 1965 act to reflect the new importance of ambient standards. Enforcement actions could be taken against dischargers violating standards on interstate waters. The federal government could initiate litigation 180 days after notifying violators. However, this provision of the Water Quality Act of 1965 was not widely

[7]Binkley et al. (1975) examine this subject in detail.

used. By the end of 1971, only 27 notifications of impending court action had been issued. A reason the courts were not used more frequently is the difficulty in proving that a particular waste source caused a violation of downstream ambient standards. Establishing this causal relationship was problematic when there were many nearby sources of waste and when the residual's behavior in natural waters was poorly understood. In these cases, alleged violators of ambient standards could easily argue that a different waste source caused the problem or that there was inadequate scientific basis for determining how the residuals were transported in waterways.

Major Shifts in Strategy during the 1970s

In amending the Federal Water Pollution Control Act in 1972, Congress introduced: (1) national water quality goals, (2) technology-based effluent limitations, (3) a national discharge permit system, and (4) federal court actions against sources violating permit conditions.

The 1972 amendments aimed to restore and maintain "the chemical, physical and biological integrity of the nation's waters." The amendments specified, as a national goal, that the "discharge of pollutants into navigable waters be eliminated by 1985." They also included an "interim goal":

> *[W]herever attainable, an interim goal of water quality which provides for the protection and propogation of fish, shellfish and wildlife and provides for recreation in and on the water [should] be achieved by July 1, 1983.*

The goal of eliminating the discharge of pollutants from waterways was criticized heavily because it did not reflect a materials balance perspective. By requiring very high percentages of removal of water-borne residuals, Congress shifted some residuals discharges to the air and land. For example, large reductions of solids in municipal wastewaters increased municipal sludge disposal problems. Another criticism centered on the high cost of "eliminating" the discharge of pollutants. Incremental residual reduction costs often increase dramatically when the percentage of removal goes beyond about 90%.

The 1972 amendments gave enormous power to the EPA administrator. They left the administrator responsible for defining "pollutants" in the context of the national goal and for making operational the various effluent restrictions which the amendments described only in general terms. The effluent limitations were specified by the administrator on an industry-by-industry basis. Because they applied nationwide, these limitations were *independent* of the particular context in which a discharge occurred.

The EPA administrator was required to set effluent restrictions that met the following general requirements of the 1972 amendments: by 1977, all dischargers were to achieve "best practicable control technology currently available" (BPT); and by 1983, all dischargers were to have the "best available technology economically achievable" (BAT). After delays caused by numerous legal challenges to the EPA administrator's effluent limitations guidelines, the BPT provisions were implemented. However, the BAT requirements were so heavily disputed that Congress modified them in the Clean Water Act of 1977.

The principal criticism of the original BAT effluent limitations was that the costs of the very high required percentage reductions in residuals would be much greater than the benefits. In defining BAT, costs were considered, but only in the general context of affordability by industry. Computations of the social benefits of stringent effluent controls were not a central factor. Congress presumed the benefits of eliminating water pollutants would be substantial. The Congress' insistence on very strict effluent limitations can also be interpreted as an effort to guarantee the rights of Americans to high-quality waters.[8]

In 1977, Congress responded to the critics of BAT by requiring it only for toxic substances. A different requirement was introduced for "conventional pollutants," such as biochemical oxygen demand and suspended solids. The effluent limitations guidelines for these pollutants were to be based on the "best conventional pollutant control technology" (BCT). Table 5.2 indicates the kinds of effluent constraints the EPA administrator set under the Clean Water Act of 1977. Illustrations of specific effluent restrictions are given above in Tables 4.6 and 4.7.

The Clean Water Act of 1977 strongly endorsed the view that water-borne toxic substances must be controlled. The act included a list of 65 substances, or classes of substances, that was used as the basis for defining toxics. This list resulted from a 1976 settlement of a legal action in which several environmental organizations sued the EPA administrator for failing to issue toxic pollutant

TABLE 5.2 Technology-Based Effluent Limitations: Examples from the Clean Water Act of 1977

Publicly Owned Treatment Works

Requirements for 85% BOD removal, with possible case-by-case variances that allow lower removal percentages for marine discharges

Industrial Discharges (bases for effluent limitations)

Toxic pollutants — BAT

Conventional pollutants — BCT; in determining required control technology, EPA is directed to consider "the reasonableness of the relationship between the costs of attaining a reduction in effluents and the effluent reduction benefits derived"

Nonconventional pollutants (pollutants that are not classified as either conventional or toxic) — BAT, but with possible case-by-case variances that allow lower degrees of treatment

[8]Velasquez (1982, p. 191) examines Congress' strict effluent limitations in the context of moral rights to a high-quality environment.

standards under the FWPCA amendments of 1972.[9] The list was subsequently refined by EPA to include 129 specific materials, referred to as "priority pollutants."

Effluent limitations required by the FWPCA amendments of 1972 (and later the Clean Water Act of 1977) formed the basis for issuing "National Pollutant Discharge Elimination System" (NPDES) permits. As explained by Quarles (1976), the permit system idea stemmed from actions taken by the Department of Justice in the late 1960s. With the support of a favorable interpretation by the U.S. Supreme Court, the U.S. attorneys relied on the 1899 Refuse Act to prosecute industrial sources of water pollution. The 1899 act, which was originally interpreted to prohibit deposits of refuse in navigable waters to keep them clear for boat traffic, was interpreted in the 1960s as applying to liquid wastes as well. In December 1970, the EPA administrator issued an executive order calling for a water quality management program using permits and penalties based on the Refuse Act of 1899. Although this program was delayed by court challenges in 1971, the Congress made it a central part of the federal strategy embodied in the FWPCA amendments of 1972.

The process of issuing an NPDES permit is discussed above in the context of Tables 4.6 and 4.7. These permits prescribe treatment requirements, construction schedules, and the timing of effluent discharges. Immediately following passage of the 1972 amendments, permits were issued by EPA. However, recognizing the responsibilities of the states in water pollution control, the 1972 amendments included provision for transferring the administration of the permit program to the states. By 1980, more than half the states issued NPDES permits directly.

The FWPCA amendments of 1972 gave the EPA substantial powers to enforce the NPDES permit program. EPA was authorized to seek court injunctions against permit violators. In addition, EPA could issue administrative orders requiring compliance with permit conditions. Failure to comply with these orders exposed the discharger to criminal sanctions and civil penalties.[10] The 1972 amendments also included provisions for "citizens suits" in federal district courts by persons "having an interest which is or may be adversely affected" by violations of discharge requirements.

The permit-based enforcement strategy used after 1972 had an advantage over the earlier enforcement process based on ambient water quality standards. Using the ambient standards, the government had to show that a particular discharge caused a violation in a downstream standard. As mentioned previously, there are difficulties in establishing these causal relations. The permit-based approach was easier to implement, since only effluent monitoring was needed to establish that a particular source was discharging more than its NPDES permit allowed.

[9]This litigation is discussed by Zener (1981, pp. 99–100).

[10]Criminal fines of $25,000/day ($50,000/day for second offenses) could be imposed on any person who "willfully or negligently" violated a permit condition or discharged without a permit. Civil penalties of up to $10,000/day could also be assessed.

Water Pollution Issues in the 1980s

Several aspects of the federal water pollution control strategy came up for administrative reassessment during the early 1980s. One item concerned the diminished authority of the states that followed passage of the federal water quality laws in the 1970s. Many industries and municipalities wanted the states to regain their traditional responsibilities as the principal administrative units in managing water quality. They felt the states would implement water pollution control laws more effectively than EPA, because the states were more familiar with local water problems. A contrary view was that state water pollution control agencies could be easily pressured by industries and municipalities, and this weakened their effectiveness as regulators. In addition, a strong federal role was appropriate because many water quality problems crossed state lines. As a consequence of the reassessment, some responsibilities assumed by EPA during the 1970s were returned to the states during the early 1980s.

Another aspect of the federal strategy that was widely debated in the early 1980s was the lack of economic efficiency in using technology-based effluent standards, especially the stringent BAT limits. Some people felt that the costs of providing BAT frequently exceeded the benefits to downstream water users. In addition, it was argued that any system of uniform effluent standards would be economically inefficient because it could not account for local circumstances that influenced the transport of residuals in water. An ambient standard could be met less expensively if effluent requirements reflected how much each discharge contributed to violating the standard. Those favoring BAT requirements recalled the ineffectiveness of the Water Quality Act of 1965 with its reliance on effluent standards tailored to individual circumstances. Supporters of BAT felt that stringent effluent constraints were essential to minimize the release of water-borne toxic materials. Whether or not the benefits of BAT were worth the high costs could not be assessed because so little was known about the full social costs imposed by discharges of toxic substances.

Controversy regarding economic efficiency also surrounded the federal subsidies for municipal treatment plants. The construction grants program budget had grown from $50 million/year in 1956 to more than $4 billion during some years in the late 1970s. The program was a prime target for federal budget cutters. The Municipal Waste Treatment Construction Grant Amendments of 1981 reduced the portion of treatment plant costs eligible for subsidies. The 1981 amendments also revised the program to give local officials a greater role in establishing priorities for federal grants.

In addition to being a period in which federal strategies were reappraised, the early 1980s was also a time when several emerging water quality problems received increased attention. One problem concerned pollution from sources other than treated and untreated wastewater discharges. Examples include lead and zinc in urban surface water flows ("runoff") following precipitation. Pesticides in agricultural runoff is another instance. The Council on Environmental Quality (1981) reported that because of such "non-point sources" of pollution,

37 states would fail to meet the interim water quality goals specified in the FWPCA of 1972.

A second widely discussed problem concerned the contamination of ground and surface waters with toxic chemicals, especially heavy metals and synthetic organic compounds such as polychlorinated biphenyls (PCBs). During the 1970s, the widespread appearance of toxic substances in water supplies and elsewhere led to the passage of four major federal laws: (1) the Safe Drinking Water Act of 1974, which instructed EPA to establish maximum safe levels of contaminants and include the levels in standards for drinking water; (2) the Resource Conservation and Recovery Act of 1976, which established a "manifest system" allowing government agencies to keep track of a unit of hazardous material from the time it left the "generator" until it was received at an approved disposal site; (3) the Toxic Substances Control Act of 1976, which regulated the production and distribution of any chemical substance presenting "unreasonable risks"; and (4) the "Superfund Act" (passed in 1980), which provided for the removal and clean up of hazardous materials released from inadequate disposal areas. These laws were not easy to implement. Many innovative regulations had to be prepared by the EPA. In addition, several federal administrative units had to coordinate their activities to devise a coherent strategy for managing toxic substances. By the early 1980s, the effectiveness of these four new laws had not been fully proven.

One other widely discussed new problem concerned the deposition of air-borne residuals on water bodies, especially lakes. Some of these residuals reach the earth when gaseous emissions of sulfur and nitrogen oxides are chemically converted to sulfuric and nitric acids and washed out of the atmosphere by rain or snow ("acid precipitation"). Evidence indicated that the dry deposition of acidic materials, for example, as particulates, was also significant. Acid precipitation and dry deposition were held responsible for the acidification of many lakes in eastern North America with consequent negative effects on fish and other aquatic organisms. The acidification of lakes was an international issue. Many Canadians felt their waters were being acidified by air-borne residuals that originated in America and were transported hundreds of miles into Canada. Acidification was not the only difficulty associated with the "fall-out" of air-borne residuals. Measurements showed that atmospheric deposition of metals and organic compounds (in addition to acids) was also causing problems in some lakes.[11] During the early 1980s, the federal government was under considerable pressure to deal with the water quality problems caused by air-borne residuals.

AIR QUALITY MANAGEMENT

There are similarities in the ways that air and water quality management programs evolved in the United States. In each case, there was little federal activity

[11]For information on the atmospheric deposition of trace metals (e.g., zinc and copper) and various organic compounds, see Council on Environmental Quality (1981, pp. 60–62).

before the 1950s. Between 1955 and 1970, the federal air quality management strategies were often modeled on water pollution control policies that had been developed a few years earlier. This occurred because the federal air and water quality laws were formulated by the same congressional committees. Parallels in federal air and water quality management strategies are elaborated below.

Early Interventions: Municipal Ordinances and Nuisance Law

The earliest programs to manage air quality regulated emissions from smokestacks using municipal ordinances. The first antismoke ordinance in the United States was issued by Chicago in 1881. It prohibited the emission of "dense smoke" from stacks and provided fines of up to $50 per day for offenders.[12]

A notable air pollution control effort was initiated by Pittsburgh around the turn of the century. In 1906, when it was still widely known as "Smoky City," Pittsburgh set up a smoke-control program based on education campaigns to improve the combustion equipment and techniques used by industries. This approach was not fully effective. City ordinances passed during the 1940s established emission standards to limit discharges of smoke, fumes, soot, and cinders. A special-purpose municipal agency administered the ordinances and monitored industrial emissions. Programs like the one in Pittsburgh have since been established in many urban areas.

Early efforts to decrease air pollution also relied on private litigation based on the law of public nuisance. Under common-law doctrine, a *public nuisance* is generally defined as "an unreasonable interference with a right common to the public."[13] Public nuisance law has been used by both individual citizens and government officials to reduce emissions of air pollutants and to force dischargers to provide financial compensation for alleged damages. Typically, a judge is required to balance the damages to injured parties against the social value of the activity causing the emission. Under these circumstances, judges make case-by-case determinations of what the emission standards should be and what pollution control devices must be installed. Thus, judges are forced to make technical decisions with little guidance to assure that similar cases are treated in similar ways. This is why nuisance law is not an effective basis for an air pollution control program.

During the 1950s, there was a shift away from nuisance law and municipal ordinances as bases for air quality management. Two factors stimulated the development of additional air pollution control strategies. One was a tragic episode at Donora, Pennsylvania. In 1948, a combination of poor atmospheric ventilation and concentrated emissions from steel mills, smelters, and other plants in and around Donora caused death to 20 people and illness to several thousand. The second factor was the growing recognition of the linkage between

[12]For a description of these early municipal programs, see Bibbero and Young (1974, pp. 1–6).

[13]This definition is used by Stewart and Krier (1978, p. 210). They elaborate on how public nuisance law has been used in the United States to control environmental quality.

automobile exhausts and photochemical smog. In the early 1950s, researchers at the California Institute of Technology identified a photochemical reaction through which hydrocarbon vapors and nitrogen oxides from auto exhausts combined to yield ozone and a number of other highly reactive organic oxidants. This type of smog, which requires sunny days, atmospheric stability, and the type of emissions produced by internal combustion engines, is a major problem in Los Angeles and many other cities.[14]

Early Federal Air Quality Laws: 1955–1967

Prior to 1955 there was no federal involvement in air quality management. With few exceptions, even the states had not seen fit to adopt air pollution control legislation. Up to that time, air quality management was viewed as a concern of city and regional agencies.

The pattern of reliance on local agencies began to change with the passage of the Federal Air Pollution Control Act of 1955 (see Table 5.3). This act established a program of federally funded research grants administered by the U.S. Public Health Service. Although the act represented an expansion of the federal role, it was a very limited one. The legislative history of the act indicates Congress' intent to restrain federal involvement and respect the rights and responsibilities of states, counties, and cities in controlling air pollution.[15]

The federal role was further extended by the Clean Air Act of 1963, which allowed direct federal intervention to reduce interstate pollution. The form of intervention followed the enforcement process in the Federal Water Pollution Control Act of 1956. This three-stage process involved (1) a conference of interested parties to gain agreements regarding necessary pollution controls, (2) a formal public hearing to be convened if the measures agreed upon at the conference were not implemented, and (3) federally initiated litigation as a last resort. Like its counterpart in the water pollution control program, this enforcement process also had an undistinguished record. Fewer than a dozen conferences were called under the 1963 act. Only one case reached the court action stage, and it required 3 years to get there. Even though it permitted federal intervention, the 1963 act continued along the lines of earlier legislation by emphasizing the role of state and local governments in air quality management. The act also continued the federally funded research and set up a program of grants to strengthen state and local air pollution control activities.

Even though scientists had linked photochemical smog to releases from automobile exhausts, a federal effort to control auto emissions was slow in coming. The Motor Vehicle Exhaust Study Act of 1960 provided funds for additional research on vehicle emissions, but it did not regulate them. The first federal restrictions on auto emissions came with the Motor Vehicle Air Pollution Control

[14]This paragraph is based on Kneese (1978); statistics describing the Donora air pollution tragedy are from Turk (1980, p. 261).

[15]This discussion of the history of federal efforts to control air quality relies on Bibbero and Young (1974, pp. 140–163) and Stewart and Krier (1978, 333–337).

TABLE 5.3 Key Federal Laws Controlling Air Pollution

Year	Title	Selected *New* Elements of Federal Strategy[a]
1955	Air Pollution Control Act	Funds for air pollution research
1960	Motor Vehicle Exhaust Study Act	Funds for research on vehicle emissions
1963	Clean Air Act	Three-stage enforcement process Funds for state and local air pollution control agencies
1965	Motor Vehicle Air Pollution Control Act	Emission regulations for cars beginning with 1968 models
1967	Air Quality Act	Federally issued criteria documents Federally issued control technique documents Air quality control regions (AQCRs) defined Requirements for states to set ambient standards for AQCRs Requirements for state implementation plans
1970	Clean Air Act Amendments	National ambient air quality standards New source performance standards Technology forcing auto emission standards Transportation control plans
1977	Clean Air Act Amendments	Relaxation of previous auto emission requirements Vehicle inspection and maintenance programs Prevention of significant deterioration areas Emission offsets for nonattainment areas
1980	Acid Precipitation Act	Development of a long-term research plan

[a]The table entries include only the *new* policies and programs established by each of the laws. Often these provisions were carried forward in modified form as elements of subsequent legislation.

Act of 1965. Based on earlier auto emission control efforts in California, the 1965 act gave the Secretary of the Department of Health, Education and Welfare authority to establish permissible emission levels for *new* automobiles beginning with the 1968 model year. The control of emissions from older vehicles was left to individual states.

Regional Ambient Standards and State Implementation Plans

In 1967, still another increment was added to the level of federal involvement in air quality management. The Air Quality Act of 1967 continued the earlier federal grants for research and for strengthening state and local programs, and it also continued the three-stage federal enforcement process. In addition, the 1967 act borrowed concepts from the Water Quality Control Act of 1965 by requiring states to develop ambient air quality standards and "state implementation plans" (SIPs) to achieve the standards. Implementation plans were to include emission requirements for controlling air pollution and a timetable for meeting the requirements. Deadlines were set for submitting ambient standards, which were to be established on a region-wide basis. If satisfactory standards were not issued on time, they would be developed by the federal government.

As a prelude to the state activity in standard setting, the Secretary of HEW was required to

1 Designate air quality control regions (AQCRs) centered around metropolitan areas.
2 Publish air quality criteria documents for: particulates, sulfur oxides (SO_x), carbon monoxide, hydrocarbons, nitrogen oxides, and photochemical oxidants. The documents were to include the latest scientific knowledge about how the above-listed pollutants influenced human health and welfare.
3 Publish control technique documents for various substances indicating the feasibility and cost-effectiveness of different emission reduction methods.

For each of the AQCRs designated by the Secretary of HEW, the *states* had to adopt ambient air quality standards using a process involving extensive public involvement and the information in the criteria and control technique documents. In addition to setting standards for these AQCRs, the states were required to indicate how the standards would be met. This information was to be part of the state implementation plan.

Although this process for standard setting and plan making appears straightforward, it required much more time to implement than most people anticipated. By April 1970, only 28 AQCRs had been designated by the Secretary of HEW, and not all of the required criteria documents and control technique documents had been issued. Because of this slow start, the states did not make great strides in setting standards and in putting together SIPs. By the spring of 1970, not one state had adopted a full-scale set of ambient air quality standards and implementation plans. This slow progress, together with the high level of public concern for environmental quality, led Congress in 1970 to redirect the federal approach for controlling air pollution.

National Ambient Standards and Federally Imposed Emission Standards

Although the Clean Air Act Amendments of 1970 continued many of the research and state aid programs established by prior legislation, several aspects of the amendments represented dramatic changes in strategy. These involved (1) the requirement that the administrator of EPA set national ambient air quality standards (NAAQS) and emission standards for selected categories of new industrial facilities and (2) the explicit delineation (by Congress) of "technology forcing" auto emission standards.

In 1970, Congress abandoned the process of setting standards on a regional basis. Instead, it required the EPA administrator to prescribe national primary and secondary ambient air quality standards for each of six "criteria pollutants," the substances for which air quality criteria documents had been required under the 1967 act. Primary standards were those which, in the judgment of the EPA administrator, were required to protect the public health. Secondary standards, which were to be at least as stringent as the primary standards, were intended to "protect the public welfare from any known or anticipated adverse effects." For example, the primary standards for sulfur oxides were to protect the population from various heart and lung diseases. The secondary SO_x standards were to protect against potential damage to vegetation.

The use of federally determined ambient standards represented a major departure from earlier reliance on state and local agencies for standard setting. States retained the authority to decide on how to achieve the NAAQS, and these decisions were to be reflected in their implementation plans. However, the states were not entirely free to make decisions since the SIPs were subject to federal approval. If a state did not comply with the 1970 Clean Air Act Amendments, it risked the loss of federal funds for air pollution control programs and the prospect of having the federal government establish its SIP.

Another manifestation of the expanded role of the federal government is the 1970 amendments' requirement that the EPA administrator issue "new source performance standards" (NSPS). These standards were to control new stationary sources categorized by the administrator as contributing significantly to air pollution. Examples of the types of facilities subjected to NSPS include portland cement plants, nitric acid plants, and municipal incinerators. In setting performance standards for each category of facilities, the EPA administrator was required to determine the "best system of emission reduction which (taking into account the cost of achieving such reduction)" had been "adequately demonstrated." These regulations were very controversial. The EPA administrator was often sued by industries who argued that the proposed NSPS were not technologically feasible or they were too costly in light of the benefits in improved air quality. Sometimes it was even charged that the proposed NSPS would cause environmental degradation, for example, when the removal of air-borne particulates would generate large quantities of solid waste.

The 1970 amendments also changed the federal approach to controlling automobile emissions. In this instance, Congress itself specified the details of what it expected in the way of reduced emissions. In frustration over the auto industry's lack of rapid progress in controlling emissions, Congress mandated a 90% decrease in hydrocarbon and carbon monoxide emissions (from 1970 levels) by 1975 and a 90% decrease in emissions of nitrogen oxides (from 1971 levels) by 1976. These reductions were technology forcing in that they went beyond what the auto industry viewed as feasible at the time the amendments were passed. Whether or not this congressional action forced the auto companies to do more effective research on controlling emissions is disputable. The original 1975–1976 standards were not attained on schedule. The Clean Air Act Amendments of 1977 relaxed the emission requirements somewhat and extended the compliance deadlines into the early 1980s. Although the modified requirements were eventually met, the early 1980s was a period of continuing debate concerning the need for additional emission cutbacks, especially for vehicles with diesel engines.

Plans for Transportation Control and Vehicle Inspection

Recognizing that restraints on stationary sources and auto emissions might not be sufficient to attain the national ambient air quality standards, Congress included a section in the Clean Air Act Amendments of 1970 requiring that SIPs include land use and transportation controls "where necessary to achieve and maintain air quality standards." The EPA administrator responded cautiously to this portion of the act since land use controls in the United States are traditionally the province of local governments. In fact, the EPA administrator only acted on this matter after he was sued. The environmental groups who initiated the litigation argued that EPA was required to issue transportation control plans (TCPs) if the NAAQS would not be met using other means.[16]

One of the first transportation control plans issued by the EPA administrator was for the Los Angeles region. In 1970, the NAAQS for photochemical oxidants were exceeded in the Los Angeles area on 250 days, and there was little hope of attaining the standard by regulating only stationary sources and auto emissions. The California SIP should have included a transportation control plan for the Los Angeles area, but it did not. Thus, EPA issued a plan that included vehicle inspections to assure the proper functioning of emission controls, retrofitting of old autos with pollution control devices, and an 82% reduction in "vehicle miles traveled" by means of gasoline rationing, car pooling, and vehicle free zones. This and other transportation control plans issued by EPA under the 1970 amendments were never implemented. The negative reaction to the federally imposed TCPs was so strong that Congress revised the TCP requirements in the Clean Air Act Amendments of 1977.

Under the 1977 amendments, any state that could not meet the primary NAAQS by 1982 could obtain a 5-year extension if it (1) implemented a program for the inspection and maintenance of auto emission control devices, and (2) revised its implementation plan to include appropriate TCPs. These transportation control plans were to be designed by the people who would be responsible for implementing them, local transportation planners and elected officials.

[16]For an account of this suit and its effects, see Stewart and Krier (1978, pp. 442–443).

The importance of vehicle inspection and maintenance is highlighted by "in-use" emissions data in a report by the National Commission on Air Quality (1981). These data indicate that large numbers of vehicles meeting applicable emission standards when they are new fail to meet those standards after a year or two on the road. Factors contributing to these failures include lack of proper maintenance, deliberate tampering with emission control systems, and fouling of catalytic converters by use of leaded gasoline. Under the 1977 amendments, states that did not implement required vehicle inspection programs could lose federal subsidies for highway and wastewater treatment plant construction. Despite this, many states had serious problems in setting up inspection programs. Difficulties resulted from a resistance to increased government regulation and concerns about the costs of inspections and follow-up repairs.

Many states developed transportation control plans in response to the 1977 amendments. These plans included the following approaches to reducing the number of vehicle miles traveled:

1 "Pooling" – Providing incentives for car and van pooling.
2 "Flex time" – Staggering work hours to reduce peak-hour traffic.
3 Priority lanes – Designating highway lanes for use only by buses and "pools."
4 Parking – Making it difficult to park, for example, by eliminating parking subsidies to employees.
5 Public transit – Developing and improving rail and bus systems.

Although transportation controls were implemented following passage of the 1977 amendments, the projected impact of these controls varied widely. Many regional transportation agencies estimated that the maximum pollution reductions obtainable by their plans would be only 5 to 8%. In contrast, EPA estimated that pollution reductions could be as high as 15 to 20% if the TCPs were "more innovative" (Seltz-Petrash, 1979).

New Sources and the Attainment of Standards

Soon after national ambient air quality standards had been promulgated, two important growth-related issues began to be debated. One concerned areas where the ambient air quality was *better* than required by the national standards. Should new sources of emissions be permitted to degrade the air quality in such areas down to the NAAQS levels? A second issue involved "nonattainment areas," regions that had not yet attained the national ambient standards. Since the standards were not met in those areas, should they be allowed to accept *any* new sources? These questions were argued extensively during the 1970s. The positions Congress took on these matters in 1977 are summarized below.

Prevention of Significant Deterioration Congress was unwilling to accept the prospect of allowing areas with air quality already above the national ambient standards to degrade to the lower limits set by the standards. Neither was it willing to exclude the possibility of having any air quality degradation in those areas. A compromise position was based on the concept of "prevention of significant deterioration" (PSD).

The 1977 amendments required that an area meeting the national ambient standards for a given air pollutant be declared a "PSD area" for that pollutant. The amendments also defined three classes of PSD areas. For each class, numerical limits indicated the maximum permissible increment of air quality degradation from all new (or modified) stationary sources of pollution in an area. Class I, which was mandatory for areas containing important national resources like the Grand Canyon, had the lowest allowable increments of degradation. For example, the allowable increase for total suspended particulates was 5 $\mu g/m^3$ measured on an annual average basis. The highest allowable changes were for Class III areas. In these cases, the total suspended particulates increment was 37 $\mu g/m^3$ measured on an annual average basis. Except for special places like national parks, each state could classify its own PSD areas.

The 1977 amendments required that state implementation plans include provisions for preconstruction review of significant new stationary sources in PSD areas. A source was considered "significant" if it would emit more than 100 tons/year. For a significant new source to be allowed into a PSD area it had to meet stringent emission reduction levels specified by the EPA administrator. In addition, the increment of air quality degradation caused by the source had to be within the numerical limits of allowable degradation available in the region. Additional requirements were imposed on new sources in PSD areas where visibility was an important issue. These added restrictions were intended to protect spectacular vistas that were threatened by air pollution in the western states.

Offsets in Nonattainment Areas The 1977 amendments also indicated that significant new sources could locate in areas that did *not* meet the NAAQS, but only if certain conditions were satisfied. The amendments required that a significant new source locating in a nonattainment area had to meet strict emission reduction requirements developed by the EPA administrator. In addition, discharges from the new source had to be more than offset by reductions in emissions from other sources in the region. After the "emission offsets" were applied, the net effect had to be reasonable progress toward meeting the NAAQS in the region. The Wickland Oil Company example in Chapter 4 illustrates the use of emission offsets (see Table 4.13).

Innovative Policies for Trading Emission Reductions

In 1979, the EPA extended the concept of emission offsets, as used in nonattainment areas, to a different context: multiple sources of air pollution generated at a single site. This extension, known as the *bubble policy,* is illustrated in Figure 5.1. The figure depicts a firm that must control releases from smokestacks at two adjacent plants. Before the bubble policy, the firm had to comply with

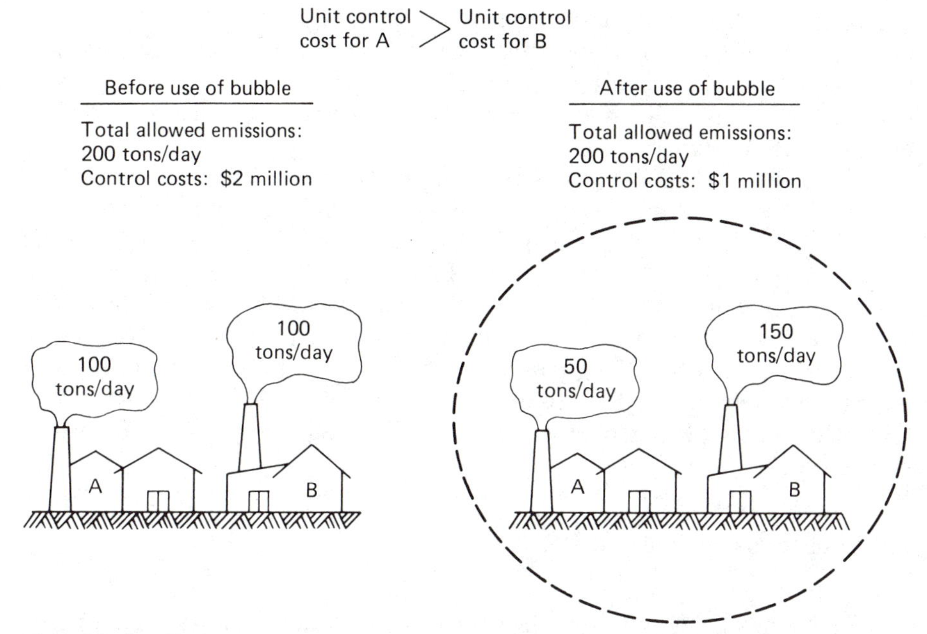

FIGURE 5.1 Reducing costs using the "bubble policy." Adapted from Environmental Protection Agency (1982).

emission standards that allowed only 100 tons/day from each plant. The total discharge was 200 tons/day. The unit cost of emission controls for plant A was much higher than that for plant B, but the emission requirements were insensitive to these cost differences. Using the bubble policy, the firm is free to decide how to reduce residuals at each plant. The only restriction is that its total discharge must be no greater than 200 tons/day. Imagine that a bubble surrounds the two plants. The policy allows the firm to make choices within the bubble, but the total discharge from the bubble is constrained. In the early 1980s, the original bubble policy was extended to include plants that were not at the same location ("multiplant bubbles").

The aim of the bubble policy was to have firms decrease the costs of achieving a particular level of residuals reduction. The policy allowed plant managers to exchange extra decreases in emissions at easily controlled sources for increases in emissions at sources that were expensive to control. Opportunities for cost savings are illustrated by a case involving the Narragansett Electric Company in Rhode Island.[17] The utility had two power plants located ¼ mile apart. Before applying the bubble policy, the sulfur dioxide releases at each plant resulted from using fuel oil with a 1% sulfur content. In applying the bubble concept, the utility planned to switch to a high-sulfur oil (2.2% sulfur) at one plant. In

[17]This description of the Narragansett Electric Company case is from the National Commission on Air Quality (1981, p. 277).

exchange, the SO_2 emissions at the other plant would be reduced by either burning natural gas or not operating when the high-sulfur fuel was used. According to EPA (1982), this application of the bubble policy would save Narragansett $3 million/year. It would also decrease overall SO_2 emissions from the two plants by 10%.

In the early 1980s, EPA took steps that encouraged firms to use the bubble policy. This involved increasing the ways a firm could use "emission reduction credits" (ERCs), reductions in emissions that exceeded those required by a state implementation plan. Under revised EPA rules, a firm could use ERCs in transactions involving either emission offsets or the bubble policy. Firms could also store ERCs in "banks" and use them at a future time. In addition, a firm could make profits by selling ERCs to other firms that needed them to meet federal and state air quality requirements. Interfirm transfers of ERCs were restricted so that ERC trading would not increase total emissions or otherwise violate state or federal laws.

Air Pollution Issues in the 1980s

The early 1980s witnessed a major reevaluation of the federal strategies for air pollution control. There was evidence that parts of the Clean Air Act Amendments of 1977 were difficult to administer and in need of streamlining. Also, the high costs of meeting Clean Air Act requirements [estimated by CEQ (1980) to have been about $22 billion in 1979] provided pressure to devise more economical strategies. The EPA rules on trading emission reduction credits illustrate the regulatory reforms introduced to save money while meeting air quality goals.

An issue that received increased attention in the early 1980s concerned unregulated toxic materials that were being discharged into the environment. The EPA administrator had authority to issue emission standards for hazardous materials under the Clean Air Act Amendments of 1970. However, as of 1981, standards had been issued for only four hazardous substances: asbestos, beryllium, mercury, and vinyl chloride. During the early 1980s, EPA attempted to improve its procedures for regulating hazardous substances, and Congress was urged to provide legislation to expedite the regulation of air-borne toxic wastes.

The early 1980s was also a time when indoor air pollution emerged as a potentially significant problem. Increasing levels of indoor air contamination were linked to the energy conservation activities of the 1970s. Efforts to reduce air leaks by means of extensive caulking and weather stripping had substantially cutback the rate of exchange between indoor and outdoor air. In some instances, air exchanges were so low that indoor air became stale. In addition, the increased use of gas stoves and unvented space heaters were linked to high indoor concentrations of carbon monoxide and nitrogen oxides. High concentrations of formaldehyde, a suspected carcinogen, were also measured in a variety of indoor settings. There was a clear need for research to gauge the public health risks associated with contaminated indoor air.[18]

[18]For a technical discussion of indoor air pollution, see Wadden and Scheff (1983).

Some of the most troublesome air quality problems of the early 1980s involved acid precipitation and dry deposition, issues introduced in the previous discussion of water pollution. Actual and potential damages to lakes and forests in much of eastern North America were of great concern in both the United States and Canada. The acidification problem was addressed by the Acid Precipitation Act of 1980 (Public Law 96-294), which set up an interagency task force to establish a long-range research program. Although the U.S. air quality legislation of the 1970s was not adequate to deal with acid precipitation and dry deposition, many U.S. officials argued that an effective federal policy could not be formulated until additional scientific research was completed. A good scientific basis for policy making was viewed as essential since the control strategies being considered involved enormous costs. A contrary perspective was that further delays in dealing with the acidification problem would have disasterous effects on numerous lakes in both the United States and Canada. Because of the strains they caused in U.S.–Canada relations and the uncertainties regarding the effectiveness of proposed control strategies, acid precipitation and dry deposition appeared as troublesome as any of the air quality issues needing attention in the 1980s.

KEY CONCEPTS AND TERMS

WATER QUALITY MANAGEMENT
- State water pollution control agencies
- Enforcement conferences, hearings, and court actions
- Construction grants for municipal facilities
- Water quality criteria and standards
- Environmental Protection Agency
- Technology-based effluent standards
- Best practicable technology
- Best available technology
- Effluent limitations guidelines
- NPDES permits
- Nonpoint sources of pollution
- Heavy metals
- Synthetic organic compounds
- Atmospheric deposition of pollutants

AIR QUALITY MANAGEMENT
- Public nuisance law
- Local ordinances controlling air quality
- Health effects of air pollution
- Auto emissions and photochemical smog
- Air quality criteria
- National ambient air quality standards
- Technology forcing auto emission standards
- New source performance standards
- Transportation control plans
- Vehicle inspection and maintenance
- Prevention of significant deterioration areas
- Nonattainment areas
- Emission offsets, banking, and bubbles
- Emission reduction credits
- Unregulated toxic air-borne residuals
- Indoor air pollution
- Acid precipitation and dry deposition

DISCUSSION QUESTIONS

5–1 Consider the three stage enforcement process introduced as part of the federal air and water quality programs in the 1950s and 1960s. In what sense could the introduction of these programs be viewed as a bold step forward? Why was the three-stage process *not* used as the central thrust of the enforcement progams employed by the EPA in the 1970s?

5–2 In the mid-1960s, federal air and water quality management strategies included the following elements: (1) establishment of ambient standards by state and/or local units of government and (2) development of "implementation plans" in which the states indicated measures to be taken to meet the ambient standards. This approach was modified fundamentally in the 1970s by the establishment of national ambient air quality standards and by the reliance on technology-based effluent standards. Why were these modifications introduced?

5–3 The following pattern was common in the 1970s: Congress sets out a general residuals management policy and an overall strategy. The EPA administrator is left to work out implementation details by publishing rules and regulations which have the force of law. What are the problems with this approach to law making?

5–4 As of 1950, the public and private resources devoted to managing wastewaters were greater than those devoted to managing air-borne residuals. Moreover, the latter were much greater than the resources devoted to controlling hazardous substances. Why do you think this was the case?

5–5 Defend the federal requirements for transportation control plans and vehicle inspection and maintenance programs. Think of alternative strategies for reducing emissions from mobile sources of air-borne residuals in areas that are violating the NAAQS for photochemical oxidants or carbon monoxide.

5–6 Emission offsets permit new discharges to enter an area that is already exceeding its ability to provide waste receptor services. Using an offset scheme, acceptable environmental quality can be maintained. What other strategies could be used to allow new growth in an area that has no additional waste receptor services to offer?

REFERENCES

Bibbero, R. J., and I. G. Young, 1974, *Systems Approach to Air Pollution Control.* Wiley, New York.

Binkley, C., B. Collins, L. Kanter, M. Alford, M. Shapiro, and R. Tabors, 1975, *Interceptor Sewers and Urban Sprawl.* Heath, Lexington, Mass.

Cleary, E. J., 1967, *The ORSANCO Story—Water Quality Management in the Ohio Valley under an Interstate Compact.* Johns Hopkins University Press for Resources for the Future, Baltimore, Md.

Council on Environmental Quality, 1980 (1981), *The Eleventh (Twelfth) Annual Report of the Council on Environmental Quality.* Council on Environmental Quality, Washington, D.C.

Environmental Protection Agency, 1982, *The Bubble and Its Use with Emission Reduction Banking.* Office of Policy Resource Management, Washington, D.C.

Freeman, A. M., III, R. H. Havemen, and A. V. Kneese, 1973, *The Economics of Environmental Policy.* Wiley, New York.

Hey, D. L., and N. H. Waggy, 1979, Planning for Water Quality: 1776–1976, *Proceedings of the American Society of Civil Engineers, Journal of the Water Resources Planning and Management Division,* Vol. 105, No. WR1 (March), pp. 121–131.

Kneese, A. V., 1978, A Commentary on Needed Changes in the 1970 Air Quality Act Amendments, *in* A. F. Freidlaender (ed.), *Approaches to Controlling Air Pollution,* pp. 433–447. MIT Press, Cambridge, Mass.

National Commission on Air Quality, 1981, *To Breathe Clean Air.* U.S. Government Printing Office, Washington, D.C.

National Utility Contractors Association, 1975, *Fulfilling a Promise—An Analysis of the 1972 Clean Water Act's Construction Grants Program.* National Utility Contractors Association, Washington, D.C.

Quarles, J., 1976, *Cleaning up America: An Insider's View of the Environmental Protection Agency.* Houghton–Mifflin, Boston.

Seltz-Petrash, A., 1979, Transportation Planners Join Battle for Cleaner Air. *Civil Engineering* **49** (11), 84–88.

Stewart, R. B., and J. E. Krier, 1978, *Environmental Law and Policy.* Bobbs–Merrill, Indianapolis.

The Conservation Foundation, 1982, *State of the Environment, 1982.* The Conservation Foundation, Washington, D.C.

Turk, J., 1980, *Introduction to Environmental Studies.* Saunders, Philadelphia.

Velasquez, M. G., 1982, *Business Ethics, Concepts and Cases.* Prentice–Hall, Englewood Cliffs, N.J.

Wadden, R. A., and P. A. Scheff, 1983, *Indoor Air Pollution: Characterization, Prediction, and Control.* Wiley, New York.

Whipple, W., Jr., 1977, *Planning of Water Quality Systems.* Lexington Books–Heath, Lexington, Mass.

Zener, R. V., 1981, *Guide to Federal Environmental Law.* Practicing Law Institute, New York.

PART THREE

ENVIRONMENTAL IMPACT ASSESSMENT

CHAPTER 6

ENVIRONMENTAL IMPACT STATEMENTS AND GOVERNMENT DECISION MAKING

During the late 1960s, many people felt that public works were significantly degrading the quality of the environment. Among the objectionable projects were highways that destroyed scenic landscapes, power plants that belched pollutants from ugly smokestacks, and water projects that drained ecologically valuable marshlands. Critics argued that the environmental consequences of these projects were not being accounted for in decision making. Rather, it was charged, government planners used only traditional decision criteria based on economic efficiency. There was much interest in forcing government agencies to seriously consider environmental factors when they formulated and evaluated alternative plans.

During the 1970s many industrialized nations established policies requiring that environmental impacts be weighed in government agency decision making. This chapter examines a widely known program, the environmental impact statement process established by the U.S. National Environmental Policy Act of 1969 (NEPA). The chapter also reviews requirements in various states and countries calling for the environmental assessment of proposed government actions.

NEPA'S OBJECTIVES AND PRINCIPAL PARTS

The fundamental aim of NEPA was to force all agencies of the federal government to integrate environmental concerns into their planning and decision making. The act's structure reveals how this was to be accomplished.

Although NEPA contains many provisions, it can be described in terms of three main parts: a policy statement, a set of "action forcing" provisions, and the creation of a new agency, the Council on Environmental Quality (CEQ). Table 6.1 summarizes these three units.

The first portion of NEPA contains a declaration of national policy. In Section 101(A), Congress declares

> *[that] it is the continuing policy of the Federal Government . . . to use all practical means and measures . . . to create and maintain conditions under which man and nature can exist in productive harmony and fulfill the social, economic and other requirements of present and future generations of Americans.*

Section 101(B) lists six additional points that clarify the "continuing responsibility" of the federal government. These items concern such things as assuring "safe, healthful, productive, and esthetically and culturally pleasing surroundings" and maintaining "wherever possible, an environment which supports diversity and a variety of choices." Section 101(B) also makes it clear that Congress used the term *environment* in a very broad sense to include social and cultural dimensions as well as physical and biological factors.

The first three items in Section 102(2) are among the principal action forcing provisions of NEPA. Section 102(2)(A) requires that specialists from a broad range of professional disciplines be involved in federal agency planning and decision making. The act requires that

> *[federal agencies shall] utilize a systematic, interdisciplinary approach which will insure the integrated use of the natural and social sciences and the environmental design arts in planning and in decision making which may have an impact on man's environment.*

TABLE 6.1 Principal Parts of the National Environmental Policy Act

Declaration of national environmental policy
- Broad policy statement — Section 101(A)
- Responsibilities of the federal government — Section 101(B)

Action forcing provisions — all agencies of the federal government shall
- Utilize an interdisciplinary approach to planning — Section 102(2)(A)
- Develop procedures to give environmental factors "appropriate consideration" in decision making — Section 102(2)(B)
- Prepare environmental impact statements — Section 102(2)(C)

Creation of the Council on Environmental Quality

Section 102(2)(B) of NEPA indicates that Congress did *not* intend to prohibit federal agencies from making decisions that had adverse environmental consequences. Instead, Congress required a balancing of effects and considerations. Section 102(2)(B) instructs federal agencies to

> *identify and develop methods and procedures . . . which will insure that presently unquantified environmental amenities and values may be given* appropriate consideration *in decision making along with economic and technical considerations. [Emphasis added]*

Section 102(2)(C) requires agencies to prepare a "detailed statement" of environmental impacts for "major Federal actions significantly affecting the quality of the human environment." This provision increased substantially the amount of information an agency considered and disseminated before it made its decisions. The content and use of these "environmental impact statements" (EISs) is considered later in this chapter.

Finally, the last part of NEPA created the Council on Environmental Quality in the Executive Office of the President. Among other things, the Council was directed to assist the president in preparing an annual environmental quality report, appraise federal agency performance in implementing the action forcing provisions of NEPA, conduct environmental studies and research, and advise the president on environmental matters.

NEPA'S ENVIRONMENTAL IMPACT STATEMENT PROCESS

The decision-making process used by federal agencies has been modified fundamentally by the EIS requirements of Section 102(2)(C). These requirements apply to virtually all federal agencies.

Many of the actions for which environmental impact statements are written involve specific federal projects, such as dams built by the U.S. Army Corps of Engineers. Many others involve decisions by federal agencies to grant permits or licenses. An example is the Nuclear Regulatory Commission's decision to give a utility a license to construct a nuclear power plant. In this case, the utility would probably need permits from several federal agencies. The agencies themselves would decide which one is to serve as the "lead agency" in preparing an environmental impact statement. Still other types of actions involve federal grants for state or local projects, such as subsidies for constructing interstate highways.

In a typical year, the majority of EISs are prepared for specific projects such as federal reservoirs and federally funded highways or housing developments. Less frequently, an EIS might involve not an individual action but a group of actions that are viewed collectively as an agency program. Examples of actions which have been the subject of "programmatic EISs" include the Forest Service's vegetation control activities in Arizona and New Mexico, and the Energy Research and Development Agency's liquid metal fast breeder reactor program.

Early Steps in the EIS Process

The CEQ publishes regulations indicating the procedures to be followed in preparing an EIS. These regulations allow agencies to identify categories of actions that are minor and unlikely to have a significant effect on environmental quality. Such actions can be "categorically excluded" from Section 102(2)(C) requirements.

For all federal actions that are not categorically excluded, a preliminary analysis is first used to determine whether anticipated impacts will be significant enough to require an EIS. If the agency finds that no significant environmental impacts are likely to occur, it follows Council on Environmental Quality (1978a) procedures for making that information public. The agency's decision not to prepare an EIS is subject to public review and may be challenged by citizens.

If the agency's preliminary analysis indicates that significant environmental impacts are likely, it must prepare an EIS and give all interested parties an opportunity to comment on its analyses and conclusions. Before initiating in-depth environmental studies, an agency determines the scope of issues to be addressed in its EIS. In this "scoping" exercise, the agency identifies individual citizens, interest groups, and other agencies likely to be interested in its environmental impact statement process. These parties are invited to participate in agency planning by making known their concerns about possible impacts. The scoping exercise also involves synchronizing the environmental analyses and the agency's decision making timetable. Environmental analyses must be scheduled so that the analytic results can be used in choosing among alternative proposals.

A formal requirement for scoping was introduced by CEQ to oppose the view that Section 102(2)(C) requires *only* the preparation of a legally defensible EIS. In the early 1970s, many agencies became so preoccupied with preparing environmental impact statements capable of withstanding legal challenges that they lost sight of the role of these documents in decision making. This occurred in reaction to the many court challenges which environmental groups raised against the adequacy of EISs in the early 1970s. A common objection was that an agency's EIS was inadequate because it either left out important impacts or predicted impacts poorly using inappropriate analytic techniques. In the face of such accusations, agencies added more and more pages of technical documentation to their EISs. Frequently the result was a legally defensible EIS that few people read completely because of its excessive length and complexity. Some critics charged that these lengthy documents were often prepared very late in an agency's decision process, and they only defended projects that had been planned without considering environmental impacts. The scoping requirement, introduced by CEQ in 1978, was intended to force agencies to integrate the EIS process into their planning and decision making.

Forecasting and Evaluating Environmental Impacts

Once the scope of an environmental impact study is estimated, an agency typically forms a team of specialists to carry out the necessary analyses. The appropriate composition of the team depends on the type of project involved. For complex projects, an ideal team might include environmental scientists, such as biologists and hydrologists, as well as engineers, social scientists, and design professionals. For a team to function effectively as a unit, there needs to be good communication among members. Communications can be fostered by having team members visit the project area together and meet frequently. Sometimes teams fail to work as a unit, and members contribute their skills to the environmental impact study without benefit of collaboration.

A principal task of those preparing an EIS is to forecast environmental impacts of alternative actions. For any particular action, an impact is defined as *the difference between the future state of the environment if the action took place and the state if no action occurred.*[1] Chapter 7 discusses the types of procedures commonly used in making predictions of future environmental conditions.

There often are many direct and indirect impacts associated with an action that requires an EIS. Direct impacts are those that follow immediately and obviously from an action or project. For example, a direct impact of a new reservoir is the inundation of land that was once dry. The indirect effects may be far removed from the new reservoir in both time and distance. For example, a possible indirect effect of a new reservoir on a coastal stream in California is the destruction of ocean beaches. This may occur long after the initial construction and result because sediments that once flowed to the ocean to replenish the beaches are trapped behind the new reservoir.

The goal of the EIS process is to have federal agencies use the forecasts of future environmental consequences to evaluate alternative actions thoughtfully. This evaluation includes environmental, economic, and engineering considerations. As elaborated in Chapter 8, the process of evaluation requires value judgments about how much environmental quality is to be "traded-off" to gain increases in economic benefits or other dimensions of human well-being. The EIS process involves many such judgments. The outcome is a recommendation by the agency regarding the action it wishes to pursue ("the proposed action" in an EIS).

Required Content of an EIS

An environmental impact statement provides a record of an agency's efforts to formulate alternative actions, predict environmental consequences, and decide on a recommended course of action. Requirements concerning the details that

[1]This widely used definition of an environmental impact is from Munn (1979, p. xvii). He clarifies much of the specialized terminology used in the environmental impact assessment literature.

must appear in an environmental impact statement have evolved gradually since 1970.

The starting point for determining the content of an EIS is Section 102(2)(C) of NEPA. It requires that an EIS include information on the following items:

- Environmental impacts of the proposed action.
- Adverse effects which cannot be avoided if the action is implemented.
- Alternatives to the proposed action.
- "The relationship between local short-term uses of man's environment and the maintenance and enhancement of long-term productivity."
- "Any irreversible and irretrievable commitments of resources which would be involved" if the proposed action were carried out.

The many judicial interpretations of NEPA have had a great influence in determining the required content of an EIS. During the 1970s there were over 1000 court actions alleging that agencies failed to comply with NEPA. Many of these charged that an agency's EIS was incomplete. The courts were thus put in the position of deciding what constituted an adequate EIS. Their judgments influenced the Council on Environmental Quality (1978a) regulations, which specify generally what an EIS should contain. The courts also influenced the more specific EIS content requirements elaborated in the NEPA procedures of individual federal agencies.[2]

According to the Council on Environmental Quality regulations, the part of an EIS that treats alternative actions constitutes "the heart of the environmental impact statement." The regulations call for more than a mere listing of the actions that the agency considered. They require a comparative analysis of the environmental consequences of the alternatives. The comparison must include the "no-action" alternative, the related activities likely to occur if the federal agency does not take positive action. Suppose, for example, that the proposed action in an EIS is an agency decision granting a license to a railroad. In this case, the no-action alternative is made up of the predictable public and private actions expected to take place if the license is not granted. This includes projected increases in freight transport by competing modes such as trucks, barges, and airplanes. The comparative analysis of alternatives is intended to give decision makers and the public an understanding of the environmental and economic costs and gains associated with each alternative.

Public and Agency Review of an EIS

Part of Section 102(2)(C) of NEPA obligates an agency preparing an impact statement to "consult with and obtain the comments of any Federal agency which has jurisdiction by law or special expertise with respect to any environ-

[2]Liroff (1976) analyzes many NEPA-related court actions brought in the early 1970s. A more recent review of cases is given by the Council on Environmental Quality (1979). Both sources were used in preparing this discussion.

mental impact involved." It also requires that copies of the EIS and the views of commenting agencies be made available to the president, CEQ, and the general public. These requirements of NEPA have yielded an elaborate process involving the circulation of the EIS in draft form, the preparation of review comments by recipients of the "draft EIS," the revision of the draft by the issuing agency and the distribution of a "final EIS" (see Figure 6.1). This review

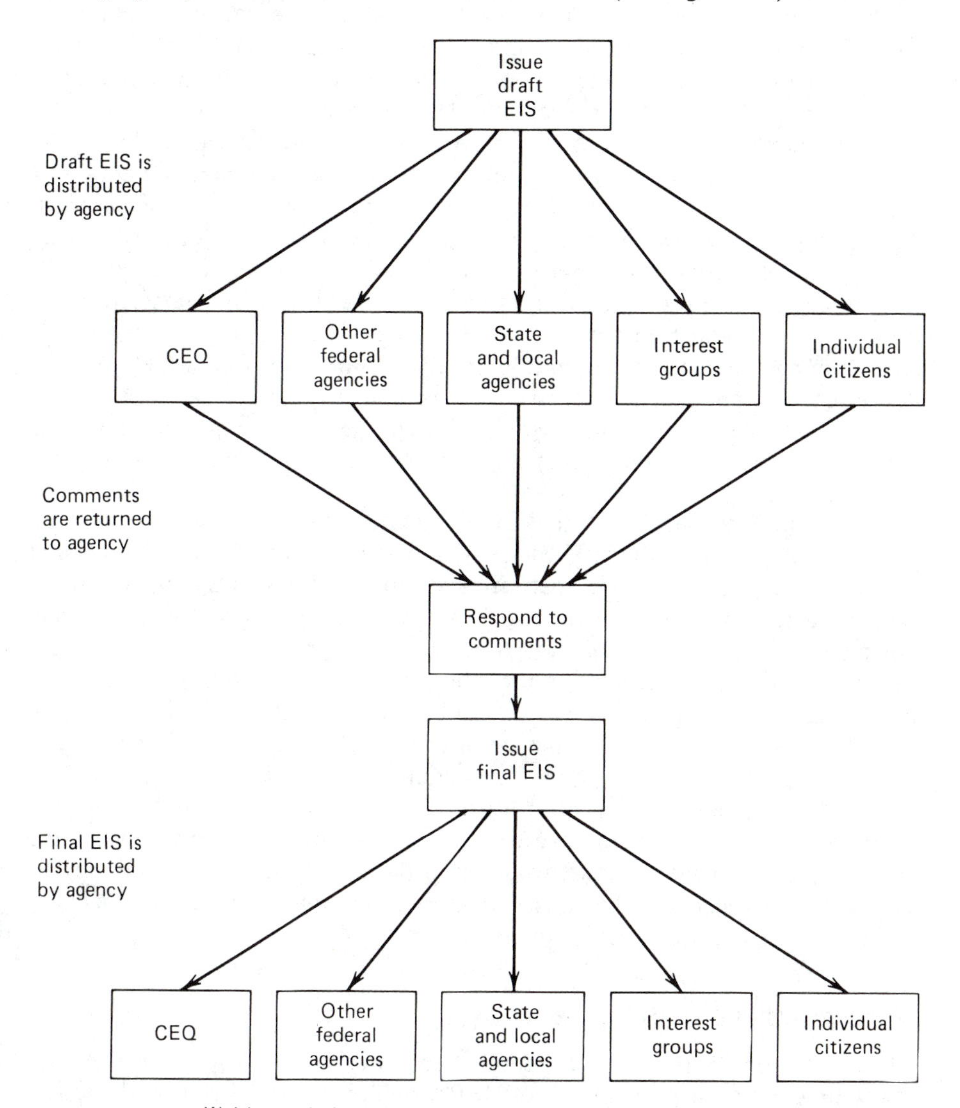

FIGURE 6.1 Process of review and comment on a draft EIS. Adapted from R. K. Jain, L. V. Urban, and G. S. Stacey, *Environmental Impact Analysis, A New Dimension in Decision Making,* 2nd ed., Van Nostrand Reinhold Co., copyright © 1981 by Litton Educational Publishing, Inc.

process provides opportunities for a full scrutiny and critique of the issuing agency's environmental analysis methods and its rationale for selecting a proposed action.

Comments resulting from the review of a draft EIS cover a wide range. Some may laud the agency's proposed action, whereas others may be disapproving. Critics might question the accuracy of facts in the draft EIS or the procedures used in forecasting impacts. They might also accuse the agency of failing to consider important environmental effects. Some commentators might request the agency to either undertake studies of actions which were not considered or abandon its proposal in favor of an alternative.

The agency issuing the EIS must respond to each criticism raised during the review of the draft. If the agency feels that a comment warrants no action, it must give a reason for its position. More substantive responses take many forms. The agency might add to its EIS by examining new alternatives or by analyzing environmental impacts not considered previously. If reviewers point out significant impacts that were omitted from the draft EIS, the agency might respond by proposing "mitigation measures." These are actions that either reduce an adverse impact or compensate for it in some way. For example, a water resource development agency might preserve an area valued as an elk habitat to "mitigate" a reservoir's inundation of an equivalent land area containing a similar habitat. In some cases, the agency might even substantially modify its proposal or abandon it entirely.

As indicated in Figure 6.1 a final EIS is issued after the agency has responded to the comments on the draft. The agency cannot implement the action proposed in the final EIS until interested parties have had a chance to see how the agency responded to the review comments. Those who are dissatisfied with the final outcome and wish to stop or modify the agency's action must, at this point, resort to traditional mechanisms for influencing administrative processes: political pressure or litigation. Pressure can be exerted, for example, by lobbying the Congress or by exerting influence through CEQ or others in the executive branch. The courts have been used frequently by private parties who seek to influence an agency's actions after a final EIS has been distributed. However, when all NEPA procedures have been complied with, there are only limited circumstances in which a court would stop the agency from implementing its proposal. As explained by Liroff (1976), these conditions involve agency decision making that is "arbitrary and capricious" in a legal sense.

Assessments of the EIS Process

Effects of the EIS process on federal agency decision making have been appraised many times. Assessments of NEPA's influence have been conducted by CEQ, individual agencies, and independent observers such as Andrews (1976) and Liroff (1976). The following points are frequently made in these assessments: although the EIS process is costly in both resources and time, it often leads to increased public participation in agency planning and to improved coordination among agencies. In addition, the process sometimes affects decisions and leads to projects that more fully account for environmental values and concerns.

Many effects of the EIS process on federal agency decision making cannot be analyzed systematically because they are unrecorded. For example, consider an agency that wants to undertake a project with only a few notable adverse impacts. Suppose the agency avoids the expense and time required to prepare a full-blown EIS by modifying the project to eliminate the few negative effects. In this case, the EIS process clearly affected the project outcome. However, there might be no official record of this influence. Subtle changes like this occur frequently and make it difficult to fully evaluate the EIS process.

Many effects of the EIS process have been overviewed in Wichelman's (1976) study of NEPA's implementation in the 1970s. His analysis includes the four phases shown in Table 6.2. The *interpretive* and *formal compliance* phases occurred during the first few years after the passage of NEPA. During these phases, the requirements of the EIS process were clarified by the courts and CEQ. Environmental planners in various federal agencies were just beginning to understand what constituted an adequate environmental impact analysis and how it might be conducted with a limited budget. During these phases, the EIS process was *not* well integrated into agency planning. Often the EIS process was simply "tacked on" after the key decisions had been made. The emphasis was on producing a legally adequate impact statement.

A number of noteworthy organizational changes took place within some federal agencies during the first few years after 1970. New organizational units were created to house the staffs responsible for EIS preparation, and people assigned to these units participated in special environmental training programs. In addition, many environmental professionals were hired, some of whom brought

TABLE 6.2 Successive Phases of NEPA Implementation[a]

Interpretive Phase

Existence of NEPA's requirements is recognized by agency leadership, but little substantive change in agency activity takes place

Formal Compliance Phase

Changes are made to comply with the act, but they are largely oriented toward meeting legal procedural requirements

Integrated Planning Phase

Changes are made to integrate NEPA implementation activities into agency decision procedures for individual actions

Programmatic Planning Phase

Changes are made to integrate environmental values into the formulation of proposed agency programs, policies, regulations, and legislation

[a]Based on information in Wichelman (1976, pp. 297–300).

outlooks and values previously absent or poorly represented in agency planning. Agency budgets were formulated with spending requirements for environmental impact statements in mind.

Wichelman argues that the continuing review provided by the courts and CEQ forced many agencies to the *integrated planning* phase by the mid-1970s. In this phase, agencies placed less emphasis on the EIS itself and more importance on making decisions based on the information generated during the EIS process. The need to integrate environmental considerations into federal agency decision making was a central theme in the CEQ regulations issued in 1978. Evidence presented by the Council on Environmental Quality (1981) suggests that this integration was, in fact, being carried out by the end of the 1970s.

The final phase in Table 6.2, *programmatic planning,* involves the integration of environmental concerns in the formulation of entire programs and policies. Judging from the small number of programmatic EISs prepared throughout the 1970s, this phase appears to be out of the reach of most federal agencies.

STATE ENVIRONMENTAL IMPACT REPORTING REQUIREMENTS

Six months after NEPA was enacted, the Commonwealth of Puerto Rico initiated the first of the "little NEPAs." These are laws modeled on NEPA and directing various state and local agencies to consider the environmental effects of their own actions. By 1979, a total of 18 states had comprehensive environmental impact reporting requirements established either by legislation or by executive or administrative orders (Table 6.3). In addition, nine states had special purpose environmental assessment requirements. An example is Nebraska's use of environmental impact reporting for state-funded highway projects.

TABLE 6.3 States with EIS Requirements[a]

States using Comprehensive Statutory Requirements[b] ***(14)***

California, Connecticut, Hawaii, Indiana, Maryland, Massachusetts, Minnesota, Montana, New York, North Carolina, South Dakota, Virginia, Washington, and Wisconsin

States using Comprehensive Executive or Administrative Orders (4)

Michigan, New Jersey, Texas, and Utah

States with Special Purpose EIS Requirements (9)

Arizona, Delaware, Georgia, Kentucky, Mississippi, Nebraska, Nevada, New Jersey,[c] and Rhode Island

[a]Based on information in Council on Environmental Quality (1979, pp. 595–602).

[b]The Commonwealth of Puerto Rico also has comprehensive statutory requirements.

[c]New Jersey has EIS programs established under a state executive order *and* special "procedural rules" for implementing state laws governing the use of wetlands and coastal areas.

For most state programs, the points that must be covered in environmental impact documents are similar to those in Section 102(2)(C) of NEPA. A few states require the consideration of additional items. For example, the California Environmental Quality Act calls for an assessment of the "growth-inducing impact" of proposed actions and a description of "mitigation measures" that could be taken to minimize adverse impacts.

Although they are similar to NEPA as far as EIS content requirements are concerned, the state programs differ from the federal program in several ways. For one thing, most states do not have an agency with wide-ranging authorities to monitor compliance and otherwise play a substantive overseer role. Arrangements for administering environmental impact reporting requirements vary from state to state. In some cases, the EIS program is managed by the state department of natural resources. Other states rely on their environmental protection agencies. Another administrative form is an environmental council or board in the governor's office, such as the Office of Environmental Quality Control in Hawaii.[3]

State programs differ greatly in terms of their applicability to private development proposals. The California statute has the most wide-ranging coverage. It applies to state-initiated actions, such as highway projects, as well as a variety of decisions made by cities, counties, and regional agencies. The local agency actions, which include the granting of building permits and zoning variances, have made the California impact assessment requirements applicable to proposals made by private parties. Environmental impact reports have been written for virtually thousands of private land development projects in California. At the opposite extreme, EIS requirements in some states apply only to a single type of activity. An example is Delaware's program requiring EISs in connection with permits issued under its Coastal Zone Act and its Wetlands Law. In other states, the environmental impact requirements are neither as pervasive as those in California nor as narrowly applied as those in Delaware. Table 6.4 provides examples of the actions covered by state EIS programs.

TABLE 6.4 Range of Actions Subject to State EIS Requirements

Decisions at state *and* local levels, including agency-initiated projects and permits, leases, licenses, and entitlements (California)[a]
Only state actions, but including state permits granted for private projects (Wisconsin)
State actions and, in some circumstances, private actions requiring decisions by lower levels of government (North Carolina)
Only state actions, but *excluding* license and permit decisions (Indiana)
Only specific types of actions, such as state-funded highway projects (Nebraska)

[a]Parentheses are used to show a state in which the indicated actions are covered. Table entries reflect conditions in the mid 1970s as reported by the Council on Environmental Quality (1974, 1976) and Burchell and Listokin (1975).

[3]Trzyna (1974) describes the many organizational arrangements used in administering little NEPAs.

Because states differ greatly in the types of actions for which environmental impact documents are required, there are big differences in the effects of the programs and in the number of documents produced. In California, which has one of the most ambitious programs, several hundred environmental impact documents might be produced in a year. In contrast, there are some states where only 5 to 10 EISs per year are prepared.

The courts have played an important role in interpreting the state counterparts to NEPA. A noteworthy instance is *Friends of Mammoth v. Mono County,* a suit filed under the California Environmental Quality Act.[4] The case concerned a "conditional use permit" that had been issued for a condominum project by the Mono County Board of Supervisors in 1971. Mono County is in eastern California near Yosemite National Park and was characterized in the California Supreme Court's opinion on the case as "one of the nation's most spectacularly beautiful and comparatively unspoiled treasures." The plaintiffs argued that Mono County's action in granting the conditional use permit would lead to significant adverse environmental impacts, and that the decision should have been accompanied by an environmental impact report. In deciding in favor of the plaintiffs, the California Supreme Court concluded that the "legislature necessarily intended to include within the operations of the act, private activities for which a government permit or other entitlement of use is necessary." This opinion was given in September 1972, and the California Environmental Quality Act was amended soon thereafter. The amendments indicated that the act's environmental impact requirements applied to a broad range of local government decisions, including actions to change zoning ordinances and to issue zoning variances and conditional use permits. The decisions of the California Supreme Court and the state legislature made the California environmental impact program a major force in integrating environmental values into local land use planning.

Another illustration of a significant court action involving state environmental review requirements is *Polygon Corporation v. City of Seattle*.[5] The Polygon Corporation applied to the city of Seattle for a permit to build a 13-story condominium. An EIS, prepared in accordance with the State Environmental Policy Act (SEPA), reported several adverse effects of the project, including obstructed views and increased traffic noise. On the basis of these anticipated effects, the Seattle superintendent of buildings did not grant the permit. The Polygon Corporation sued the city of Seattle arguing that SEPA required only that an environmental impact statement be prepared and circulated. The corporation felt that SEPA alone did not confer the authority to deny a permit on grounds related to the act's environmental policy objectives. In rejecting the Polygon Corporation's analysis, the Washington State Supreme Court argued that "SEPA confers on the City . . . the discretion to deny a building permit appli-

[4]The presentation of the Friends of Mammoth case is based on Rodgers (1976, pp. 67–68).

[5]A more complete description of the Polygon Corporation case is given by the Council on Environmental Quality (1978b, p. 405).

cation on the basis of adverse environmental impacts disclosed by an EIS." This decision by the Washington Supreme Court supports the view that state laws modeled on NEPA may go beyond procedural issues concerning the preparation of EISs and actually force the integration of environmental values into decision making by state and local governments.

Although approximately half of the states have EIS requirements of one form or another, the effectiveness of the state programs has been highly variable. Many programs have been hampered by inadequate funding for the preparation and review of environmental documents. States that have made substantial commitments to their EIS programs have the *potential* for using them effectively in an array of public and private decision processes. However, adequate funding does not assure program effectiveness. As Wandesforde-Smith (1981) has documented, even in a state as committed to environmental impact reporting as California, the results have not been uniformly effective.

ENVIRONMENTAL IMPACT ASSESSMENT REQUIREMENTS OUTSIDE THE UNITED STATES

The same factors that motivated the enactment of NEPA also led other countries to develop requirements for environmental impact assessments (EIAs) during the 1970s. Chief among these factors was the public's concern over environmental degradation caused by large-scale projects such as highways, dams, mines, and electric power plants. Countries that developed environmental impact assessment requirements include Australia, Canada, Japan, Thailand, and some members of the European Economic Community.

Many countries paid careful attention to the influence of NEPA on federal agency planning in America. Of special interest was the role of the U.S. courts in enforcing EIS requirements and increasing the ability of citizens to affect agency decisions. Many countries wanted to avoid giving such power to the courts and allowing citizens to observe the internal decision processes of administrative agencies. Thus, EIA programs were often structured to preserve the discretionary powers of government authorities. Environmental impact assessment requirements were generally established either administratively using existing authorities or by means of new legislation. Events in the Federal Republic of Germany and Australia demonstrate these approaches.

West Germany's "Principles" for EIA

During the early 1970s, officials in the Federal Republic of Germany debated the merits of passing a law such as NEPA. They decided that requirements for environmental impact assessments of federal actions could be introduced more effectively using a "cabinet resolution," a form of administrative guidance that applied to the various agencies of the federal government.[6]

[6]The discussion of EIA in the Federal Republic of Germany relies on an analysis by Kennedy (1981).

In 1975, a cabinet resolution entitled "Principles for the Environmental Impact Assessment of Federal Actions" recommended procedures to be followed in conducting assessments. The "Principles" defined the content of an EIA and indicated the circumstances in which an assessment should be conducted. A large number of federal projects and programs were potentially covered by the Principles.

If a federal agency decided to conduct an EIA, the Principles guided the overall exercise. They required that the environmental consequences of proposed federal actions be predicted and evaluated. The Principles also urged authorities to find ways of avoiding or reducing adverse environmental effects. Environmental considerations were to be balanced against other public and private concerns in reaching final decisions.

Because the Principles took the form of a cabinet resolution and not a law, they were not enforceable by the courts. Unlike individuals and citizens' groups in the United States, those in West Germany had no mechanism to force an agency to conduct an environmental impact assessment. The individual federal ministries decided whether an EIA was required in any particular circumstance, and they had considerable latitude in making their determinations. For example, several ministries felt that federal actions regulated by existing pollution control and land planning statutes were exempted from meeting EIA requirements. This provided an "escape clause" that was sometimes used by agencies to avoid implementing the Principles.

In studying the effects of the Principles in the 3½ years after they were issued, Kennedy (1981) found no visible modifications in agency planning. The only change he saw was that several federal ministries issued internal memoranda recognizing the existence of the Principles. In addition, the Ministry of Transportation issued a statement explaining why the Principles did not apply to its activities. Kennedy concluded that the Principles had not led to an open decision process that allowed outsiders to judge whether the agencies were making decisions that were more sensitive to environmental values.

Because the internal workings of federal ministries in West Germany are not open to public view, it is difficult for observers to comment definitively on changes effected by the Principles. From the evidence available to him, Kennedy concluded that West Germany was far from integrating environmental factors into agency decision making. A response to this criticism was that, since government officials are committed to it, EIA will gradually become a routine component of decision making by federal agencies. The evidence suggests that the process will be slow without something more forceful than voluntary compliance to make the federal ministries apply the Principles.

Australia's Legislated "EIS Process"

In 1974, the Australian Parliament passed "The Environmental Protection (Impact of Proposals) Act," EP(IP)A. The act included provisions for "public environmental inquiries" to make government officials and the public more aware of the environmental issues at stake in a particular circumstance. Only two inquiries were held in the 1970s, both times for decisions related to mining. The EP(IP)A also established an EIS process for projects undertaken by the national (commonwealth) government in Australia. The process involved preparing a draft EIS, circulating it for public comment, and issuing a final EIS before the project was implemented. Although it appears similar to the EIS process used in the United States, the administration of the EIS process under EP(IP)A is different from that of its American counterpart.[7]

A key distinction between the processes used in Australia and America concerns the circumstances under which an EIS is required. In America, any federal agency action can potentially require an EIS, and citizens may legally challenge an agency's decison not to prepare one. In Australia, the requirements are more complex. The minister responsible for a proposed action (the "action minister") is the only one who can decide that an EIS is required. If it appears that a proposal will have significant impacts, the action minister may refer it to the federal Department of Environment. The environmental department then offers its opinion on whether an EIS may be required and refers the matter back to the action minister. If the Department of Environment has judged that an EIS *may* be required, then the action minister uses criteria specified in the procedures for implementing the EP(IP)A to determine whether an EIS should be prepared. This process leaves much room for administrative discretion. For example, the action minister has two opportunities to exercise discretion: once in making the initial referral and once in acting on the opinion of the environmental department. Hollick (1980) reports that between 1974 and 1980 only about 55 impact statements were prepared under EP(IP)A. In the corresponding time period, the number of impact statements prepared under NEPA was well over 5500.[8]

The comparatively small number of EISs prepared in response to the Australian law reflects both the power of the federal ministers to exercise discretion and the political climate that prevailed in Australia in the late 1970s. The EP(IP)A was passed originally by a labor government which was committed to a federal role in environmental management. In 1975, it was replaced by a conservative government that wanted to leave environmental management in the hands of the states. The conservative government could accomplish this without repealing EP(IP)A because the federal ministers controlled how the act was implemented. The ministers could not be sued by citizens' groups in the same way that federal agencies in the United States were sued for failing to implement NEPA. The High Court of Australia did not uphold the right of citizens to sue the commonwealth government for failure to comply with the procedures of EP(IP)A.

As in America, the EIS process in Australia takes place at both the national

[7]The discussion of Australian EIS requirements is based on reports by Hollick (1980) and Formby (1981).

[8]This figure is based on the Council on Environmental Quality's (1979, p. 589) estimate that about 11,000 EIS's were prepared between January 1, 1970 and December 31, 1978. In 1979, another 1400 EISs were issued (Council on Environmental Quality, 1980, p. 384). Thus, about 1250 EISs were prepared in an average year during the 1970s.

and the state levels. The six states in Australia have considerable power and play an important role in environmental management. As of 1980, all states had published procedural guidelines for environmental impact assessments. The guidelines differed markedly from one state to the next. According to Formby (1981), highly industrialized New South Wales had the most extensive program involving EIAs and public environmental inquiries of any government in Australia. In contrast, the EIA program in the much less industrialized state of Tasmania had not been used widely.

As of 1980, the environmental impact assessment programs at the commonwealth and state levels in Australia were operating largely at the discretion of government authorities and achieving mixed results. Although some administrations were not implementing their environmental impact procedures frequently, Australia still had a more vigorous approach to EIA than many other countries. When major physical development projects were proposed, environmental impact assessments were often used effectively to raise the level of consideration given to environmental factors. In Formby's (1981) words, a "creeping environmentalism" had set in. The various EIA programs initiated during the 1970s were gradually raising the environmental awareness of private developers, government officials, and the public at large.

KEY CONCEPTS AND TERMS

NEPA's OBJECTIVES AND PRINCIPAL PARTS
- Statement of national environmental policy
- Action forcing provisions
- Mandate for interdisciplinary planning
- "Appropriate consideration" of environmental values
- Environmental impact statement
- Council on Environmental Quality

NEPA'S ENVIRONMENTAL IMPACT STATEMENT PROCESS
- Lead agencies
- Actions requiring an EIS
- The "programmatic EIS"
- Scoping process
- "No-action" alternative
- EIS review procedures
- Comments on draft EIS
- Issuance of a final EIS
- Court actions under NEPA

STATE ENVIRONMENTAL IMPACT REPORTING REQUIREMENTS
- NEPA as a model
- Differences among state requirements
- EISs for private development projects

ENVIRONMENTAL IMPACT ASSESSMENT REQUIREMENTS OUTSIDE THE UNITED STATES
- Administratively established EIA
- Administrative discretion
- Enforceability of EIA requirements
- Public environmental inquiries

DISCUSSION QUESTIONS

6–1 Give four illustrations of actions that are subject to NEPA requirements and do not involve the construction of a facility by a federal agency. At least one example should involve a project constructed by a private party.

6–2 Suppose you were working in the office of a federal agency and your principal task was to prepare environmental impact statements for actions proposed by your office. How would you explain the most important aspects of your job to a visitor from a foreign country?

6–3 To what extent do you think the requirements for review and consultation on environmental impact statements have led to the attainment of NEPA's policy goals? Would you be in favor of minimizing those requirements as a way of reducing the cost of complying with NEPA? Justify your position.

6–4 How would you characterize the role of the courts in the implementation of NEPA? Why were the courts called upon so frequently to litigate cases involving agency compliance with NEPA?

6–5 The term *cumulative environmental impacts* is often used when a collection of minor effects from different actions combine to cause a significant environmental impact. For example, a project to line a short section of a stream with concrete may cause no notable effects on aquatic life. However, the combined effects of a dozen such projects could create major disturbances to fish and other species. How might a programmatic EIS be used to consider cumulative impacts? What other approaches can be used to account for cumulative impacts?

6–6 To what extent do the little NEPAs differ from each other and from NEPA? In what sense is the California Environmental Quality Act more far reaching than NEPA?

6–7 Suppose you are called upon to advise the ministry of environment in a western European country on whether or not it should advocate the adoption of a law like NEPA. What advice would you offer based on what you know about experiences with NEPA and little NEPAs in the United States? How would you modify your response if the ministry was in a developing country?

6–8 Compare the federal environmental impact assessment requirements in the United States, West Germany, and Australia. Indicate how differences in administrative discretion and court enforceability account for the differences in the implementation of EIA programs in these countries.

REFERENCES

Andrews, R. N. L., 1976 *Environmental Policy and Administrative Change: Implementation of the National Environmental Policy Act.* Lexington Books, Lexington, Mass.

Burchell, R. W., and D. Listokin, 1975, *The Environmental Impact Handbook.* Center for Urban Policy Research, Rutgers University, New Brunswick, N.J.

Council on Environmental Quality, 1974, *The Fifth Annual Report of the Council on Environmental Quality,* U.S. Government Printing Office, Washington, D.C.

Council on Environmental Quality, 1976, *The Seventh Annual Report of the Council on Environmental Quality.* U.S. Government Printing Office, Washington, D.C.

Council on Environmental Quality, 1978a, Regulations for Implementing the Procedural Provisions of NEPA. *Federal Register* **43,** 55, 978–56,007.

Council on Environmental Quality, 1978b, *The Ninth Annual Report of the Council on Environmental Quality.* U.S. Government Printing Office, Washington, D.C.

Council on Environmental Quality, 1979, *The Tenth Annual Report of the Council on Environmental Quality.* U.S. Government Printing Office, Washington, D.C.

Council on Environmental Quality, 1980, *The Eleventh Annual Report of the Council on Environmental Quality.* U.S. Government Printing Office, Washington, D.C.

Council on Environmental Quality, 1981, *The Twelfth Annual Report of the Council on Environmental Quality.* U.S. Government Printing Office, Washington, D.C.

Formby, J., 1981, The Australian Experience, in T. O'Riordan, and W. R. D. Sewell (eds.) *Project Appraisal and Policy Review,* pp. 187–225. Wiley, Chichester, England.

Hollick, M., 1980 Environmental Impact Assessment in Australia: The Federal Experience. *Environmental Impact Assessment Review* **1** (3), 330–336.

Kennedy, W. V., 1981, The West German Experience, in T. O'Riordan, and W. R. D. Sewell (eds), *Project Appraisal and Policy Review*. pp. 155–185. Wiley, Chichester, England

Liroff, R. A., 1976, *A National Policy for the Environment, NEPA and Its Aftermath*. Indiana University Press, Bloomington.

Munn, R. E. (ed.), 1979 *Environmental Impact Assessment, Principles and Procedures,* 2nd ed. Wiley, Chichester, England

Rodgers, J. L., Jr., 1976, *Environmental Impact Assessment, Growth Management and the Comprehensive Plan.* Ballinger, Cambridge, Mass.

Trzyna, T. C., 1974, *Environmental Impact Requirements in the States: NEPA's Offspring,* report prepared for Office of Research and Development. U.S. Environmental Protection Agency, Washington, D.C.

Wandesforde-Smith, G., 1981, The Evolution of Environmental Impact Assessment in California, in T. O'Riordan, and W. R. D. Sewell (eds.), *Project Appraisal and Policy Review,* pp. 47–76. Wiley, Chichester, England.

Wichelman, A. F., 1976, Administrative Agency Implementation of the National Environmental Policy Act of 1969: A Conceptual Framework for Explaining Differential Response. *Natural Resources Journal* **16** (2), 263–300.

CHAPTER 7

APPROACHES TO FORECASTING ENVIRONMENTAL IMPACTS

Following the passage of the National Environmental Policy Act, federal agencies needed to develop techniques for analyzing environmental impacts. As a result, government agencies, consultants, and university and private research groups produced an avalanche of what are termed *environmental assessment methods.* The volume of literature produced is indicated by Canter's (1979) review of assessment procedures, which includes about 175 references to different approaches. An extensive bibliography is given by Clark, Bisset and Wathern (1980).

Impact assessment methods can be divided into two broad categories corresponding to basic planning activities: forecasting and evaluation. Forecasting consists of predicting the environmental impacts of alternative actions and is considered in this chapter. Evaluation, the subject of Chapter 8, is the process of putting relative values on different impacts and establishing a preference ordering among alternative plans. Using these definitions, value judgments about whether predicted impacts are good or bad are part of evaluation, not forecasting.

AIDS TO IMPACT IDENTIFICATION

Much of the early literature on environmental assessment procedures did not emphasize techniques for forecasting and evaluating impacts. Rather, it identified the types of impacts typically associated with a particular kind of project or activity. Information of this sort was used to help organize environmental impact studies and to suggest topics for further investigation.

The "checklist" is a simple aid to impact identification. Many federal agencies in the United States have produced checklists to help their staffs review draft EISs prepared by other agencies. The "Environmental Impact Statement Guidelines" developed by the Region X Office of the Environmental Protection Agency (1973) provide an example. These guidelines include separate checklists for highways, dredging and spoil disposal, land management, airports, water resource developments, nuclear power plants, and pesticide use. Each list indicates the issues and analyses that EPA reviewers expect to see in an EIS involving a project (or activity) of a given type.

Some checklists denote only broad categories of impacts that may be associated with a particular project type. An example is the list developed by Arthur D. Little, Inc. (1971) to aid in identifying the environmental effects of transportation projects. It includes eight categories of impacts and a much larger number of subcategories. For instance, one of the eight categories is "soil erosion"; it contains two subcategories, "economics and land use" and "pollution and siltation." This particular list emphasizes that a transportation project may cause impacts during any one of several phases: planning and design, construction, and operations. For instance, under "planning and design," the checklist includes impacts associated with land speculation that may occur in anticipation of a new transportation facility. It also considers the dislocation of families that sometimes occurs when property is purchased to construct transportation projects.

Information useful in identifying environmental effects is also given in surveys of literature on impacts for a given project type. Examples are the reviews for transportation facilities by Bigelow-Crain Associates (1976) and for land development projects by Berns (1977). The literature reviews often catalogue information on environmental changes caused by past projects. Sometimes they include case studies demonstrating particular effects. For a given type of impact, the reviews often refer to documents describing commonly used forecasting procedures.

JUDGMENTAL APPROACHES TO FORECASTING

Forecasting plays a central role in environmental impact assessment. Indeed, an environmental impact is generally defined as a projected change in the value of one or more measures of environmental quality. Forecasts are typically made for a number of alternative plans, including the no-action alternative which consists of related events likely to occur if the agency pursues none of *its* alternatives. The no-action forecasts provide a base against which the environmental conditions for other alternatives can be compared.

The most frequently used approaches to forecasting impacts rely heavily on expert opinion. In this context, an "expert" is someone with special knowledge useful in forecasting. Thus, for example, a real estate agent might be considered an expert in predicting the land use changes induced by a proposed highway.

More generally, the term *expert* refers to a person with extensive academic training and practical experience relevant to the forecasting task. An expert's familiarity with impacts caused by similar projects in analogous settings commonly plays a key role in forecasting.

To illustrate how expert opinion is used in predicting environmental impacts, consider a proposal that involves filling a marsh to construct a large housing project. Suppose a biologist with extensive knowledge of marsh ecosystems is asked to forecast the biological impacts of the project. The biologist would employ standard scientific references and field investigation methods to characterize the project area from a biological perspective. If time were available, he or she might search for information on the observed effects of housing projects in similar settings. The forecast would consist of an opinion based on the above-mentioned information and the biologist's understanding of how marsh ecosystems function and respond to disturbances of the type proposed.

The judgments of specialists are also commonly used in forecasting the *social* impacts of projects. McCoy's (1975) approach to predicting the effects of a highway proposed for the Georgetown neighborhood in Lexington, Kentucky, is illustrative. The city highway department and the mayor's office were consulted for details regarding the proposed highway. Statistical descriptions of Georgetown were obtained from the U.S. census, and field visits provided further information about local residents and their activities. A questionnaire was administered to find out how residents spent their time, how they perceived their neighborhood, and how much they knew about the proposed highway and the process of relocating their residences. In addition, the literature on projects having social impacts on urban neighborhoods was reviewed. All of this information provided the basis for McCoy's projections of how the highway would affect the residents of Georgetown.

Individual specialists make forecasts of environmental impacts by working either independently or in small groups. The advantage of a group is that the experts can build on each other's ideas and thereby provide a more useful forecast. However, there are some potential difficulties in using ordinary meetings as a format for conducting group forecasting exercises. One is that a few vocal individuals may dominate the deliberations of a group. Another possible problem is that some specialists may feel pressured to give opinions in conformity with others in the group.

The "Delphi method" is one of the many procedures developed to increase the effectiveness of experts in making forecasts as a group.[1] It avoids the problems above by not relying on group meetings. The method obtains the opinions of experts by means of mail questionnaire surveys. Throughout the exercise, the anonymity of individuals providing responses to the questionnaires is preserved. Several iterations (or "rounds") of the questionnaire survey are used

[1]Armstrong (1978) and Porter et al. (1980) discuss numerous techniques to elicit forecasts from groups of experts. The procecures are sometimes referred to collectively as "judgmental" (or subjective) forecasting methods.

to give individuals a chance to revise their previous forecasts based on the judgments of others in the group. Opinions are shared by means of statistical summaries of responses from preceding rounds. The summaries are mailed out with each successive round of the questionnaire. Statistical measures, such as the median and the interquartile range of the forecasts, are employed in these summaries to reduce group pressure to arrive at a concensus forecast.

The Delphi method was used by Cavalli-Sforza et al. (1982) to forecast changes in land use resulting from each of three transportation projects in San Jose, California. They organized a group of 12 experts including transportation planners, engineers, economists, city officials, and representatives of citizens' groups. Each specialist received a questionnaire requesting numerical estimates of future population, housing, employment, and transit use in each of several zones impacted by the proposed projects. The results from the first-round questionnaire were summarized and mailed to each of the original respondents with a request that they revise their previous projections where appropriate. Experts whose forecasts deviated significantly from the median values were asked to rationalize their responses, and their explanations were included anonymously as part of the between-round summary information. This entire process was repeated three times. Although Cavalli-Sforza and her colleagues experienced difficulties in maintaining the enthusiasm of participants to complete the long and complex questionnaire, the land use forecasts finally obtained appeared quite reasonable. However, because the projections were for the year 1990, it is impossible to comment on their accuracy.

PHYSICAL MODELS IN FORECASTING

Physical models are small-scale three-dimensional representations of reality. They have been used to make predictions for thousands of years and are familiar, in some form, to almost everyone. A centuries-old example is the architect's model of a proposed building.

In addition to depicting individual buildings, physical models can show how the appearance of an entire landscape or cityscape would change with the addition of a new project. This is demonstrated by the model of downtown San Francisco developed by the Environmental Simulation Laboratory at the University of California, Berkeley (see Figure 7.1). Using special camera equipment, films could be made showing what a pedestrian would observe in walking through various parts of the city. The model was applied in forecasting the visual effects of proposed high-rise office buildings in San Francisco. Scale models of the buildings were set within the model of San Francisco, and films were made to portray views with the new office buildings in place. Models like this have simulated the visual impacts of proposed projects in a variety of settings. They have also been used to estimate how proposed high-rise buildings would cast shadows and thereby decrease sunlight on city streets.

Physical models can also serve in predicting the impacts of projects on water bodies. Models are especially useful in analyzing tidal estuaries because the

FIGURE 7.1 Berkeley Environmental Simulation Laboratory's model of downtown San Francisco. (Photograph by Kevin Gilson. Courtesy of the Environmental Simulation Laboratory, University of California, Berkeley.)

complex mixing of fresh and salt water that occurs in these water bodies makes it difficult to apply other forecasting techniques. Physical models of estuaries have provided forecasts of how navigation projects and port facilities influence water surface elevation, current velocity, salinity, and waste dispersion characteristics. Predictions have also been made of the effects on estuaries of reducing fresh-water inflows, such as occurs when an upstream reservoir is developed for water supply purposes.

Estuary models generally include an array of electrical and mechanical devices to represent the effects of tides and the inflow of fresh waters. Instruments are included for measuring water depth, velocity, temperature, and the concentrations of salinity and various dyes used to simulate wastewater discharges. Figure 7.2 shows a small portion of the Corps of Engineer's model of Chesapeake Bay, the largest estuary in the United States. This model is built to a horizontal scale of 1 ft = 1000 ft and occupies about 9 acres. It is scaled such that 1 year of real time can be simulated in 3.65 days.

The design of estuary models is based on scientific principles. However, these principles alone are not sufficient to guarantee an accurate representation of reality. Extensive measurements of water circulation and quality characteristics in a real estuary must be obtained to "verify" the model of that estuary. This is a time-consuming process in which model features, such as the roughness of

FIGURE 7.2 Portion of the physical model of Chesapeake Bay. (Courtesy of the U.S. Army Corps of Engineers.)

the model's surface, are changed gradually until the behavior of the model replicates that of the estuary. Models can be adjusted to provide accurate forecasts of how physical changes within an estuary influence water surface elevation and velocity. However, many authorities feel these models should not be relied on for quantitative predictions of changes in water quality variables.[2]

[2]The limitations of physical models of estuaries are elaborated by Tracor, Inc. (1971, p. 494).

Small-scale physical models are also used to forecast how air-borne residuals will be transported after they are discharged. Figure 7.3 shows results from an air pollutant dispersion modeling study using a wind tunnel at an EPA research laboratory. Model dimensions were determined using physical principles and the pollutant dispersion patterns in the model provided a reasonable approximation to corresponding patterns in the atmosphere. The model used fans to simulate wind conditions and tracers to represent air-borne residuals. Electronic instruments measured the velocities and concentrations of emissions downwind of the discharge point.

Figure 7.3 indicates how the concentration distributions (or "plumes") of residuals are projected to vary under different combinations of stack height and building width. The plume associated with the narrow building is almost the same as the plume with no building at all. However, the wide building causes a significant amount of "plume downwash," the descending movement of air-borne residuals downwind from the discharge. The results shown qualitatively in the figure were confirmed by quantitative measurements of tracer concentrations at different points downwind of the smokestack. This type of wind tunnel experiment can serve to forecast concentrations of air-borne residuals emitted from smokestacks near collections of buildings.

FORECASTING WITH MATHEMATICAL MODELS

Mathematical models are also used to predict environmental impacts. These "models" are constructed from combinations of algebraic or differential equations. They are generally based on either scientific laws or statistical analyses of data, or both.

Predicting Bacterial Concentrations Using Scientific Principles

The law of conservation of matter is the foundation for most models used in forecasting water and air quality impacts. This principle states that the mass of a substance leaving a specified volume of water (or air) must equal the inflow of the substance plus any production and minus any decay or change in storage within the volume unit. In other words,

$$\text{Outflow} = \text{inflow} + \text{production} - \text{decay} - \text{change in storage}$$

Use of this "mass balance" concept in model building is demonstrated below for coliform bacteria, a commonly employed indicator of water pollution by human feces. Although coliform bacteria are usually not harmful in themselves, they signal the possible presence of microorganisms capable of causing diseases such as typhoid fever and cholera.

A model to forecast the coliform concentration (in number of organisms per unit volume of water) downstream of a municipal wastewater discharge is developed by writing the mass conservation equation for bacteria in a specified

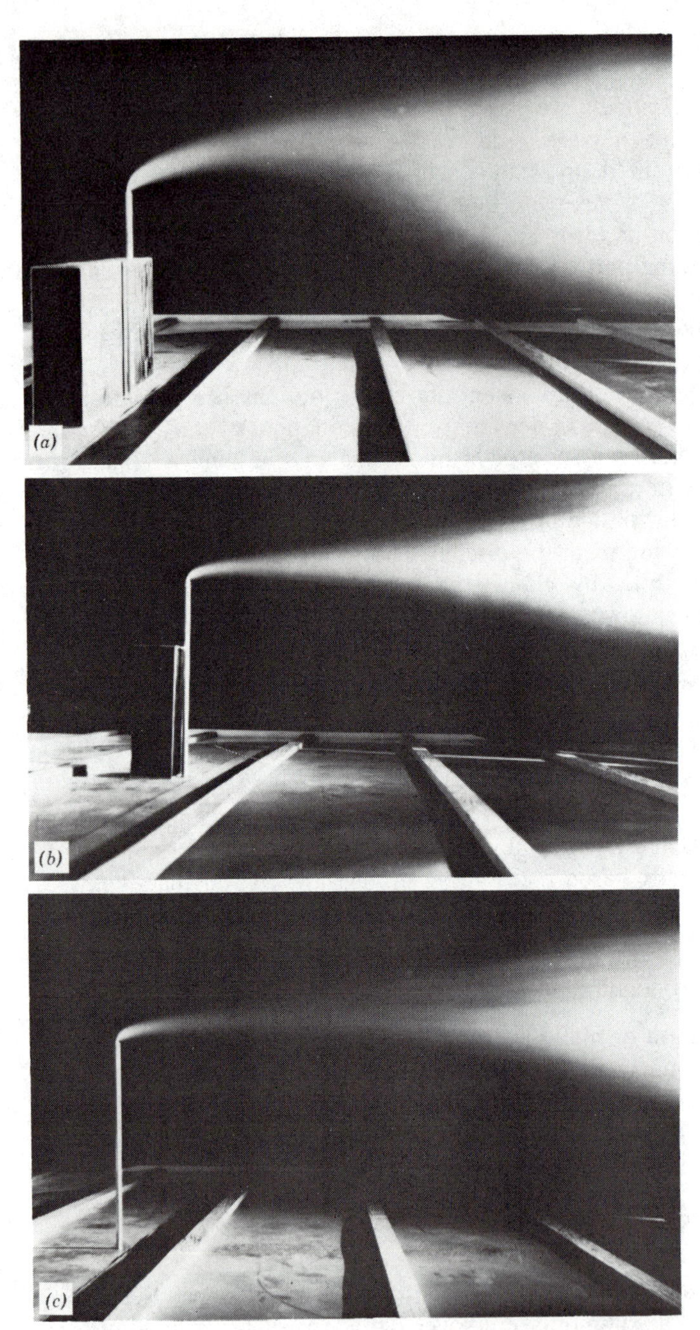

FIGURE 7.3 EPA model studies of air pollutant dispersion. Reprinted with permission from *Atmospheric Environment,* Vol. 10, W. H. Snyder and R. E. Lawson, Jr., "Determination of a Necessary Height for a Stack Close to a Building—A Wind Tunnel Study," Pergamon Press, New York. (Photos supplied courtesy of the Fluid Modeling Facility, Environmental Protection Agency, Research Triangle Park, North Carolina.) (*a*) Building width twice its height. (*b*) Building width one third its height. (*c*) Stack without building.

volume unit within the stream. For simplicity, suppose the stream flow and the wastewater discharge are both constant. Under these "steady-state" conditions, there is no change in storage of bacteria within the volume unit. The only other component needed to build the forecasting model is a statement about the way bacteria are transformed within the stream. Field studies indicate that changes in the number of bacteria follow a regular pattern which can be described mathematically. Many representations can be used, depending on how accurately the real bacterial transformation process is to be simulated. A very simple formulation, and one that is often considered adequate for making forecasts, assumes that the number of bacteria will decrease according to a first-order reaction as the water flows downstream. The phrase *first-order reaction* is a shorthand way of saying that the *rate* of bacterial die-off is proportional to the number of bacteria present.[3]

Figure 7.4 summarizes components of the model building process. A typical volume unit located x length units downstream of the point of discharge is selected. Since the velocity of streamflow is a constant (u), this volume unit can also be described as having traveled for t time units downstream of the discharge. "Travel time" is related to distance and velocity using

$$t = \frac{x}{u} \qquad \textbf{(7-1)}$$

A mass balance equation is written for the bacteria entering and leaving the typical volume unit. As shown by Thomann (1972), if the in-stream bacterial transformation process is represented as a first-order decay reaction, the solution to the differential equation representing the mass balance is[4]

$$N(t) = N_0 e^{-kt} \qquad \textbf{(7-2)}$$

where

t = time of travel downstream of the discharge point
$N(t)$ = concentration of coliform at time t
N_0 = concentration of coliform just below the point of discharge ($t = 0$)
k = bacterial die-off coefficient

[3]The term *first order* is used because the bacterial decay process is described with a first-order differential equation.

[4]This formulation presumes that the physical transport of bacteria is accomplished only by the average motion of the stream ("advection"). There is no transport of bacteria due to diffusion or dispersion. The differential equation for this case is

$$u\frac{dN}{dx} + kN = 0$$

subject to the initial condition: $N = N_0$ at $x = 0$. For simplicity, equation 7-2 shows the solution to this equation with time of travel substituted for (x/u).

The Physical Conditions

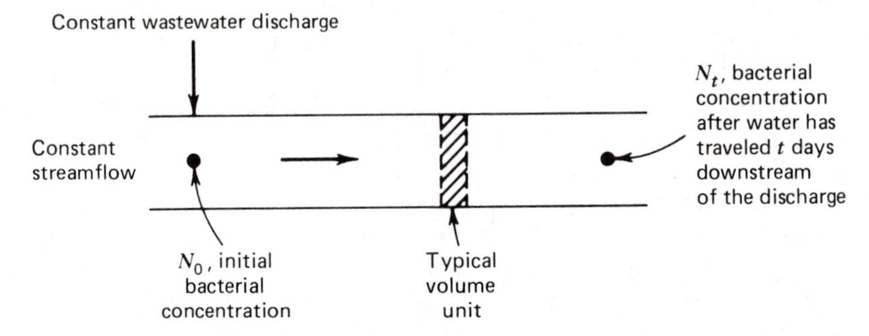

Relationship between travel time and distance

With a constant stream velocity (u), travel time (t) and distance downstream from the discharge (x) are related by

$$t = \frac{x}{u}$$

Mass balance for typical volume unit

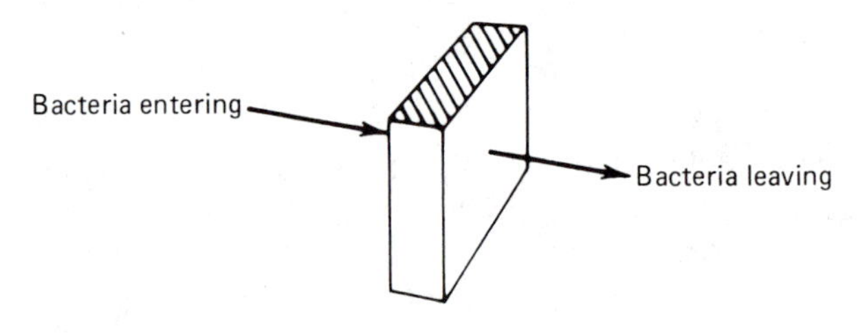

Number of bacteria leaving = Number of bacteria entering − Bacterial die-off

Representation of bacterial die-off process

First-order reaction: die-off *rate* is proportional to number of bacteria present, where k is the die-off coefficient.

Resulting forecasting model

$$N_t = N_0 e^{-kt}$$

FIGURE 7.4 Modeling bacterial concentration in a stream.

Equation 7-2 can serve to predict how downstream bacterial concentrations will change in response to either a new discharge of wastewater or an increased treatment of an existing waste source. Before the equation can be used to make forecasts, the bacterial die-off coefficient must be determined. Numerous empirical studies of bacterial decay in streams provide the basis for estimating k. In a typical field investigation, samples are retrieved from several locations downstream of a municipal waste source under conditions when both the streamflow and the wastewater discharge are approximately constant. Laboratory measurements of the coliform concentration are made for each sample. The laboratory data provides N_t for several values of t including $t = 0$. The observed values of log (N_t/N_0) versus travel time are plotted and a straight line is passed through the data points. The die-off coefficient is estimated based on the slope of the line. Figure 7.5 is a plot of coliform concentration data for the Tennessee River below Chattanooga. The figure demonstrates the use of a formula for computing k from the slope of the line passing through data points for the first several days of travel time. Because the points trace out a nonlinear curve beyond a travel time of about 5 days, the assumption of a first-order bacterial

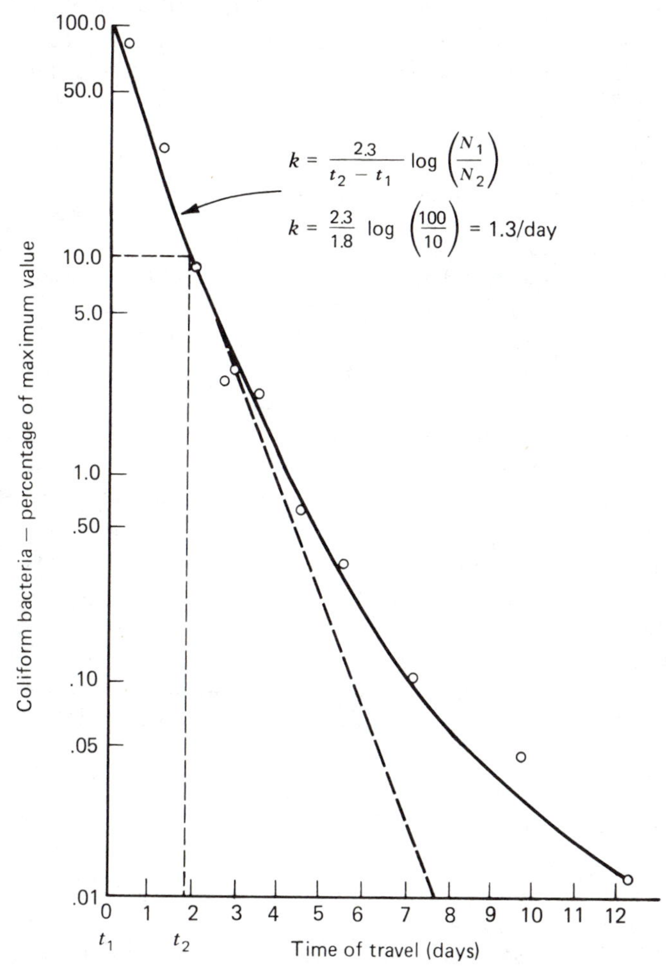

FIGURE 7.5 Estimation of bacterial die-off coefficient: Tennessee River below Chattanooga. From *Systems Analysis and Water Quality Management* by R. V. Thomann. Copyright © 1972, Environmental Science Services Division. Environmental Research and Applications, Inc., New York. Used with the permission of McGraw–Hill, New York.

decay process is not valid for t greater than 5. The entire die-off process may be represented with two straight-line segments. In this case, a separate value for the die-off coefficient is calculated for the period beyond the fifth day.

A typical application of equation 7-2 is to predict the effect of disinfecting a municipal wastewater discharge with chlorine. Studies determining bacterial die-off coefficients on similar streams can be used to estimate k. Existing coliform concentration at the point of discharge is obtained by direct measurement. The expected percentage reduction of bacteria due to chlorination provides a basis for computing N_0, the bacterial concentration at the discharge point *after* disinfection. Once k and N_0 are determined, equation 7-2 is used to calculate values of bacterial concentration at selected downstream locations. Step-by-step instructions for carrying out the calculations are given by Hydroscience, Inc. (1971). They also show how to estimate bacterial concentrations resulting from a completely new wastewater discharge.

A Statistical Model for Predicting Carbon Monoxide Levels

Statistical forecasting models are often used when there is insufficient scientific knowledge to describe the transport and transformation of residuals mathematically. In developing a statistical model, intuition, previous studies, and scientific theories are used to postulate a relationship among what are considered to be the key variables. The assumed relationship contains constants ("model parameters") that are estimated based on measured values of the variables.

The statistical approach to forecasting is illustrated by Tiao and Hillmer's (1978) model for predicting carbon monoxide (CO) concentrations at a particular monitoring site 25 ft from the San Diego Freeway in southern California. Their modeling effort built on previous empirical studies that identified variables affecting CO in the ambient air. Some of these variables concerned traffic flow, especially traffic speed and volume, and others described local meteorological and topographic conditions.

Based on an analysis of factors likely to influence CO at their site, Tiao and Hillmer identified two principal variables, traffic density and wind speed. They postulated that carbon monoxide concentrations were related to these variables using

$$C_t = a + MD_t \exp[-b(W_t - W_0)^2] \quad \textbf{(7-3)}$$

where

C_t = CO concentration at the monitoring site for hour t (parts per million)

D_t = traffic density on the adjacent freeway for hour t (vehicles per hour)

W_t = wind speed in the direction perpendicular to the freeway for hour t (miles per hour)

The constants, a, b, M, and W_0, are determined using past measurements of the model variables, C_t, D_t, and W_t. The parameter a represents the background CO concentration that exists independent of the freeway traffic. The exponential term in the equation accounts for the transport of carbon monoxide caused by atmospheric turbulence. Carbon monoxide emissions from motor vehicles on the freeway are reflected in M, a proportionality constant that is multiplied by traffic density.

Once the form of the equation relating the variables had been assumed, the next step involved estimating the parameters, a, b, M, and W_0. This was done by collecting hourly measurements of C_t, D_t, and W_t for a 2½–year period beginning in June 1974. A preliminary analysis of these data indicated that hourly CO concentrations at the monitoring site were significantly influenced by both the season and the day of the week. Therefore the measurements of C_t, D_t, and W_t were divided into groups corresponding to different seasons and days of the week, and these groupings were maintained in all subsequent stages of model building.

To see how model parameters were estimated, consider the set of data for Sundays during "summer" (June through October). Hourly measurements of C_t, W_t, and D_t, during the 5-month period beginning in June 1975 were used to determine the parameters for the version of equation 7-3 applying to summer Sundays. A standard statistical procedure, the method of least squares, was used to compute a, b, M, and W_0. The procedure is called *least squares* because it yields model parameters that give the smallest value to the sum of the squares of the errors. An *error* is defined as the difference between estimated and observed values of C_t for a specific hour and Sunday within the 5-month measurement period. Consider, for example, the hour from noon to 1 P.M. on the first Sunday in June. The *estimated* value of C_t is computed by substituting into equation 7-3 values of wind speed and traffic density measured at that particular hour and day. The *observed* value of C_t is the carbon monoxide concentration recorded at the monitoring site for that precise hour. The differences in estimated and observed CO concentrations are computed for each summer Sunday hour for which there are measurements in the 5-month period. They are then squared and added together. Values of a, b, M, and W_0 are selected such that the sum of the squared errors is a minimum.

Applying the method of least squares to the data for summer Sundays in 1975, the following parameter values were computed: $a = 1.79$, $b = 0.013$, $M = 0.019$, and $W_0 = 2.54$. These values were substituted into equation 7-3 to provide a basis for estimating CO concentrations at the monitoring site during summer Sundays. Figure 7.6 provides an indication of how well this model represents the hourly variations in carbon monoxide. The points labeled "actual CO" represent the average hourly CO measured at the site during Sundays in the summer of 1975. The solid line represents CO concentrations estimated with equation 7-3. The values of D_t and W_t used in estimating CO with equation 7-3 were also average hourly values based on all of the data for Sundays in the 5-month summer period. As indicated in Figure 7.6, the statistical model replicates the observed data quite closely.

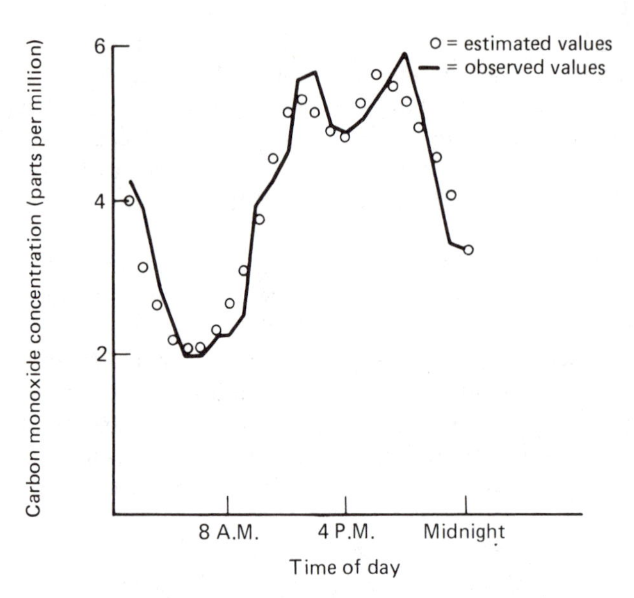

FIGURE 7.6 Estimated versus observed values of carbon monoxide for summer Sundays in 1975 at the San Diego Freeway monitoring site. Reprinted with permission from G. C. Tiao and S. C. Hillmer, *Environmental Science and Technology,* Vol. 12, pp. 820–828. Copyright © 1978, American Chemical Society.

Once the model parameters have been specified, equation 7-3 can be used to predict how changes in traffic density influence carbon monoxide at the monitoring site. However, this equation applies only to summer Sundays at that particular site. For other time periods and physical locations, the four model parameters must be reestimated using measurements appropriate to the new site and time period. Because the form of equation 7-3 is not based on physical laws, it may not provide a suitable representation of how C_t, D_t, and W_t are related in settings different from the San Diego Freeway site. The lack of transferability from one location to another is a principal shortcoming of statistical models. An important advantage is that a forecasting equation applicable to particular conditions can often be obtained, even though scientific theories relating the variables are incomplete.

FORECASTING MODELS BASED ON "SOFT INFORMATION"

The previous cases involved models to predict concentrations of bacteria in streams and carbon monoxide in the atmosphere. Because these variables can be measured, it is possible to determine whether the models' forecasts are accurate. Although it is customary to develop forecasting models for variables that can be determined solely from quantitative observations, some modeling experts find this too restrictive. They argue that the intuitive feelings that people have about ambiguous qualitative concepts such as "quality of life" should not be ignored when formulating government plans and policies. They also believe that mathematical procedures can be used to convert personal judgments into useful forecasts.

Several mathematical models have been developed to make quantitative forecasts on the basis of "soft information," subjective opinions, and judgments about the relationships among poorly defined variables. One of these is called "KSIM," a model developed by Kane et al. (1973) to help people refine their intuitions and opinions and make them more useful in a policy-making context.

The KSIM model was used effectively in a workshop convened to assess impacts associated with a possible U.S. deep-water port system for imported oil.[5] Workshop participants had diverse backgrounds and were knowledgeable about the subject. They formulated a workshop problem statement by spelling out basic assumptions and articulating a number of "what if" questions that they would try to answer. A typical question was, "What if crude oil imports were cut off?" Participants identified what they felt were the key variables and put them in the form required by the KSIM approach: each variable was scaled to range from a minimum of 0 to a maximum of 1. "Initial conditions," values of the variables at the beginning of the forecasting exercise, were also set. The five variables agreed upon by the participants were energy use, domestic energy supply, domestic investment in new energy sources, reliance on traditional sources of energy, and public confidence in the government. For each variable the meaning of the values of 0 and 1 were explained in qualitative terms.

The next step involved development of a "cross impact matrix," a table with the above-mentioned five variables arrayed down the first column and across the top row.[6] For each pair of variables in the matrix, workshop members decided whether a relationship existed, and if it did, what its "strength" was on a scale from -3 to $+3$. For example, participants felt that the linkage between domestic investment in new energy sources and domestic energy use was positive with a strength of $+2$. In other words, investments in new supplies of energy would lead to an increase in energy utilization. Each number in the cross-impact matrix represented the *perceptions* of the workshop participants regarding how two variables in the matrix related to each other.

The KSIM model was introduced after the numbers in the cross-impact matrix were specified. The model consists of a particular set of differential equations that is *assumed* to govern the interactions between the variables.[7] Details of the model are unimportant here. The key point is that the same general equations

[5]This discussion of the application of KSIM to analyze policies for deep-water port development is from Mitchell et al. (1975, pp. 145–154.)

[6]The cross-impact matrix in KSIM is different from those in the "cross-impact matrix method" reported by Gordon and Hayward (1968). Their matrices have elements that represent the extent to which the probability of one "event" is enhanced (or inhibited) if a second event occurs. These probabilities are employed to provide information about which sequences of future events are the most likely to occur.

[7]The reasoning used in picking the particular set of equations used in KSIM is given by Kane et al. (1973).

are employed in every application of KSIM. In any particular case, parameters in the equations are estimated using the cross-impact matrix and the information on initial conditions.

Solutions to the differential equations that make up KSIM consist of values of the variables over time. An illustrative result from the deep-water port workshop is shown in Figure 7.7. By examining the solutions, workshop participants were able to refine their initial impressions about cross impacts. Participants in a typical KSIM exercise often continue to revise the cross-impact matrix until the solutions they obtain are plausible and compatible with their intuitions.

At this point in the workshop, the cross-impact matrix was enlarged to accomodate "policy interventions," and the KSIM model was solved again. Information on the impacts of interventions took the form of curves like those in Figure 7.7.

The workshop exercise was *not* used exclusively as a means of producing forecasts. As Mitchell et al. (1975, p. 142) put it, "in KSIM sessions it is often the process (exchange of views, surfacing of issues, identification of concerns, and so on) that is more important than the product."

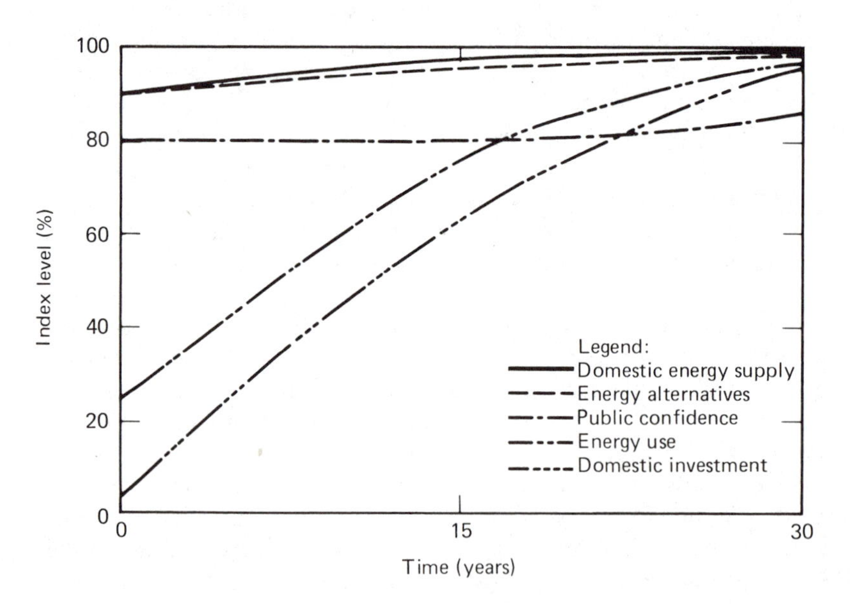

Consider: Initial projected variable interactions of the national deep water port system

Implications: Domestic investment increases steadily to near maximum.
Energy use increases steadily to near maximum.
Public confidence is stable.
Domestic energy supply increases to maximum.
Energy alternatives increase to maximum.

FIGURE 7.7 Sample results from the KSIM Deep Water Port Workshop. From Mitchell et al. (1975).

KSIM is clearly not a forecasting model in the usual sense. Its virtue is in showing the *logical consequences* (over time) of the *model users' intuitions and assumptions* about interactions among the variables. In KSIM, the forms of the equations are set. They are not modified to fit a specific problem. The initial conditions and the numerical entries in the cross-impact matrix are changed in successive applications of the model to help workshop participants refine their intuitions. Since the variables cannot be measured, the accuracy of forecasts cannot be tested by making comparisons with observed values.

Mathematical models using soft information have been disputed, especially by those who expect forecasting models to be based on empirical data and scientific theories. Critics have argued that relationships among variables must be established scientifically and tailored to each problem setting. In addition, comparisons between predicted and actual outcomes are necessary to gauge a model's worth.

Those defending models based on soft information feel that a quantitative comparison with observed data is not the only way to establish a model's validity. Instead, they argue, a forecasting model should be judged in relation to other possible means of making predictions. For example, an alternative to the KSIM projections in the deep-water port study above consists of verbal descriptions of future scenarios. Value judgments and personal opinions would be employed in generating the scenarios, but chances are that these judgments and opinions would not be made explicit. A defense of KSIM is that it requires a clear delineation of judgments and assumptions. Models developed from soft information are also justified by their ability to employ whatever information exists about relationships, even if that information is scanty and of marginal quality. Traditional modeling approaches may not accommodate some important information because it is too incomplete to be acceptable in a precise modeling exercise.[8]

Models using soft information are different in kind from those developed from physical laws and statistical analyses. They are not meant to provide rigorous, quantitative predictions of the type given by equations 7-2 and 7-3. Models such as KSIM are more appropriately viewed as tools for sharpening the qualitative discussions of the possible impacts of government actions. When considered as aids to communication, models using soft information arouse much less controversy.

MODEL CALIBRATION AND VALIDATION

Before concluding this introduction to mathematical models, consider the distinctions between model validity and model "calibration." A model is *calibrated* by using empirical data to estimate its parameters. The process of determining

[8]Justifications for mathematical models based on soft information are elaborated by Forrester (1968, Chapter 3). For a more complete discussion of Forrester's use of such models, and the controversy that his work has generated, see Pugh (1977).

the four parameters in equation 7-3 to predict carbon monoxide provides an illustration. Once a model has been calibrated, it can be used to make forecasts of the variable in question under future conditions. However, little can be said about the accuracy of those forecasts unless the model has been validated.

A model is *validated* by conducting tests to see if its forecasts are likely to be close to the outcomes eventually observed. To test a model, it is necessary to compile a set of observations of the model variables that is different from the data used in calibration. An independent data set is needed to avoid the logical inconsistency of both formulating *and* testing a model using the same measurements.

A valid forecasting model is one that consistently yields predictions that are "reasonably close" to what occurs in reality. However, it is often impracticable to test a forecasting model by comparing its predictions with observed outcomes. One reason is that the expense of gathering the necessary data is frequently prohibitive. Often, what little empirical data can be gathered must be used to estimate the model parameters. Validation is particularly difficult when long-range forecasts are involved. Typically there is little enthusiasm for conducting studies 10 or 20 years after a model was used to see how the forecasts turned out. Sometimes it is not even meaningful to compare forecasts with actual outcomes. This occurs, for instance, when the invention of a new technology leads to events that could not possibly have been anticipated in making the original forecasts.

In addition to comparing predicted and observed effects, there are several other ways of gauging a model's worth. Some experts consider a forecasting model valid if it "behaves" in reasonable ways. For example, an air pollution model can be tested by examining predicted air quality levels under various assumptions about government pollution control regulations. A measure of the model's validity is given by whether these predictions are plausible and consistent with the modeler's intuitions. Because forecasting models are frequently used to aid decision making, some experts feel that validity questions should be examined only in the specific setting in which a model is to be used. For example, in some decision-making contexts, a model that gives forecasts within $\pm 100\%$ of actual outcomes may be considered adequate and, in some sense, valid. Such a wide range of predicted outcomes would be unsatisfactory if the results were not useful to decision makers. These different conceptions of model validity must be considered in interpreting the results from modeling studies.

KEY CONCEPTS AND TERMS

AIDS TO IMPACT IDENTIFICATION
- Checklists
- Impacts of project phases
- Literature reviews by project type

JUDGMENTAL APPROACHES TO FORECASTING
- Expert opinion
- Impacts of past projects
- Delphi method

PHYSICAL MODELS IN FORECASTING
- Modeling visual impacts
- Estuary models
- Wind tunnels

FORECASTING WITH MATHEMATICAL MODELS
- Mass balance equation
- First-order reaction
- Statistical models
- Model parameters
- Method of least squares
- Soft information
- KSIM and cross impacts
- Model calibration
- Model validity

DISCUSSION QUESTIONS

7-1 Give an example of a physical model that can be used to forecast environmental impacts. Try to select an illustration different from those mentioned in Chapter 7. Indicate the types of variables that your illustrative model would forecast. Also indicate the types of projects or actions that could be examined with the model.

7-2 Suppose you were asked to use the Delphi method to predict the land use changes that would accompany alternative reservoir projects. What kinds of perspectives would you try to have represented among the experts that would perform the forecasts?

7-3 Phosphorous and nitrogen are nutrients that support the growth of algae in streams. The distribution of these nutrients is sometimes modeled mathematically by assuming that they decay over time according to a first-order reaction. Suppose the original concentration of total nitrogen just below a wastewater discharge is 8 mg/l.

(i) Draw a graph showing total nitrogen versus time of travel if the decay coefficient is assumed to be 0.5/day.

(ii) Forecast the effects of providing wastewater treatment that reduces the concentration of total nitrogen below the point of discharge to 6 mg/l.

7-4 The use of KSIM and other mathematical models with soft variables can be attacked easily by those accustomed to more traditional approaches to developing models. Indicate the criticism you would expect to be advanced against KSIM. What kind of a defense could be mounted?

7-5 Indicate a circumstance in which you might use a mathematical model with soft variables to forecast environmental impacts.

REFERENCES

Armstrong, J. S., 1978, *Long-Range Forecasting: From Crystal Ball to Computer*. Wiley, New York.

Arthur D. Little, Inc., 1971, *Transportation and Environment: Synthesis for Action—Impact of National Environmental Policy Act of 1969 on the Department of Transportation,* Vol. I. Arthur D. Little, Inc., Washington, D.C.

Berns, T. D., 1977, The Assessment of Land Use Impacts, in J. McEvoy, III and T. Dietz (eds.), *Handbook for Environmental Planning, The Social Consequences of Environmental Change,* pp. 109–161. Wiley, New York.

Bigelow-Crain Associates, 1976, "State and Regional Transportation Impact Identification and Measurement," report prepared for National Cooperative Highway Research Program. Transportation Research Board, National Research Council, Washington, D.C.

Canter, L. W., 1979, *Water Resources Assessment—Methodology and Technology Sourcebook.* Ann Arbor Science Publishers, Ann Arbor, Mich.

Cavalli-Sforza, V., L. Ortolano, J. S. Dajani, and M. V. Russo, 1982, *Transit Facilities and Land Use: An Application of the Delphi Method,* Report IPM-15. Department of Civil Engineering, Stanford University, Stanford, Calif.

Clark, B. D., R. Bisset and P. Wathern, 1980, *Environmental Impact Assessment, A Bibliography with Abstracts,* Mansell, London.

Environmental Protection Agency, 1973, *Environmental Impact Statement Guidelines,* revised edition. Region X Office, Seattle, Wash.

Forrester, J. W., 1968 *Principles of Systems,* Text and Workbook, 2nd preliminary ed. Wright-Allen, Cambridge, Mass.

Gordon, T. J. and H. Hayward, 1968, Initial Experiments with the Cross-Impact Matrix Method of Forecasting. *Futures* **1** (2), 100–116.

Hydroscience, Inc., 1971, *Simplified Mathematical Modeling of Water Quality,* report prepared for the Environmental Protection Agency. Washington, D.C.

Kane, J., I. Vertinsky, and W. Thomson, 1973, KSIM: A Methodology for Interactive Resource Policy Simulation. *Water Resources Research* **9** (1), 65–79.

McCoy, C. B., 1975, The Impact of an Impact Study, Contribution of Sociology to Decision-Making in Government. *Environment and Behavior* **7** (3), 358–372.

Mitchell, A., B. H. Dodge, P. G. Kruzic, D. C. Miller, P. Schwartz, and B. E. Suta, 1975, *Handbook of Forecasting Techniques,* IWR Report 75-7. U.S. Army Engineer Institute for Water Resources, Ft. Belvoir, Va.

Porter, A. L., F. A. Rossini, S. R. Carpenter, and A. T. Roper, 1980, *A Guidebook for Technology Assessment and Impact Analysis.* Elsevier/North Holland, New York.

Pugh, R. E., 1977, *Evaluation of Policy Simulation Models: A Conceptual Approach and Case Study.* Information Resources Press, Washington, D.C.

Thomann, R. V. 1972, *Systems Analysis and Water Quality Management.* Environmental Research and Applications, Inc., New York (reissued by McGraw–Hill, New York).

Tiao, G. C., and S. C. Hillmer, 1978, Statistical Models for Ambient Concentrations of Carbon Monoxide, Lead and Sulfate Based on LACS Data. *Environmental Science and Technology* **12** (7), 820–828.

Tracor, Inc., 1971, *Estuarine Modeling: An Assessment,* report prepared for the Environmental Protection Agency, Washington, D.C.

CHAPTER 8

METHODS AND PROCESSES FOR EVALUATING ENVIRONMENTAL IMPACTS

The evaluation (or rank ordering) of alternative proposals involves much more than environmental issues. Political, technical, and economic factors must be considered along with environmental impacts in making evaluations. Although this chapter touches upon these other concerns, it does so only to put the role of environmental issues in context. The emphasis throughout is on planning and decision making in the public sector.

The term *evaluation* is sufficiently vague to warrant elaboration. It has been defined by Lichfield, Kettle, and Whitbread (1975, p. 4) as "the process of analyzing a number of plans or projects with a view to searching out their comparative advantages and disadvantages and the act of setting down the findings of such analyses in a logical framework." They stress that evaluation is *not* decision making. Instead, it assists decision making by highlighting the differences between alternatives and providing information for subsequent deliberation.

Value judgments are made so frequently that it is sometimes difficult to identify evaluation as a distinct planning activity. Consider four common planning tasks: (1) identifying problems and goals, (2) formulating alternative proposals, (3) forecasting impacts, and (4) evaluating alternatives. The delineation of problems and goals is no less a value judgment than the choice to analyze some proposals in detail and not others. Even though value judgments are made implicitly throughout planning, it is useful to distinguish the *systematic* consideration of impacts in ranking alternatives with the term *evaluation*. The techniques for ranking are commonly referred to as *evaluation methods*.

Before introducing specific techniques, some general observations are in order. As planning proceeds, one of the four planning tasks listed above may be emphasized at some time. However, the four tasks are *not* carried out in a lockstep, sequential order. Determinations are continually being made and revised regarding the nature of the planning problem, the alternatives to be considered, and so on.

Observe also that more than one set of values is relevant in choosing among alternative public sector proposals. Within democratic nations, it is commonly accepted that the values of *all* individuals who may be affected by public decisions should be considered. Adopting this perspective, evaluation also includes the process of identifying different segments of the public and ascertaining their feelings and opinions about alternative plans. Thus, in addition to techniques for organizing information to assist in ranking alternatives, evaluation methods include procedures for determining how individuals and groups value alternative public actions.

Although there are literally hundreds of evaluation methods, there is little agreement among experts about which are best. This chapter introduces procedures that are either widely discussed in the public sector evaluation literature or commonly used in practice, or both. Because the public can play an important role in evaluating alternatives, the chapter also presents techniques for involving citizens in planning.

ISSUES IN MULTICRITERIA EVALUATION

The simplest circumstance for ranking alternatives occurs when there is only one decision criterion and all impacts are measured in the same units. Suppose, for example, the rule for choosing among alternative projects is the maximization of net economic benefits. In this case, the difference between total benefits and costs (in monetary units such as dollars) provides an index of a project's merit and a basis for project selection.

When environmental factors play a role in ranking alternatives, they are frequently considered as one of several criteria for decision making. The criteria may be measured in units as different as dollars and number of residences relocated. Such measures are "incommensurable." They are not readily compared or easily combined into a single index of a proposal's overall worth.

An Example Involving Alternative Reservoir Sites

Consider a hypothetical situation with two alternative water resource projects and two evaluation criteria. Plan A is to develop a reservoir that yields high economic benefits by reducing the likelihood of flooding to homes downstream of the dam (see Figure 8.1). However, it also inundates an important wildlife refuge area. Plan B locates the reservoir farther upstream. It destroys a smaller portion of the wildlife refuge, but it leaves the downstream homes subject to

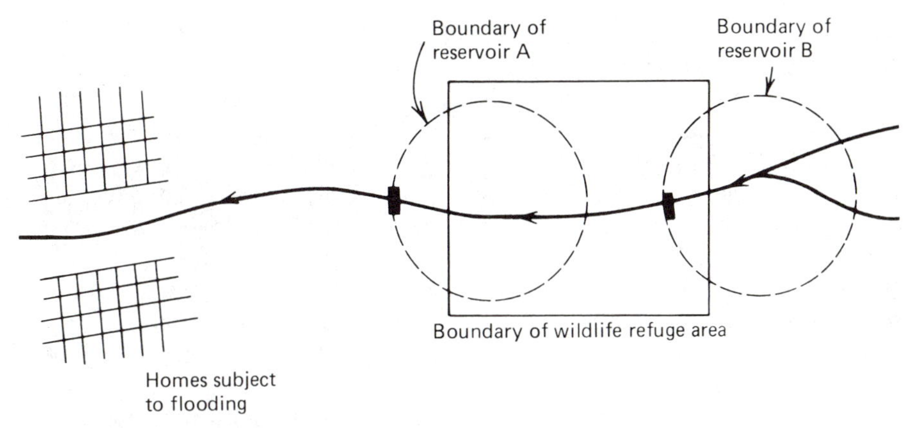

FIGURE 8.1 Two alternative reservoir projects.

higher potential flood damages. This example is much simpler than real situations. Typically, there are many more alternative projects, and numerous other impacts pertinent to decisions about which proposal to adopt.

Table 8.1 summarizes the hypothetical reservoir selection problem. It contains information on the two criteria to be used in ranking the alternative plans: (1) maximization of the "present value" of net economic benefits, measured in dollars;[1] and (2) minimization of acreage of the wildlife refuge area destroyed by inundation. Even assuming these are the only relevant criteria, there is still no obvious way to decide which project to implement. The criteria are expressed in dissimilar units, and there is no agreed-upon technique for manipulating the evaluative information to yield a final decision.

Evaluative Factors and Weights

It is convenient to view the ranking of alternative public actions in terms of *evaluative factors.* These are the goals, objectives, concerns, and constraints that

TABLE 8.1 Effects of Alternative Reservoir Proposals

	Plan A (Downstream Site)	Plan B (Upstream Site)
Present value of net economic benefits ($\$10^6$)	97	85
Area of wildlife refuge inundated (acres)	5000	1000

[1]The present value concept is used because costs and benefits occur at different future times. A dollar in year t is worth more than a dollar in year $t + \Delta t$, since it can be invested to yield interest during the time period, Δt.

various decision makers and segments of the public consider important in ranking alternatives. In the reservoir example, the evaluative factors are net economic benefits and area of wildlife refuge disturbed. Since plan ranking is strongly influenced by the choice of evaluative factors, there frequently is controversy over who should define them. In public sector planning there are three principal sources of evaluative factors: institutions, community interactions, and technical and scientific judgments. The discussion below considers ways that planners delineate evaluative factors from each of these sources.

Often, there are individuals potentially affected by a proposed government project who cannot participate directly in the plan evaluation process. As a matter of convenience, these people are referred to as the *nonlocal public*. Their goals and concerns are expressed *institutionally* at the national, state, and local levels in various laws, policies, and programs. Their feelings may also be reflected in the policy statements of various groups such as the chamber of commerce and the Sierra Club. For example, suppose in the reservoir selection problem that a state law prohibits the destruction of significant portions of wildlife refuge areas. This law allows citizens far from the project site to have an influence on the choice of a reservoir. In fact, such a law might be sufficient to force the selection of the upstream reservoir, since it has much less effect on the refuge area. Planners identify institutional sources of evaluative factors by communicating with government officials and interest group representatives, and by examining relevant laws, policy statements, and regulations.

Planners can interact directly with members of the *local public* to help define problems from a local perspective and identify issues which individuals consider important in choosing among alternative proposals. To accomplish this, planners may provide information describing the problems and possible actions, as they understand them, and the impacts associated with these actions. Members of the local public can then provide "feedback" to planners about their own perceptions of the problems and what they consider important in ranking alternatives.

Evaluative factors are also based on technical or scientific judgments which members of the public may neither appreciate nor recognize. For example, planners may deem it important to maintain the habitats of certain species in the interests of a project area's long-term ecological stability.

Sometimes planners must translate evaluative factors into technical terms that can be used to guide the formulation of alternatives, impact analysis, and evaluation. For example, the desire of citizens to maintain trout fishing in a local stream may be converted by planners into numerical constraints on stream temperatures and concentrations of dissolved oxygen.

The relative importance of different evaluative factors is often discussed in terms of *weights*. Sometimes weights are indicated explicitly, as in the assertion "maintaining the stream as a trout fishery is twice as important as the need to preserve ten acres of open space." More frequently, project selections are made through a concensus building process in which weights are not stated. The concept of weights is still helpful in these cases, even though the weighting is implicit. Suppose, in the reservoir example, the upstream project was selected in order to save 4000 acres of wildlife refuge. The cost in terms of net economic benefits foregone is $12 million (see Table 8.1). It could be argued that by selecting the upstream reservoir, an unstated weighting took place. The decision makers implicitly valued the refuge area at a minimum of $3000 acre (the total net benefits foregone divided by the acres saved).

In thinking about evaluation, it is helpful to ask the following questions: whose evaluative factors and weights are relevant and how should they be determined and used in ranking alternatives? This chapter examines various evaluation techniques in terms of both evaluative factors and weights.

EXTENSIONS OF BENEFIT–COST ANALYSIS

The benefit–cost analysis (BCA) approach to choosing among alternatives has its foundations in utilitarian philosophy and economic theory. Practical application of BCA was fostered by water resource development laws passed by the U.S. Congress in the 1930s. Since then, the benefit–cost approach has been used to evaluate many different types of projects and programs.[2]

Limitations of Traditional BCA

The economic benefits of a public project are measured by the willingness of individuals to pay for the project's outputs. The costs of a project are the gains that must be sacrificed by using the "inputs" (resources such as land and labor) necessary to implement the project. Economists view costs as opportunities foregone, potential gains given up by not using the resources in other ways. If a project's inputs and outputs are traded in competitive markets, the market prices are used to compute the project's costs and benefits. Frequently, competitive markets do not exist. For example, in the reservoir problem of Figure 8.1, flood protection is not sold in competitive markets. When suitable market prices are unavailable, costs and benefits are estimated using procedures that rely heavily on the professional judgments of individuals performing the BCA. For a given situation, different analysts may choose different methods to estimate costs and benefits. Critiques of the resulting decisions are common. Government planners have been accused of making judgments that favor project development by overestimating benefits and underestimating costs.

The use of BCA to rank alternatives presupposes that economic benefits and costs adequately represent all of the significant effects. There are no weights involved. All effects are measured in monetary units such as dollars, and they are simply added together. (Another way of saying this is that the weights are zero for evaluative factors different from economic benefits and costs.) When

[2]Eckstein (1958) explains the origins of BCA and how its practical application was fostered by federal water resource legislation. For a general introduction to BCA, see McCallister (1980).

maximization of net economic benefits is used exclusively to rank alternatives, a narrow, government-oriented perspective for defining the public interest is adopted. This is because the procedures for cost and benefit estimation are carried out largely by the staff of the agency proposing the action. The estimates can be made without involving the public directly. Although BCA has played an important role in public sector planning, the results are *not* generally used as the exclusive basis for decision making. More commonly, BCA results are treated as one element in a broader political process aimed at yielding widely acceptable decisions.

BCA is of limited usefulness in aiding decisions that result in significant environmental impacts. An important reason is that BCA does not systematically consider impacts that cannot be described appropriately in monetary terms. This is illustrated by the reservoir example in Figure 8.1. A BCA clarifies the economic gains that must be foregone to minimize the area of the wildlife refuge that is inundated. However, a BCA does not provide a basis for ranking the two plans. A ranking might be possible if acres of wildlife refuge could be valued in monetary terms. There are significant methodological problems in making such estimates. Moreover, many people are philosophically opposed to evaluating biologically important areas in terms of dollars.

Another shortcoming of BCA is its failure to account for equity considerations. The emphasis is on aggregate economic effects, not on which groups gain and lose if a project is implemented. This is demonstrated by supposing that Plan B in Figure 8.1 provides more flood protection to low-income families than Plan A. In this instance, a traditional BCA indicates that Plan A is preferred since it has higher net economic benefits. If the interests of low-income groups were to be favored, Plan B might be a better choice. This income redistribution issue is not treated when only net economic benefits are examined. The limitations of BCA have stimulated efforts to modify the approach and make it applicable to a wider range of evaluation problems.

Extending BCA to Consider Multiple Objectives

Important early work in broadening BCA was carried out by Marglin (1962) and Maass (1966). They argued that it was often inappropriate to use a single objective, maximization of net economic benefits, because government programs were frequently intended to serve "multiple objectives." To evaluate them systematically, *classes of benefits* should be defined to correspond to each of the objectives.

The illustrative cases discussed by Marglin and Maass often involve two objectives: maximization of net income to a nation as a whole and maximization of net income to a particular region or group of citizens. The *national income net benefits* are defined in the same way as the net benefits included in a traditional BCA. Analysis of the second objective requires the introduction of "income redistribution net benefits," the net income that flows to the particular region or group singled out for special treatment. For each project under consideration, both national income and income redistribution net benefits are computed. Proposals may then be ranked using a weighted sum of the contributions to each objective. A variation of this is to maximize contributions to one objective, subject to the requirement that a minimum contribution to the second objective be provided.

Using the weighted sum of objectives technique, the evaluative factors considered in ranking alternatives are the objectives themselves. Although Marglin and Maass often included only the national income and income redistribution objectives in demonstrating their ideas, they recognized that other objectives might need to be considered. In any particular situation, they felt that precise objectives should be specified by appropriate public decision makers.

Determination of the weights to be used in computing the weighted sum of objectives has been a subject of continuing debate. Much of the large amount written on the subject of weights is, in Steiner's (1969, p. 48) words, "almost entirely theoretical and assertive." The empirical work that has been done relates mainly to tradeoffs between national income and income redistribution. It uses past choices and tax data as bases for inferring the weights that appear to have governed past decisions. However, many analysts question the suitability of using weights implied by previous choices to decide what the weights should be.

Some social scientists have suggested that policymakers should articulate the relative importance of different objectives whenever a new government program is established. Agency analysts could then translate this information into weights that could be employed in evaluating alternative proposals.[3] This approach has not been widely used. Few legislators seem inclined or able to articulate specific weights on objectives like income redistribution and environmental quality.

Since the late 1960's, there has been substantial activity in a part of mathematical optimization theory that treats the multiobjective evaluation issues examined by Marglin and Maass. This branch of theory, known as "multiobjective programing," concerns the limited set of multicriteria evaluation problems that can be described fully by mathematical equations and inequalities. The appendix to this chapter contains an example that introduces multiobjective programming.

TABULAR DISPLAYS AND THE SUM OF WEIGHTED FACTOR SCORES

A number of commonly used evaluation methods rely on tabular displays of information. Table rows generally correspond to evaluative factors in one form or another. Columns correspond to the alternative projects under consideration. A few types of table entries are commonly used. In some cases, an entry consists of a brief description of how a particular project is likely to influence a given evaluative factor. In other cases, an entry is a numerical score characterizing the effects of an alternative on a factor.

[3]This is a procedure advocated by Freeman (1970) for programs administered by federal agencies in the United States.

Some idea of the extent to which tables are used is given by Canter's (1979) analysis of 28 environmental impact statements for wastewater management proposals. Twenty of these EISs included table of the type described above. Most of these 20 impact statements relied on numerical scores to characterize the effects of alternatives.

Table Entries Based on Ordinal Scales

Numerical table entries often result from comparative analyses of how alternatives influence an evaluative factor. Rank ordering is sometimes used as the basis for assigning scores. Suppose, for example, that five alternative highway routes are ranked in terms of the noise they will cause in a nearby residential area. The route that causes the least noise is assigned a "1", the route causing the second to the least amount of noise is assigned a "2", and so on. Another approach to making comparative observations involves categories, for example, "positive effect, no impact, and negative effect." Sometimes these categories are assigned arbitrary numerical values, such as $+1$, 0, and -1.

A typical display to aid in ranking alternatives is shown in Table 8.2. It was used in evaluating several wastewater management plans (labeled arbitrarily as R, S, T, U, and V) for North Monterey County in California. Rank ordering was used to score the alternatives in terms of costs and energy consumption. For most evaluative factors, however, the alternatives were rated using four categories: adverse, beneficial, problematic (unknown or open to question), and none.

The entries in Table 8.2 are based on *ordinal scales*. Only information reflecting a qualitative comparison or ordering of alternatives is given. Nothing is implied about the magnitude of the differences among alternatives. For example, asserting that one alternative has a beneficial impact on air quality and another has an adverse impact does not say anything about *how much* of a difference exists between the impacts of the two alternatives.

Sum of Weighted Factor Scores

A common approach to summarizing evaluative information uses the sum of weighted factor scores for each alternative. First, each proposal is assigned a score for each evaluative factor. All scores must be within the same numerical limits. For example, scores might vary from 1 to 10, with 10 representing the best alternative. Weights are assigned to indicate the relative importance of each factor and they are used in computing a weighted sum. This approach is illustrated by a case involving land use planning for the city of Palo Alto, California.

In the early 1970s, Palo Alto engaged Livingston and Blayney, a firm of planning consultants, to examine the suitability of alternative types of land development for the foothills at the edge of the city. Livingston and Blayney's (1971) procedure for evaluating alternative land use plans centered around nine evaluative factors. In their view, these factors encompassed all of the consid-

TABLE 8.2 Summary Evaluation of Alternative Treatment and Disposal Plans—North Monterey County, California EIS[a].

	Alternatives[b]					
Potential Impacts	**R**	**S**	**T**	**U**	**V**	**No Action**
Physical/Biological Impacts						
Archaeological resources	P	P	P	P	P	N
Air Quality	A	A	A	A	A	A
Soils and crops	N	P	P	P	P	N
Agricultural practices	N	P	P	P	P	N
Seismic risks	A	A	A	A	A	A
Groundwater quality	N	B	B	B	P	A
Surface water quality	B	B	B	B	B	A
Monterey Bay water quality	B	B	B	B	B	A
Water supply and reuse	N	B	B	B	B	N
Public Health—water contamination	B	B	B	B	P	N
Public Health—land contamination	N	A	A	A	A	N
Energy consumption in treatment and disposal of wastewater (rank)[c]	2	6	4	4	3	1
Aesthetics	B	B	B	B	B	A
Land use changes	N	A	P	A	A	N
Salinas River biota	A	P	A	A	A	N
Salinas River lagoon biota	B	B	B	B	B	N
Marine biota	B	B	B	B	B	N
General construction impacts	A	A	A	A	A	N
Economic Impacts						
Construction cost (rank)	3	5	4	6	2	1
Operating cost (rank)	2	6	3	5	3	1
Local cost (rank)	3	5	4	6	2	1
Overall cost (rank)	3	5	4	6	2	1
Social Impacts						
Growth inducement-accommodation	A	A	A	A	A	N
Local acceptance	A	P	P	P	P	A

[a]Adapted from Canter (1979). Reprinted by permission of Information Resources Press, Arlington, Virginia, copyright © 1979. The original version of this table is from the "Final Environmental Impact Statement and Environmental Impact Report, North Monterey County Facilities Plan," Vol. I, issued by the EPA and Monterey Peninsula Water Pollution Control Agency, San Francisco, August 1977.

[b]*Key*: B, beneficial; A, adverse; P, problematic (unknown or open to question); N, none.

[c]Comparative ranking from most acceptable (1) to least acceptable (6) alternative.

erations (aside from "political factors") significant in ranking the alternatives. The consultants used professional judgment to assign each factor a weight indicating its importance relative to other factors. The nine factors, along with their weights (in parentheses) are as follows: 10-year cost (5); 20-year cost (5);

social impact (8); transportation needs (3); ecological impact (5); fire hazard (2); visual impact (5); geologic impact (3); and hydrologic impact (2). For each factor, the consultants assigned an ordinal score reflecting the influence of each land use plan. Scores ranged from 1 (worst) to 5 (best). For each plan, a weighted sum was computed by multiplying the score for each factor times the associated factor weight and adding up the products. The results were used in making recommendations regarding alternative land use proposals.

Evaluation procedures using a sum of weighted factor scores have one clear advantage over many other methods: they are not difficult to implement. In a typical application, problems in determiing which impacts are significant and how they are valued are settled by the collective judgments of the planning analysts. Once the scores and weights are assigned by the analysts, only simple arithmetic is needed to compute the sum of weighted scores for each plan.

An extensive reliance on the value judgments of planners is sometimes cited as a weakness of the sum of weighted factors approach. Planners often exercise great control over the selection of factors, the scoring of alternatives for each factor, and the assignment of weights. Critics argue that the choices of factors and weights made by planners may be inconsistent with the views of persons affected by proposed plans. A response to this criticism is to involve the public in determining factors and weights, and this is sometimes done in practice.

The validity of processes typically used to select weights has also been challenged. As shown by Hobbs (1980), the common practice of assigning weights based on an "importance scale" from 1 to 10 does not accurately reflect the preferences of those choosing the weights. Theoretically rigorous procedures for determining weights are described by Hobbs, but they are often hard to implement.

Another shortcoming is that the sum of weighted factor scores does not adequately inform decision makers. The factor scores and weights are somewhat arbitrary. In addition, when these numbers are aggregated into a single index, much useful information about impacts is buried.[4] Tradeoffs among alternatives are not illuminated, and the value judgments made by the planning analysts are not revealed. Critics of the weighted factors approach suggest that simple prose descriptions of the main impacts of alternatives would more clearly highlight tradeoffs and require planning analysts to make fewer value judgments. A problem with this suggestion is that the amount of information involved in describing the numerous impacts typically associated with several alternatives can be overwhelming.

The Goals–Achievement Matrix

Hill's "goals–achievement matrix" represents an extension of the logic behind the previously mentioned tabular displays and sums of weighted factor scores.

[4]The process of computing weighted sums has also been criticized because the ordinally scaled numbers that are multiplied and added are not defined to permit manipulation using ordinary arithmetic operations. Elliott (1981) discusses the way alternative scales of measurement are used in creating indexes for ranking alternatives. Hobbs (1980) takes up this point in connection with scales used for assigning weights.

His approach to scoring the effect of an alternative on an evaluative factor uses goals and objectives as the basis for defining benefits and costs. According to Hill, benefits indicate progress toward desired community objectives while costs are retrogressions from these objectives.[5]

The goals–achievement approach explicitly considers the incidence of benefits and costs. It requires identification of the various groups of individuals or establishments that may be affected by a particular proposal. Weights are assigned to indicate how much each goal is valued by the groups. The use of weights that reflect how different groups view the *same* goal distinguishes Hill's approach from the weighted factors scores technique. The goals–achievement method relies on a second set of weights to indicate the overall importance of one goal relative to another. These "community weights" are similar to the weights used in computing a sum of weighted factor scores.

Hill proposes two distinct ways of organizing data to assist decision makers in ranking proposals.[6] One involves presenting only information on scores and weights without attempting to compute an overall index of a plan's worth. This is illustrated in Table 8.3, which compares two transportation plans (A and B) in terms of "accessibility" (ease of travel between two points) and community disruption. Both an "uptown group" and a "downtown group" will be affected by the proposals. Plan A increases accessibility for the uptown group, while decreasing it for the downtown group. Plan B has the opposite effect. Also, Plan A disrupts the uptown neighborhood, but it has no impact downtown. In contrast, Plan B has no influence on the uptown neighborhood, but it causes disruption downtown. These impacts are translated into scores on an ordinal

TABLE 8.3 Hill's Goals–Achievement Matrix: An Example[a]

	Goal 1 Accessibility			Goal 2 Community Disruption		
Community weights on goals	2			1		
	Group Weights (Goal 1)	Plan A	Plan B	Group Weights (Goal 2)	Plan A	Plan B
Uptown group	3	+1	−1	3	−1	0
Downtown group	1	−1	+1	2	0	−1
Extent of goals achievement		+2	−2		−3	−2

[a]This table's format is adapted from Hill (1968).

[5]The discussion here, which is based on Hill (1967, 1968), uses the terms *goals* and *objectives* interchangeably in the interests of simplicity. Hill (1967, p. 22) makes the following distinctions: a goal is "an end to which a planned course of action is directed." In contrast, an objective denotes a goal that "is believed to lead to another valued goal rather than having intrinsic value in itself."

[6]Hill's approach has been reviewed critically by McCallister (1980) and Litchfield, Kettle, and Whitbread (1975); it has been applied widely in land use planning exercises in England.

scale: +1 = positive effect, 0 = no effect, and −1 = negative effect. A goals–achievement matrix (Table 8.3) summarizes the information.

Hill's second approach to presenting information on scores and weights uses indexes to show how well the goals are achieved by each plan. The indexes are computed in the same way as sums of weighted factor scores, except that account is also taken of how the different groups weigh each goal. The computations are explained in two parts. One involves the "extent of goals achievement" by a given plan for a particular goal. For each goal, this is calculated as a sum of products of group weights multiplied by the ordinal scores representing the effects of the plan. Consider the extent to which Plan A achieves the accessibility goal. For the uptown group, the weight (3) is multiplied by the score representing Plan A's effect on accessibility (+1) to yield +3. The product of the downtown group's weight (1) and score (−1) is −1. Summing the products for both groups yields +2, a measure of how well Plan A meets the accessibility goal. For all other combinations of goals and plans, the extent of goals achievement is computed in the same way. Results are at the bottom of Table 8.3.

The second part of the calculation determines the "weighted index of goals achievement." This is done by multiplying the previously computed extent of goals achievement values and the community weights shown at the top of Table 8.3. For Plan A, the weighted index of goals achievement is the sum of products of the Plan A entries in the last row of Table 8.3 multiplied by the appropriate community weights,

$$(2)(+2) + (1)(-3) = +1$$

A similar computation for Plan B yields a weighted index of −6.

In considering the incidence of effects, the goals–achievement approach requires that weights be obtained for both the community as a whole and for individual groups within the community. Although Hill does not recommend a specific procedure for determining weights, he does suggest the following possibilities:

1 *The decision makers may be asked to weigh objectives and their relative importance for particular activities, locations, or groups in the urban area.*
2 *A general referendum may be employed to elicit community valuation of objectives.*
3 *A sample of persons in affected groups may be interviewed concerning their relative valuation of objectives.*
4 *The community power structure may be identified, and its views on the weighting of objectives and their incidence can be elicited.*
5 *Well-publicized public hearings devoted to community goal formulation and valuation can be held.*
6 *The pattern of previous allocations of public investments may be analyzed in order to determine the goal priorities implicit in previous decisions on the allocation of resources (Hill, 1967, p. 25).*

Many of Hill's suggestions involve the public directly. The following discussion of public involvement techniques elaborates on the numerous ways of identifying the goals of different groups.

INVOLVING THE PUBLIC IN EVALUATION

Many people feel that the selection of evaluative factors and weights should not be left exclusively in the hands of professional planners. Several arguments support the view that citizens should be allowed to influence public sector planning directly. Some of these refer to laws and regulations that mandate public involvement in planning. Others concern the rights of citizens in a democracy and the view that only those affected can place values on social and environmental impacts. Public involvement is considered a legitimate component in government agency planning in the United States and many other countries. Some agencies use the term *public* broadly to include individual citizens and interest groups, elected and appointed officials, and governmental administrative units that may have an interest in the actions being planned.

Commonly Used Public Involvement Techniques

The public's contribution to planning government projects is not limited to just evaluating alternatives. Many agencies involve the public in all planning tasks, including the formulation of alternatives and the prediction of impacts. No single approach is relied on for including the public in planning. Often several techniques are put together in a program tailored to meet the objectives of a particular planning study.

Most public involvement programs include meetings, and several meeting types are commonly used (see Table 8.4). The public hearing is the most rigid. A hearing officer generally governs the proceedings and a stenographer makes a *verbatim* transcript. Presentations are formal and there is little interaction among participants. Large group meetings can be much less formal than hearings. However, it is difficult for many citizens to contribute directly in large

TABLE 8.4 Meeting Types Commonly Used to Include Citizens in Planning[a]

Public hearings
Large public meetings
- Official presentation followed by question period
- Panelists offering alternative viewpoints
- Informal "town meeting" structure

Large public meetings utilizing small group discussions
Public workshops
Informal small group meetings

[a]Based on information in Creighton (1981).

assemblies unless provisions are made to break up into small groups for part of the time. Workshops are generally used to have individuals focus on a specific planning task. The least structured meeting types are small, informal get-togethers. These are sometimes held in private homes and they provide an opportunity for citizens and agency planners to exchange ideas in a casual setting.

Public involvement programs rarely involve only meetings. Written materials and coverage in the "mass media" are useful in providing information *to* the public. Public displays and exhibits, especially when staffed by planners to answer questions, are also valuable. Different techniques are used to obtain information *from* the public. Interviews and mail questionnaires can be effective for this purpose. By randomly selecting persons to be interviewed or surveyed, it may be possible to obtain a representative sample of opinions and feelings.

Table 8.5 lists public involvement methods that do not rely on meetings. The last technique listed, advisory groups, is different from those discussed above. Group members engage in a two-way communication with the agency performing a planning study. Agencies often form advisory groups to provide a convenient way of communicating informally with individuals representing a range of interests. Advisory groups can assist an agency in formulating and evaluating alternatives, and in designing other methods to include citizens in planning. For example, an advisory group might help prepare an information brochure that explains the motivations for an agency's planning study.

The design of an effective public involvement program requires both skill and effort. In the United States, many agencies have developed sophisticated instruction manuals to help their staffs avoid common mistakes and take advantage of the experience gained by others.[7]

TABLE 8.5 Public Involvement Techniques Not Based on Meetings

Providing information to the public

Reports, brochures, and information bulletins

Mass media coverage
- Press releases
- Radio and television "talk shows"
- Documentary films

Public displays and exhibits

Obtaining information from the public

Interviews

Mail questionnaires

Establishing two-way communications

Advisory groups (also called task forces and citizens' committees)

[7]The *Public Involvement Manual* prepared by Creighton (1981) provides detailed instruction in the use of numerous techniques for including citizens in agency planning.

Integrating Public Involvement into Planning: A Case Study

A shortcoming of many public involvement exercises conducted by federal agencies in the United States is their lack of integration with other planning and decision-making activities. Sometimes the information provided by citizens is not considered systematically, and public involvement is viewed as just another hurdle to be jumped along the route to implementing an agency's project.

A planning study carried out by the U.S. Army Corps of Engineers, San Francisco District Office ("the district"), demonstrates how a public involvement program can be organized to assist both agency planners and citizens in dealing with important problems. The study, which was carried out in the 1970s, concerned flooding on San Pedro Creek in Pacifica, California, a small coastside community south of San Francisco.[8] The district was committed to involving the public in each of four planning tasks: (1) identifying the water-related problems and needs of the San Pedro Creek area, (2) formulating alternative plans to deal with flooding and other water problems, (3) forecasting the impacts of the various proposals, and (4) evaluating the alternatives. The district felt that citizens should be given opportunities to express their opinions throughout all stages of the planning investigation.

At the outset, the district identified numerous offices and agencies for inclusion in their public involvement program. Among these were the Pacifica City Council, the city manager, and the state and federal fish and wildlife agencies. Local residents living either along the creek or in the flood plain were also to be involved in the study. Individuals who often played an important role in Pacifica's community affairs were interviewed to determine which citizens and groups might be interested in the district's study. Those questioned initially were identified by a review of back issues of local newspapers. The initial interviews generated the names of other people who should be contacted.

Having determined which individuals, groups, and agencies would be involved in the planning investigation, the district delineated specific public involvement program objectives. A primary goal was to have the public informed on all aspects of the San Pedro Creek study. This required that citizens be given details on the district's perception of the water-related problems in the San Pedro Creek area. The public also needed information about possible plans to deal with those problems and the impacts of the alternative plans. Another of the district's objectives was to have two-way communications with the public. This required that citizens have opportunities to react to the district's ideas and proposals.

The San Pedro Creek study was to be carried out over a 2-year period. To meet its public involvement program objectives over so long a time period, the district had to use several techniques. Not everyone with an interest in the San Pedro Creek study would either need or want to be involved on a continual basis over a 2-year period. Many individuals and groups would be content if they were only consulted when the district was about to make a key decision.

[8]Wagner and Ortolano (1976) provide a detailed account of the planning process used in the San Pedro Creek flood control study.

The district formed a "citizens' advisory committee" to maintain regular communication with at least one public entity. The committee consisted of five Pacifica residents selected by the city council. Collectively, they represented the people likely to have the greatest interest in the outcome of the San Pedro Creek investigation. These included local homeowners, merchants in a shopping center within the flood plain, and local environmental groups. The citizens' advisory committee provided information throughout the study. It also helped design other elements of the district's public involvement program.

To facilitate a two-way information flow between the district and various segments of the public, a "citizen information bulletin" was prepared a few months after the study began. A questionnaire to be returned to the district was inserted in the bulletin. Both the bulletin and questionnaire were mailed to about 1200 citizens and officials. The bulletin described the district's preliminary ideas about the San Pedro Creek flooding problem, possible alternative actions, and the likely impacts of those actions. The questionnaire considered the same topics and provided a convenient opportunity to comment on and supplement the district's preliminary concepts.

A "public workshop" on San Pedro Creek flood problems was held a few weeks after the bulletins and questionnaires were distributed. It was run informally by the citizens' advisory committee using a three-part format. First, the participants met as a whole to hear general remarks about the planning study and the purpose of the workshop. After that, the participants divided into small group discussions which were led by committee members. Finally, the participants were reassembled for an exchange of information about what occurred in the small groups.

The workshop gave people a chance to react to the district's preliminary ideas and to suggest additional factors that should be considered in formulating and evaluating alternative flood control plans. During the year following the public workshop, the district completed preliminary engineering, economic, and environmental studies for several proposals. Although it had met monthly with the citizens' advisory committee during this period, the district felt a need for additional communication with the public. It wanted feedback on whether all important evaluative factors were considered in the economic and environmental impact studies. The district also wanted to know how different individuals and groups weighed the evaluative factors and how they would rank the alternatives which the district had examined.

To provide this second opportunity to communicate with all segments of the public, another citizen information bulletin and questionnaire were prepared. Because the second bulletin summarized results from studies that had been completed since the public workshop, it was more detailed and elaborate than the first. The distribution of the second bulletin and questionnaire was coordinated with a meeting of the Pacifica City Council that focused on the San Pedro Creek flooding problems. Based on the information in the bulletin and presentations by the district, the city council developed its own ranking of the district's proposals. The city council's evaluation was later used by the district in judging which action should be recommended for implementation.

During the San Pedro Creek study, the public provided the district with much useful information. The public comments offered insights into which factors local residents considered important in evaluating alternative plans. For example, after learning of some preliminary flood control proposals, many Pacifica resident's expressed concern over the creek-side vegetation that would be destroyed. The district responded by formulating a plan that would reduce the flood problems without destroying the valued vegetation. An example of public involvement in evaluating alternative plans is given by the city council meeting on the San Pedro Creek problem. The ranking of alternatives resulting from that meeting had an important influence on the district's decision making.

The district's public involvement program helped yield a flood control plan with which both Pacifica and the Corps of Engineers were pleased. Even though there was no dispute over the final proposal, the plan was not implemented. This unsettling outcome resulted because the city of Pacifica was unable to generate its share of the total project costs.

ENVIRONMENTAL MEDIATION

In many cases, even when there is extensive citizen involvement in planning, government agency decisions are objectionable to some interests. If there is great dissatisfaction, those adversely affected may sue in an effort to modify or halt the agency's final plan. In the United States during the 1970s, many citizens' groups used the courts for this purpose.

There are several reasons why judicial review is often inappropriate as a mechanism for resolving environmental disputes between citizens' groups and agencies. Court actions are costly, and citizens adversely affected by an agency proposal may have difficulty raising the required funds. In addition, litigation can involve much time, sometimes years, before a settlement is reached. Furthermore, courts frequently are not inclined to rule on whether an agency "made the right choice." Even when the points of contention between citizens and an agency concern substantive issues such as increased traffic congestion, the litigation may revolve entirely around procedural (due process) questions. The court's resolution of the environmental dispute may, in such cases, be unrelated to the basic conflicts existing among those affected by the agency's proposal.[9]

"Environmental mediation" emerged in the 1970s as an innovative means for resolving environmental disputes. The following definition, used by the Office of Environmental Mediation at the University of Washington in Seattle, illuminates the mediation process:

> *Mediation is a voluntary process in which those involved in a dispute jointly explore and reconcile their differences. The mediator has no authority to impose a settlement. His or her strength lies in the ability to assist the parties*

[9]An elaboration of the points in this paragraph is provided by Lake (1980, pp. 20, 44–49).

in resolving their own differences. The mediated dispute is settled when the parties themselves reach what they consider to be a workable solution.[10]

Efforts by the Office of Environmental Mediation to resolve a dispute over a proposed extension of Interstate-90 (I-90) into downtown Seattle clarify the definition of environmental mediation. The dispute was over a proposal advanced in 1975 by the Washington State Department of Highways to build a 10-lane extension to I-90. The proposal was labeled the "4-2T-4" plan since it involved 4 automobile lanes in each direction and 2 lanes reserved for transit systems. Supporters of the plan included residents of two nearby communities, Bellevue and Mercer Island, who could commute to Seattle more easily on the I-90 extension. Opponents included the city of Seattle and various citizens' groups. They feared that the project would increase urban sprawl and have negative effects on air quality and noise. Two governmental units were eager to find a compromise version of the I-90 extension: the local county (King County) and the county wide transit agency (Metro). They feared that without a compromise, the Seattle region would lose the opportunity to obtain sizeable federal subsidies for dealing with its regional transportation problems. In addition, delays in implementing the proposed $500 million project were estimated to cost $140,000/day. The State Highway Commission, the ultimate decision maker in this case, recognized that to receive available state and federal highway funds, a proposed extension needed the support of each local jurisdiction. There was substantial motivation to avoid a court battle and resolve the conflict over the 4-2T-4 plan quickly.

In March 1976, the governor of Washington appointed two mediators from the Office of Environmental Mediation. The two individuals had previously mediated environmental disputes in the Seattle area and were known for their impartiality. Participants in the negotiations included elected officials from the three cities and King County and representatives from Metro and the State Department of Highways. The State Highway Commission's ability to implement any negotiated agreement was assured, since the commission was represented in the mediation process by the State Highway Department. In addition to initiating discussions and facilitating negotiations among formal participants, the mediators kept the general public informed and maintained communications with various citizens' groups. For example, they arranged to have some formal negotiation sessions open to the public. Individual citizens and groups could also influence the mediation process by working through their elected representatives.

After several months, a formal "memorandum agreement" was ratified by each of the governmental units participating in the negotiations. It called for an eight-lane facility, with special features for carpools and transit systems. In addition, portions of the route would be covered in order to minimize environ-

[10]This definition, and the following discussion of the mediation effort concerning I-90 in Seattle, is based on Cormick and Patton (1980). See Talbot (1983) for a more detailed description of the I-90 case.

mental impacts. The various parts of the agreement were specified as being acceptable *only as a total package.* This encouraged continued cooperation, since each body's support for the new proposal was predicated on the implementation of *all* features in the memorandum agreement.

Since the mid-1970s, there have been dozens of applications of environmental mediation in the United States. Because many of them have involved citizens directly, mediation is sometimes confused with citizen involvement in planning. Environmental mediation is not a citizen participation technique; it is a process for making decisions. The goal is to resolve environmental conflicts by negotiating a formal agreement that can be implemented.

Environmental mediation is in its formative stages. There are many unsettled questions concerning the funding for the qualifications of mediators and the selection of participants in the negotiations.[11] In addition, there is a danger that the process will be abused, for example, by providing a delaying tactic for participants unwilling to engage in "good-faith bargaining."[12] However, in light of the successful applications since 1975, mediation appears to be a promising supplement to court actions in resolving environmental disputes.[13]

[11]For an elaboration of these and other unsettled questions, see the article by Susskind (1981) and the two discussions that follow it.

[12]A description of potential shortcomings of environmental mediation is given by Lake (1980, pp. 71–73).

[13]Talbot's (1983, p. 97) analysis of six environmental mediation efforts led him to conclude that "mediation is a supplement, rather than an alternative, to legal action in environmental disputes." He argues that the threat of impending litigation provided the impetus for mediation in the six cases he analyzed.

APPENDIX TO CHAPTER 8

MULTIOBJECTIVE PROGRAMMING

This appendix introduces concepts from multiobjective programming[14] that are useful in ranking alternative projects. Consider a hypothetical situation in which a government agency must decide on how two resources, water and fertilizer, are to be distributed between two farms. Each farm grows a different crop and has different requirements per acre for water and fertilizer (see Table 8.6).

The government must decide how much new land to bring into production on farms 1 and 2. In algebraic terms, it must select X_1 and X_2, where X_1 represents the number of acres on farm 1 that the government will provide with needed quantities of water and fertilizer. The variable X_2 is interpreted in the same way.

The range of possible government decisions can be represented in algebraic terms. Based on the information in Table 8.6, the government's decision (X_1, X_2) must satisfy the following conditions:

$$\text{limit on available water—}3X_1 + 2X_2 \leq 100$$

and

$$\text{limit on available fertilizer—}X_1 + 2X_2 \leq 50$$

In addition, neither X_1 nor X_2 can be less than 0. Both farmers are assumed to have more land than can be brought into production by using all the water and fertilizer available.

TABLE 8.6 Data for Multiobjective Programming Example

	Water		Fertilizer	
	Quantity	Units	Quantity	Units
Farm 1 Requirements	3	Acre-feet/acre	1	100 lb/acre
Farm 2 Requirements	2	Acre-feet/acre	2	100 lb/acre
Total Resources available to be allocated	100	Acre-feet	50	100 lb

[14]Readers with a background in calculus and matrix algebra are referred to mathematical introductions to this subject by Cohon (1978), Zeleny (1982), and Goicoechea, Hansen, and Duckstein (1982).

The mathematical inequalities can be represented on a graph whose axes are X_1 and X_2, the decisions regarding how many acres to plant on each farm. This is shown in Figure 8.2. All points in the quadrilateral ABCD represent combinations of X_1 and X_2 that are feasible in that they satisfy the inequalities representing the limits on water and fertilizer. They also meet the conditions that X_1 and X_2 be nonnegative.

The shaded area in Figure 8.2 represents an infinite number of feasible plans. Evaluative factors must be introduced to determine which plans are preferred.

For illustrative purposes, assume the government has two major concerns. One centers on its desire to maximize contributions to net national income. These contributions per acre of land brought into production on farms 1 and 2 are represented by the coefficients Π_1 and Π_2, respectively. To make the arithmetic simple, assume that $\Pi_1 = \Pi_2 = 1$. Using these definitions, the largest net national income occurs when X_1 and X_2 are chosen to maximize

$$Z_1 = \Pi_1 X_1 + \Pi_2 X_2 = X_1 + X_2$$

Figure 8.3 contains parallel lines representing different constant values of Z_1. If these lines were superimposed on the "feasible decisions" in the shaded area of Figure 8.2, the point B would be seen to represent the plan with the largest value of Z_1. This plan is identified by solving the two equations in Figure 8.2 simultaneously. It has $X_1 = 25$ acres, $X_2 = 12.5$ acres, and a value of $Z_1 = 37.5$.

The government's second objective is based on the political difficulty it will face if it appears to favor one farm over the other in allocating water and fertilizer. The government's aim, in this regard, is to minimize the difference between the acreage brought into production on each farm. This "equity ob-

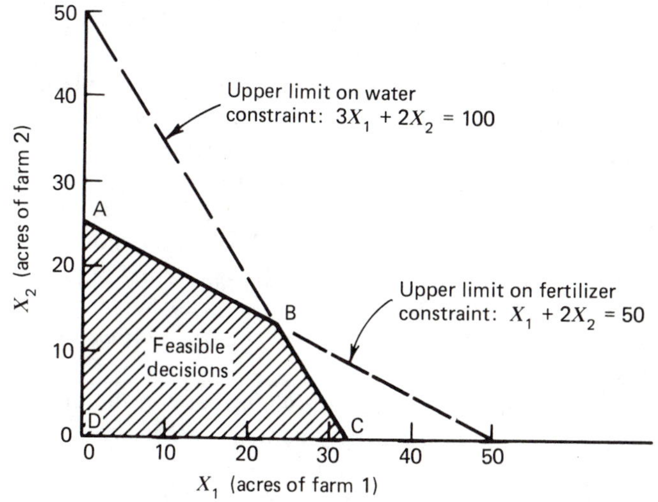

FIGURE 8.2 Set of feasible governmental decisions.

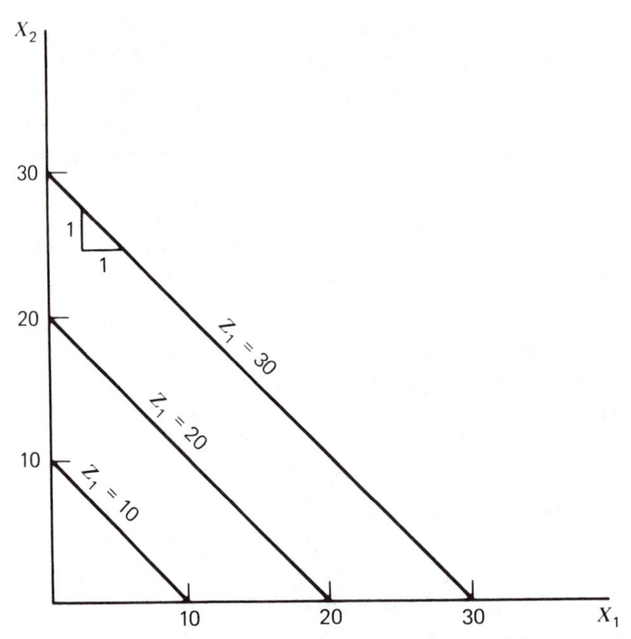

FIGURE 8.3 Constant values of the net national income objective.

jective" may be represented mathematically as the minimization of the absolute value[15] of the difference between X_1 and X_2,

$$Z_2 = |X_1 - X_2|$$

As indicated in Figure 8.4, constant values of Z_2 are straight lines with a slope of 45°. The minimum of Z_2 occurs for points on the line through the origin.

The concept of "dominance" in multiobjective programming is demonstrated by investigating the best (or "optimal") solutions if only equity is considered in reaching a decision. All points on the line $Z_2 = 0$ and within the space of feasible decisions are optimal with respect to the equity objective (see Figure 8.5). The other objective, net national income, is increased by making X_1 plus X_2 as large as possible. On the basis of national income, it can be argued that with $Z_2 = 0$, point E in Figure 8.5 is the only feasible solution that should be considered in selecting a plan. Point E is at least as good as all others in terms of the equity objective, and it yields the largest contribution to net national income of all the solutions with $Z_2 = 0$. In other words, plan E *dominates* all other feasible solutions having $Z_2 = 0$.

Only points on the line EB in Figure 8.5 should receive further consideration, because all other feasible plans are dominated. Any point on EB is preferable

[15]It is not adequate to use the simple difference between X_1 and X_2, since it could be minimized by making $X_1 = 0$ and $X_2 = 25$, the largest feasible value of X_2 (see Figure 8.2). This is why the absolute value of the difference is used.

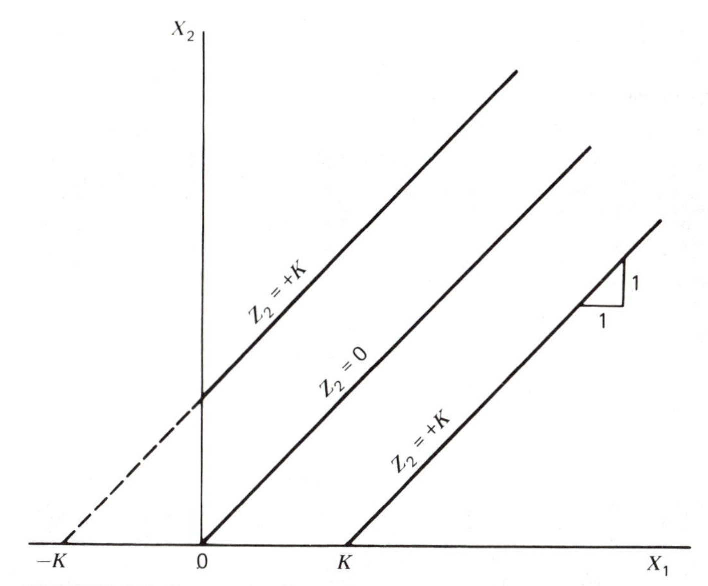

FIGURE 8.4 Constant values of the equity objective.

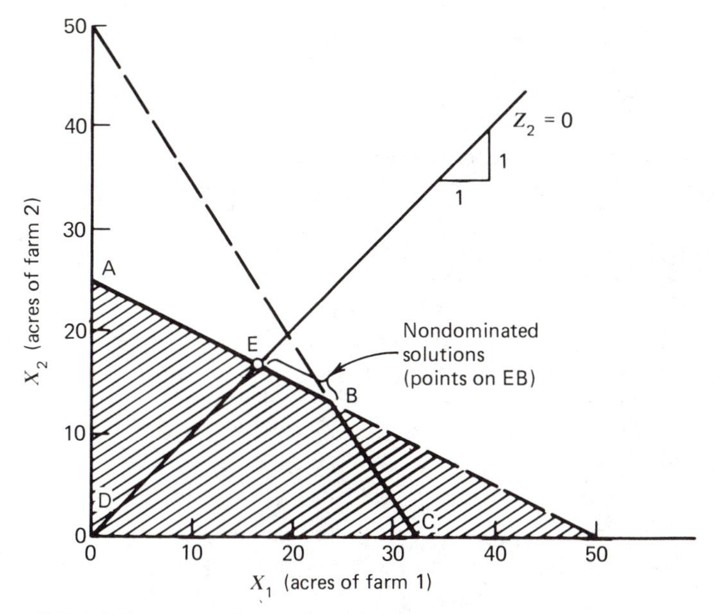

FIGURE 8.5 Nondominated solutions on a plot of X_1 versus X_2.

to feasible solutions along the 45° line passing through it. The reasoning used in establishing the dominance of plan E applies here also. Consider, for example, a 45° line passing through point B. All feasible solutions on that line have the same value of Z_2 as point B. However, solution B is preferable since it has the highest value of net national income. All plans on the line segments AE and

BC are dominated by points E and B, respectively. Those along AE are worse than solution E in terms of both the equity and national income objectives. Similarly, points along BC perform worse than plan B.

To further analyze this decision problem, it is convenient to transform points on the line EB from the plot containing the "decision variables" (X_1, X_2) in Figure 8.5 to a graph whose axes represent the objectives (Z_1, Z_2). Solution E has $X_1 = 16.67$ and solution B has $X_1 = 25$. By inspection of Figures 8.2 and 8.5, all points on EB must satisfy

$$X_1 + 2X_2 = 50$$

for X_1 between 16.67 and 25. Solving this equation for X_2 yields

$$X_2 = ½ (50 - X_1)$$

Table 8.7 shows values of X_1 and X_2 for points along line EB in Figure 8.5. Corresponding values of Z_1 and Z_2 are also indicated. Observe that the first row in Table 8.7 represents solution E. The combinations of Z_1 and Z_2 in Table 8.7 are plotted on a graph with axes Z_1 and Z_2 in Figure 8.6. The Z_2 axis is constructed such that the equity objective is improved by moving from the bottom to the top of the graph. This is done by plotting the differences in areas brought into production in the reverse of the usual order.

The dominance concept focuses the attention of decision makers on a subset of the feasible solutions, namely, the nondominated ones in Figure 8.6. The figure demonstrates how much sacrifice must be made in the equity objective

TABLE 8.7 Examples of "Non-dominated" Solutions[a]

X_1	X_2 $X_2 = ½ (50 - X_1)$	Net National Income Objective $Z_1 = X_1 + X_2$	Equity Objective $Z_2 = \|X_1 - X_2\|$
16.67	16.67	33.3	0
17	16.5	33.5	0.5
18	16	34	2
19	15.5	34.5	3.5
20	15	35	5
21	14.5	35.5	6.5
22	14	36	8
23	13.5	36.5	9.5
24	13	37	11
25	12.5	37.5	12.5

[a]The table contains only a subset of the nondominated solutions because it considers only ten values of X_1. The variable X_1 can take on an infinite number of values in the interval from 16.67 to 25.

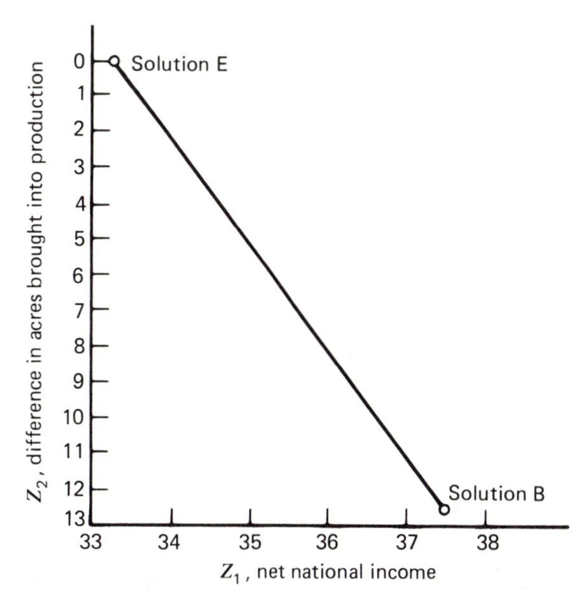

FIGURE 8.6 Nondominated solutions on a plot of Z_1 versus Z_2.

to increase net national income. However, the figure does not indicate which of the nondominated solutions should be chosen. This cannot be determined without information about how the objectives are weighted relative to each other.

If the selection of X_1 and X_2 is to be made by a *single* decision maker, there are many procedures that can assist in making the final choice. An illustrative approach involves the "isopreference curves" shown in Figure 8.7. In general, these curves are derived for a specific decision maker in a particular context. Isopreference curves are defined such that a decision maker would be equally pleased with any of the combinations of objectives represented by points on a given curve. In Figure 8.7, for example, the decision maker's preference for a solution providing $(\overline{Z}_1, \overline{Z}_2)$ is exactly the same as his or her preference for a solution giving $(\overline{\overline{Z}}_1, \overline{\overline{Z}}_2)$. To derive an isopreference curve, an analyst asks the decision maker a structured sequence of questions intended to reveal acceptable tradeoffs among objectives.[16] Using this procedure, the analyst seeks more than just the weights on the two objectives. A complete representation of a particular decision maker's preferences for selected objectives is sought. For the hypothetical curves in Figure 8.7, the levels of preference increase on curves that are further from the origin in the north-easterly direction. The "best compromise solution" is the point on the set of nondominated solutions that lies on the highest isopreference curve.

[16]Applications of the procedure for developing isopreference curves are presented by Goicoechea, Hansen, and Duckstein (1982, pp. 118–140).

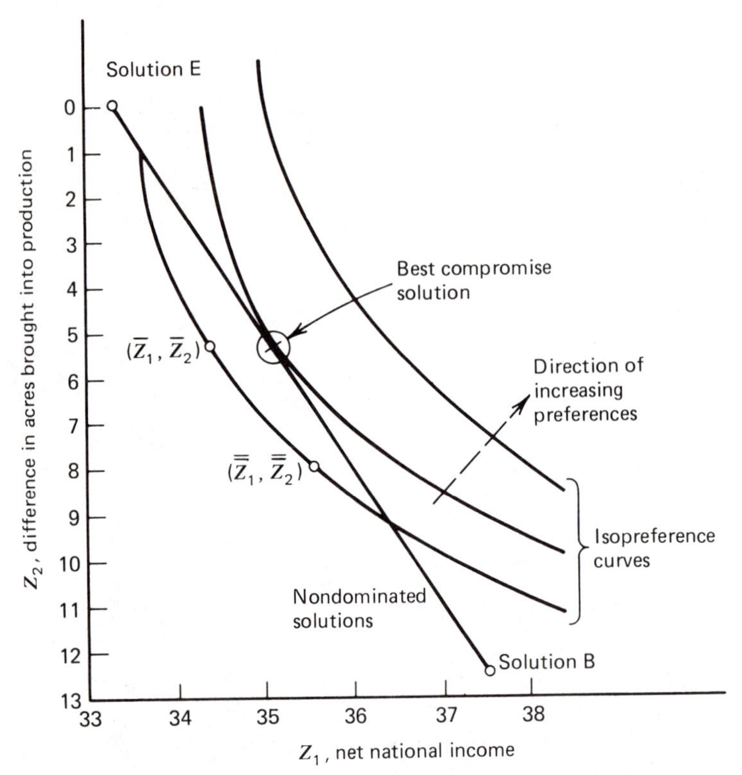

FIGURE 8.7 Use of isopreference curves to derive a "best compromise solution."

Mathematical programming methods to assist in choosing among nondominated solutions are not widely used in public sector planning. One reason is the procedures often concern the preferences of only a single individual. In public sector planning, there are usually many decision makers involved. Another explanation for the limited use of these mathematical techniques is that they are often too complex to be readily understood by decision makers.

KEY CONCEPTS AND TERMS

ISSUES IN MULTICRITERIA EVALUATION
- Four common planning tasks
- Incommensurate effects
- Sources of evaluative factors
- Weights reflecting tradeoffs

EXTENSIONS OF BENEFIT–COST ANALYSIS
- Limitations of BCA
- Benefits defined in terms of objectives
- Multiple objectives
- National income net benefits
- Income redistribution net benefits
- Weighted sum of objectives
- Bases for estimating weights

TABULAR DISPLAYS AND THE SUM OF WEIGHTED FACTOR SCORES
- Ordinal factor scores
- Bases for scoring
- Goals–achievement matrix

INVOLVING THE PUBLIC IN EVALUATION
- Identifying segments of the public
- Including citizens in planning activities
- Alternative meeting types
- Two-way information flows
- Information bulletins
- Surveys and questionnaires
- Advisory groups
- Litigation to settle disputes
- Environmental mediation

MULTIOBJECTIVE PROGRAMMING
- Net national income objective
- Equity objective
- Nondominated solutions
- Isopreference curves
- Best compromise solution

DISCUSSION QUESTIONS

8-1 A suburban community is considering several alternative land development schemes for a parcel of undeveloped land within its boundaries. The following hypothetical information represents results from a study by a consultant to the city's planning department. The ratings shown vary from 5 (best) to 1 (worst).

Compute the sum of weighted factors scores for each alternative and note the plan with the highest sum. What happens to the rank ordering of the

		Ratings for Alternative Proposals		
Factors	**Weights**	**Plan 1**	**Plan 2**	**Plan 3**
Net contribution to city tax base	10	5	3	1
Increased traffic congestion	5	1	3	5
Adverse visual effects	2	1	2	5
Attainment of low-cost housing goals	5	3	5	1
Adverse effects on school crowding	8	3	5	1

three proposals if the column of weights is changed from (10,5,2,5,8) to (5,10,5,5,8)? Criticize this approach to evaluation. What defense could be offered in response to such criticism? Would you consider using this method if you were a consultant to a city planning department? Justify your position and present a different way to rank alternatives if you find the above procedure unsatisfactory.

8-2 Suppose you are to advise a highway agency on what it can do to ensure that citizen participation in the planning and evaluation of its proposals is constructive. What advice would you offer regarding (1) the timing of citizen involvement activities and (2) techniques used to involve the public in planning?

8-3 List three reasons why planners working for a federal agency like the U.S. Army Corps of Engineers might not be eager to involve the public in their planning. What arguments could be used to counterbalance the three reasons you offered?

8-4 The city of Fulopville has a problem regarding future use of its land adjacent to San Giralamo Bay ("the Baylands"). A creek that flows through the Baylands to the Bay carries storm runoff from a residential section. This area floods periodically when high tides in the bay retard the drainage of creek flows. The county flood control engineers have proposed to create retention basins in the heart of the Baylands to temporarily store creek drainage during floord periods. Based on a BCA of five flood control plans the engineers believe they have the optimal solution.

The engineers' proposal has elicited strong objections from two interest groups. One group has called for a complete renovation of the Baylands to create a park with tennis courts, barbeque pits, parking lots, and marinas. They recognize that flood control is important but feel it should not limit recreational use of the Baylands. The second group feels that flood control should be accomplished, but without altering the Baylands. They also oppose the park concept since water fowl in the Baylands would be driven off by large numbers of people "recreating." They argue that the Baylands, in its natural form, is important to the ecological integrity of the Bay.

Assist the city's decision makers by providing a critique of the evaluative criteria used by the engineers. What actions might the decision makers take to resolve the dispute?

8-5 Consider the example defined by Table 8.6 in which it is necessary to find X_1, and X_2, the number of acres of land on farms 1 and 2 which are to be brought into production. Suppose, in addition to the constraints on water supply and fertilizer, there is a third constraint on labor, namely,

$$3X_1 + 3X_2 \leq 105$$

This means each acre (on each farm) requires three units of labor, and the total labor force is limited to 105 people. (The odd-looking numbers are used to make the answers easy to obtain.)

(i) Prepare a graph of the set of feasible decisions in this modified problem.

(ii) Make a graph of the nondominated solutions in the "objective space" (the counterpart to Figure 8.6). The objectives for this case are the same as in the example in the appendix.

REFERENCES

Canter, L. W., 1979, *Environmental Impact Statements on Municipal Wastewater Programs.* Information Resources Press, Washington, D.C.

Cohon, J. L. 1978, *Multiobjective Programming and Planning.* Academic Press, New York.

Cormick, G. W. and L. Patton, 1980, Environmental Mediation: Defining the Process through Experience, *in* L. M. Lake, (ed.). *Environmental Mediation: The Search for Concensus.* Westview Press, Boulder, Colo.

Creighton, J. L., 1981, *The Public Involvement Manual.* Abt Books, Cambridge, Mass.

Eckstein, O., 1958, *Water Resource Development, the Economics of Project Evaluation.* Harvard University Press, Cambridge, Mass.

Elliott, M. L., 1981, Pulling the Pieces Together, Amalgamation in Environmental Impact Assessment. *Environmental Impact Assessment Review* **2** (1), 11–37.

Freeman, A. M., III, 1970, Project Design and Evaluation with Multiple Objectives, *in* R. H. Haveman, and J. Margolis (eds.), pp. 347–363. *Public Expenditures and Policy Analysis.* Markham, Chicago, Ill.

Goicoechea, A., D. R. Hansen, and L. Duckstein, 1982, *Multiobjective Decision Analysis with Engineering and Business Applications.* Wiley, New York.

Hill, M., 1967, A Method for the Evaluation of Transportation Plans. *Highway Research Record* No. 180, 21–34.

Hill, M., 1968, A Goals-Achievement Matrix for Evaluating Alternative Plans. *Journal of the American Institute of Planners* **34,** 19–28.

Hobbs, B. F., 1980, A Comparison of Weighting Methods in Power Plant Siting. *Decision Sciences* **11** (4), 725–737.

Lake, L. M. (ed.), 1980, *Environmental Mediation: The Search for Concensus.* Westview Press, Boulder, Colo.

Lichfield, N., P. Kettle, and M. Whitbread, 1975, *Evaluation in the Planning Process.* Pergamon, Oxford.

Livingston and Blayney, Inc., 1971, *The Foothills Evironmental Design Study: Open Space vs. Development.* Final Report to the City of Palo Alto prepared by Livingston and Blayney, City and Regional Planners, San Francisco, Calif. unpublished.

Maass, A., 1966, Benefit-Cost Analysis: Its Relevance to Public Investment Decisions, *in* A. V. Kneese, and S. C. Smith (eds.). *Water Research,* pp. 311–328. Johns Hopkins University Press for Resources for the Future, Inc., Baltimore, Md.

Marglin, S. A., 1962, Objectives of Water Resource Development: A General Statement, *in* A. Maass, et al. (eds.), *Design of Water-Resource Systems,* pp. 17–87. Harvard University Press, Cambridge, Mass.

McAllister, C. M., 1980, *Evaluation in Environmental Planning.* MIT Press, Cambridge, Mass.

Steiner, P. O., 1969, *Public Expenditure Budgeting.* The Brookings Institution, Washington, D. C.

Susskind, L., 1981, Environmental Mediation and the Accountability Problem. *Vermont Law Review* **6** (1), 1–47.

Talbot, A. R., 1983, *Settling Things, Six Case Studies in Environmental Mediation.* The Conservation Foundation, Washington, D.C.

Wagner, T. P., and L. Ortolano, 1976, *Testing an Iterative, Open Planning Process for Water Resources Planning,* Report No. 76–2. U.S. Army Engineer Institute for Water Resources, Ft. Belvoir, Va.

Zeleny, M., 1982, *Multiple Criteria Decision Making.* McGraw–Hill, New York.

PART FOUR

LAND USE AND THE ENVIRONMENT

CHAPTER 9

ENVIRONMENTAL CONSIDERATIONS IN LAND DEVELOPMENT: INSTITUTIONAL ISSUES

Recently, increased attention has been given to the relation between land use and environmental quality. Table 9.1 lists adverse environmental effects that often result when land is developed for residential and commercial uses. As the

TABLE 9.1 Some Adverse Environmental Effects of Residential and Commercial Development

Air Quality

Air-borne residuals from space heating
Emissions from motor vehicle traffic induced by development

Water Quality

Soil erosion caused by construction activities
Residuals in storm runoff over newly paved areas
Domestic sewage from new population centers

Noise

Noise from motor vehicle traffic induced by development

Biological Resources

Direct loss and disruption of wildlife habitats

Visual Resources

Loss of scenic views, for example, by covering hillsides with subdivisions.

table indicates, there are both obvious and subtle impacts of land development worthy of serious consideration.

In the United States, decisions regarding the use of privately held land have traditionally been made at the local level. A typical land development process is initiated by individual or corporate landowners with a concept for developing a particular parcel. Before development can proceed, a proposed project must be shown to conform with land use policies and controls. The most common form of control is the local zoning ordinance.

Local governments have demonstrated only a limited ability to account for the environmental impacts of land development. Many cities have not given these impacts a high priority. Environmental concerns have often been overshadowed by the single-minded pursuit of growth to create new tax revenues. Cities that try to preserve environmental quality by restricting growth are often unsuccessful because zoning ordinances are flexible and easily modified. Powerful land developers can frequently gain exceptions to zoning ordinances or have them changed. In addition, efforts of cities that limit their own growth are sometimes offset by neighboring cities that promote growth. Many land development impacts such as traffic congestion and air pollution have little respect for municipal boundaries and often need to be controlled using region-wide efforts.

The limited effectiveness of local governments in restraining development has led to new schemes for growth management. Although some innovative approaches have taken place at the local level, many represent the expansion of federal and state roles in land management. Because land use regulations change frequently and vary markedly in different areas, a comprehensive and current review of new land management efforts cannot be presented. This chapter introduces several interesting approaches that have been tried in the United States.

FEDERAL CONTROLS ON LAND USE

For many years, the U.S. government has controlled the use of federally owned lands which constitute roughly one third of the nation's total area. A small fraction of these public lands consists of national parks, monuments, and wildlife refuges. Approximately one quarter of them are national forests. The majority of federal land, roughly 60% of the total, is under the jurisdiction of the Bureau of Land Management. The many uses of these public lands include outdoor recreation, wilderness preservation, livestock grazing, and mineral and timber production. In addition to establishing constraints on use, the agencies responsible for federally owned land exert controls by collecting user fees and issuing permits and leases.[1]

Several long-standing federal programs have had *indirect* effects on the use of privately owned land. Environmental factors were not considered system-

[1]This discussion of federal controls on land use decision making relies on information presented by Moss (1977).

atically in the design of these programs. Noteworthy examples are federal grants subsidizing the development of "physical infrastructure," major public projects such as highways and municipal sewer systems. These facilities often have a dramatic effect on *where* homes and factories are located. A report of the Council on Environmental Quality (1976) refers to such facilities as "growth shapers" because of their influence on patterns of land development. Another noteworthy example of the indirect federal influence on land use concerns programs that encourage private home ownership. These include federally insured home mortgage programs and the allowance of mortgage interest payments as deductions from taxable income. Federal subsidies to homeowners have fostered the development of suburban single-family "tract housing" as opposed to high-density, inner-city dwelling units. By encouraging a sprawling low-density-type housing pattern, federal subsidies have contributed to a reliance on private automobiles for commuting to and from work. This has been a major cause of air pollution and traffic congestion in many U.S. cities.[2]

Federal Environmental Legislation Affecting Land Use

Several new federal policies and programs influencing land use have been established since 1960. Three of these have been discussed at length in previous chapters: the National Environmental Policy Act, the Clean Water Act, and the Clean Air Act. Illustrations of how these laws affect land development are given below.

NEPA influences land use by requiring environmental impact statements for federal actions on growth-shaping projects. Examples include federal subsidies for highways and municipal wastewater treatment facilities and permits for nuclear power plants. The Council on Environmental Quality's (1978) regulations for implementing NEPA instruct agencies to predict how their actions influence land use. In addition, the regulations require consideration of whether proposed agency actions conform with local land use plans and policies.

The Clean Water Act affects decisions regarding the location of industrial facilities. The act's influence is applied through the National Pollution Discharge Elimination System permit program. It is also exerted via requirements that firms pretreat their wastes before emitting them into municipal sewer systems constructed with federal funds. This pretreatment is intended to prevent troublesome industrial waste from reducing the effectiveness of municipal treatment processes. The size of some communities is also influenced by the Clean Water Act. A city that receives a federal grant for wastewater treatment plant construction must document that the subsidized plant will not encourage new population growth and thereby add to existing water quality problems. In addition, the federal government can prevent new hookups to municipal sewers if a city is not meeting its obligations under the act.

The Clean Air Act can also have substantial effects on land use. Its precon-

[2]For numerous additional examples of how federal policies and programs indirectly affect the use of private land, see Glickman (1980).

struction review requirements for major stationary sources of air-borne residuals can influence where industries locate. This is illustrated in Chapter 4 by Wickland Oil Company's effort to locate a terminal in the San Francisco Bay area. Portions of the Clean Air Act requiring transportation control plans for selected metropolitan areas may alter land use in the long run. These plans often strive to reduce the number of vehicle miles traveled in trips from residences to work places. Eventually, they may affect decisions about where new residences and work places are located in relation to each other.

Other Recent Federal Programs Affecting Land Use

Three other recent federal laws that have notable effects on land development are the Coastal Zone Management Act, the National Flood Insurance Act, and the Wild and Scenic Rivers Act.

Coastal Zone Management Congress passed the Coastal Zone Management Act of 1972 to assist states in developing institutions, plans, and procedures to manage land use in coastal areas. Congressional action was prompted by the severe pressures being exerted on coastal ecosystems. Sources of these pressures included municipal and industrial waste discharges in coastal waters, offshore oil extraction, marina development, and estuary and wetland filling to create land for development.

The Coastal Zone Management Act is administered by the National Oceanic and Atmospheric Administration, an agency within the U.S. Department of Commerce. The act defines "coastal states" to include not only those that border on the oceans or other salt water bodies but also those that border on the Great Lakes. Under the definitions in the act, 30 states and four U.S. territories are eligible for grants. States obtain federal funds for the protection of coastal resources using a two-stage process. In the first stage, states receive money from the secretary of commerce to help develop their coastal management plans. In the second stage, states obtain funds to assist in implementing federally approved plans. There are no sanctions against states that fail to prepare adequate coastal plans. The principal incentive for state action under the act is the availability of federal grants.

Flood Plain Zoning In 1968, Congress added a new component to its strategy for coping with flooding problems by passing the National Flood Insurance Act. Before that time, the principal federal strategy for dealing with flooding involved building dams and levees, and clearing, straightening, and widening stream channels to reduce the likelihood of flooding. Implementation of these measures, which led people to feel secure about building in flood plains, was often followed by intense development in flood prone areas. When the rare flood that exceeded the design capacity of the structural works arrived, the results were sometimes catastrophic. In such cases, the federal government paid twice: once for the construction of the flood control measures and once for disaster relief for those who built in "protected" flood plains and were later washed out by a rare, severe flood.

The National Flood Insurance Program (which was strengthened by the Flood Disaster Prevention Act of 1973) works as follows. Communities participating in the program adopt regulations consistent with federal criteria for land uses within areas having "special flood hazards." Individuals with property in these areas can obtain flood insurance from licensed insurance agents and brokers. Although the insurance policies are written by the private insurance industry, the federal government subsidizes the rates and establishes the limits of available coverage.

There are substantial economic penalties for communities and individuals in designated flood hazard areas who do not participate in the National Flood Insurance Program. These concern eligibility requirements for various types of loans and federal assistance. The penalties have encouraged widespread participation. By the late 1970s, more than 17,000 local units of government had either adopted the land use regulations needed to participate in the flood insurance program or indicated their intention of doing so. Typical regulations consist of "flood plain zoning ordinances" to control the type of building that takes place in flood hazard areas.[3]

Wild and Scenic Rivers Another recent land use related program is the one established by the Wild and Scenic Rivers Act. This 1968 law set up a program to protect selected rivers and their adjacent lands. To become part of the National Wild and Scenic Rivers System, an area must possess "remarkable scenic, recreational, geological, fish and wildlife, historic, cultural or other similar values. . . ." Rivers included in the system are to be preserved in a free-flowing condition. By 1980, portions of 19 rivers were covered by the act.

Two kinds of protection are afforded by the Wild and Scenic Rivers Act. First, funds have been appropriated by Congress for the outright purchase of lands in designated areas. Second, the act imposes stringent constraints on the development of water projects which may affect the flow of officially classified wild and scenic rivers.

LAND USE REGULATION AT THE STATE LEVEL

On several occasions in American history, individual states have engaged in deliberate land use planning.[4] More often, however, state influence on land use

[3]This discussion of flood insurance is based partially on information presented by Kusler (1980, pp. 22–26).

[4]Linowes and Allensworth (1975) provide an historical account of the role of the states in land use planning.

has been an indirect consequence of programs to develop infrastructure, especially road networks. These programs have often had a dramatic effect on patterns of land development.

With few exceptions, traditional state level planning for either land use or infrastructure did not include systematic efforts to consider environmental factors. Major efforts to integrate environmental concerns into state land use planning and decision making occurred only recently. Several illustrations have already been mentioned. One involves the "little NEPAs," state laws mandating environmental impact assessments for actions taken by state (and sometimes local) agencies. In some states, for example in California, these laws have had a significant effect on land development. Other examples concern portions of the Clean Air Act and the Clean Water Act that influence land use. Many state air and water quality agencies play a large role in administering parts of these federal acts.

Increasingly, states are recognizing that serious environmental problems may result when land use decisions are based on the narrow perspectives of local governments. Since 1960, many states have introduced regulatory programs that summon a broader perspective and include a concern for environmental values.

Statewide Planning and Comprehensive Controls

The most forceful state programs constraining land use require that private developers comply with statewide restrictions. This approach is *not* widely practiced. Linowes and Allensworth (1975) indicate that only a handful of states have land use control programs of significance. Hawaii and Vermont have especially ambitious programs. Other states with strong land management regulations include Maine, Florida, and Oregon.

State-Level Zoning in Hawaii The Hawaii zoning program was initiated in 1961 in response to the adverse effects of urban sprawl in the Honolulu area and the need to limit the conversion of agricultural land to urban uses. Hawaii's program relies on a State Land Use Commission appointed by the governor. The commission classifies *all* public and private lands into one of four types of districts: urban, rural, agricultural, or conservation.

The State Land Use Commission exercises exclusive power only in conservation districts. Authority is shared in each of the other district types. For urban districts, locally established zoning regulations provide the principal land management tool. However, the boundaries of these districts are established by the commission. For agricultural and rural districts, counties control land development by issuing "use permits." The commission can prohibit specific proposals by vetoing county permit decisions.

Although Hawaii's zoning program and the State Land Use Commission have been criticized by both conservationists and developers, parts of the state land management effort have been successful. The conversion of agricultural land to urban uses has not been eliminated, but patterns of urban growth have been effectively guided. Based on an assessment of Hawaii's zoning, Healy and Rosenberg (1979, p. 186) indicate that it made "urban expansion far more compact and orderly than it would have been without the law."

There are unique aspects of Hawaii's institutional arrangements that limit the zoning program's utility as a model for other states. The history of Hawaii is characterized by a high degree of political control at the state level. In addition, land ownership patterns in Hawaii are unusual. It has been reported that over 85% of the land in Hawaii is owned by less than 100 individuals, corporations, trusts, or the government, with about half of this being privately held. Once the owners of very large land areas gave support to the new land use control concept, Hawaii was able to implement its ambitious statewide zoning program.[5]

Vermont's Development Permits Another way of controlling land use from the state level is to require permits for major land development projects. This approach was initiated by Vermont in 1970 to control the adverse environmental impacts of large projects, especially vacation-home developments.[6]

The Vermont Environmental Control Act (passed in 1970) indicates that major land development plans can not be implemented until a permit has been issued by a district commission. There are nine such commissions, each composed of three citizens appointed by the governor. The act applies to both private and public projects. Examples of proposals subject to the law's permit requirements are (1) subdivisions of more than 10 lots, (2) housing developments involving more than 10 units, and (3) commercial or industrial projects on more than 1 acre. This 1-acre limit is raised to 10 acres if the proposed project is within a town having a local zoning ordinance and subdivision regulations.

Before granting a permit, a district commission must find that a proposed development will not result in any of the following: unreasonable soil erosion or highway congestion, undue water or air pollution, or an excessive burden on existing water supply or on the ability of a municipality to provide governmental services. In addition, the proposal must not have an undue negative effect on any rare and irreplaceable natural areas or on the scenic beauty of an area. District commissions can only issue permits for proposals that conform with appropriate local, regional, and statewide plans.

Healy and Rosenberg (1979) analyzed aspects of the first 8 years of the act's implementation. Among other things, they reported that (1) a very high percentage of the 3000 or so applications for permits made during this period was eventually granted; (2) permits were almost always issued subject to conditions; and (3) the most common conditions (in a random sample of permits) concerned air and water pollution, erosion, traffic congestion, and aesthetics. Except for traffic congestion, these same factors were also the most frequent bases for denying permits. Nearly all permits included conditions aimed at enhancing environmental quality.

[5]The information in this paragraph is from Linowes and Allensworth (1975, p. 61).

[6]This discussion of the Vermont permit program relies on Healy and Rosenberg (1979).

Controls on Selected Resources and Activities

Although state land management programs sometimes involve farreaching controls like those in Hawaii and Vermont, a more common approach is either to protect special resources (for example, coastal zones) or to regulate a particular activity such as strip mining. Table 9.2 indicates the extent to which states have imposed these types of restrictions.[7]

The most frequently used special purpose controls are tax assessment policies aimed at conserving certain types of land use. For example, a tax program may be structured to ease the economic burdens of farmers who wish to keep their land for agricultural purposes. One such burden is the high tax that must often be paid by farmers whose land has great potential for conversion to residential or commercial use. High taxes can result if assessments are based on the market value of land (which may be very high for land near urban areas) as opposed to the value of land for farming. Many states have provided relief by taxing farmers based on the value of their land for agricultural purposes. In exchange for low tax, the land must be kept in agricultural use for a specified period.

Many states also have programs for controlling strip mining and the reclamation of mined lands. Typical programs require mining companies to obtain permits stipulating the land reclamation that will be needed. In some cases, performance bonds must be posted to guarantee that the reclamation will be carried out.

Power plant siting controls are also common at the state level. They take many different forms. In some states, such as Vermont, a nuclear power facility cannot be built without the approval of the legislature. In others, permits to construct power plants above a certain size must be obtained from a state agency.

TABLE 9.2 Selected State Programs for Controlling Specific Resources and Activities[a]

Type of Program	Number of States with Program in 1974
Land use tax incentives	30
Surface mining	27
Power plant siting	23
Wetlands management	16
Flood plain management	15
Coastal zone management	14

[a]Based on information in Linowes and Allensworth (1975, pp. 30–31).

[7]Popper (1981) provides a detailed account of single-purpose state land use controls established under the following laws enacted in the 1970s: Maryland's Power Plant Siting Act, Pennsylvania's Surface Mining Conservation and Reclamation Act, and California's Coastal Zone Conservation Act.

Criteria for granting permits often require a demonstration that the proposed facilities will be "compatible" with the environment.

As noted in Table 9.2, numerous states manage the use of flood plains, wetlands, and coastal areas. Many flood plain management programs involve the states in assisting local governments to meet land use criteria established under the federal flood insurance program. Wetlands protection efforts often consist of stringent controls (implemented via permit requirements) on dredging, filling, and other activities within or adjacent to marshes, swamps, and other wetlands. Development restraints and permit requirements are also common elements in coastal zone management programs.

Creation of Regional and Interstate Agencies

In most circumstances, a state's power to regulate the use of private property is delegated to cities and counties. In unusual cases, as in Hawaii and Vermont, this power is exercised directly through state land use controls. Another way of implementing state regulatory authority is through state-created regional institutions with the power to restrict land use. Table 9.3 lists five widely dis-

TABLE 9.3 Examples of Regional Organizations with Land Use Control Authority

Name	State(s)	Year of Formation	Examples of Control Authorities
Adirondack Park Agency	New York	1971	Creation of "zones" and use of permit program to control major projects in selected zones
Hackensack Meadowlands Development Corporation	New Jersey	1968	Zoning of Meadowlands and use of broad powers to preserve open space and build facilities
Twin Cities Metropolitan Council	Minnesota	1967	Influence on growth by providing sewer systems and approving plans of infrastructure agencies, cities and counties
San Francisco Bay Conservation and Development Commission	California	1965	Use of permits to control filling and development of San Francisco Bay
Tahoe Regional Planning Agency	California/ Nevada	1969	Bistate agency with permit program to control development in Lake Tahoe basin

cussed regional entities that can influence land use. They have two things in common. First, the perspectives of these regional organizations are broader than those of the cities and towns within their jurisdictions. Second, environmental issues played an important role in the formation of each of the entities.

Of the five organizations listed in Table 9.3, three have a similar overall approach: The Adirondack Park Agency, the San Francisco Bay Conservation and Development Commission, and the Tahoe Regional Planning Agency. Each engages in a plan development exercise and relies on a permit system to implement the plan. There are, of course, differences in the extent to which the entities are sensitive to the views of cities and counties within their jurisdictions. Likewise, the entities differ in the degree to which they adhere to their own plans.

The Hackensack Meadowlands Development Corporation and the Twin Cities Metropolitan Council follow unique strategies. The Hackensack Meadowlands Development Corporation was created to deal with 20,000 acres of marshlands and waterways (within a few miles of Manhattan) that had been used as a solid waste disposal site for 45 years. The area has 14 cities within its boundaries. The mandate to the corporation was to turn the acreage into recreational, commercial, and industrial parks and to enforce strict environmental regulations in the process. As a consequence, the powers and activities of the Hackensack Meadowlands Development Corporation are more sweeping than those of the other organizations in Table 9.3. The Twin Cities Metropolitan Council's strategy for guiding regional growth has relied heavily on its control over selected infrastructure systems. The council's greatest powers concern the provision of sewer systems, but it also has an influence on decisions concerning the location of parks and transportation facilities.[8]

GROWTH MANAGEMENT AT THE CITY AND COUNTY LEVELS

Since the mid-1960s many cities and counties have tried to cope with extraordinarily high rates of population growth. Some cities doubled their populations in less than a decade. This rapid growth is often accompanied by negative effects such as the loss of valuable open space, decreased air quality, and overloaded wastewater treatment plants. Often, rapidly growing communities also experience financial problems in trying to expand their public services to accommodate the new growth. The result is that traditional tools for managing growth such as comprehensive plans, zoning ordinances, and subdivision regulations have been supplemented in many cities and counties by innovative strategies.

Weaknesses of Traditional Land Use Controls

The expression of local land use policies often consists of a comprehensive plan that is implemented by means of zoning ordinances and subdivision regulations.

[8]More detailed descriptions of the organizations listed in Table 9.3 are given by Moss (1977) and by Schnidman, Silverman, and Young (1978).

In addition to controlling the type and intensity of land use, a typical zoning ordinance includes detailed "dimensional controls" such as maximum building heights and minimum lot sizes. Subdivision regulations assure that subdivisions are compatible with each other in terms of such things as the dedication of public streets, parks, and school sites, and the provision of public facilities.

If a proposed land development is consistent with applicable zoning ordinances and subdivision regulations, the requisite building permit is issued. If the proposal is inconsistent with the zoning, the developer can try to obtain an exception to the ordinance, formally termed a "variance." Alternatively, the developer can appeal directly to the locally elected governing body, such as the city council or county board of supervisors.

To examine shortcomings of the traditional land development process, consider the relationships between comprehensive plans and zoning ordinances. Since the 1920s, zoning ordinances have played the key role in controlling land use in the United States. In many instances, the ordinances preceded the development of a comprehensive plan. Sometimes zoning ordinances were applied for years in the complete absence of a plan.

Despite its widespread use, zoning has a number of serious disadvantages.[9] Although it can restrict the types of permissible projects, zoning is not a good mechanism for controlling how quickly development occurs. In addition, it is ineffective for communities that want to grow outward gradually from their already built-up areas. This deficiency of zoning leads to "leap frog development," the process of bypassing vacant land near a city's center and developing less expensive land on its outskirts.

Another limitation of zoning ordinances is that powerful development interests can often get around them. Delafons (1969) reports on the ease with which variances have been granted and the frequency with which ordinances have been changed in response to pressures from developers. Also, since comprehensive plans are not always required in developing zoning ordinances, much zoning has occurred without thorough planning. For all these reasons, cities and counties have increasingly turned to other mechanisms of land use control.[10] One of the many recently developed schemes is based on numerical quotas.

Numerical Quotas on Residential Development

The population of Petaluma, California, increased from 19,000 in 1965 to 31,000 in 1973. The adverse environmental and social consequences of this rapid growth were dramatic. To gain control over its rate of growth, Petaluma established an ordinance in 1973 putting a quota on the maximum number of residential

[9]Delafons (1969) provides an account of the history of zoning in the United States, including a discussion of its advantages and reasons for its popularity.

[10]Numerous supplements to zoning are described in the four-volume set, *Management and Control of Growth* [edited by Scott, Brower, and Miner (1975) and Schnidman, Silverman, and Young (1978)]. Not all accounts of these new approaches are favorable. For a critique of nontraditional growth management programs, see Frieden's (1979) assessment of local growth constraints in California during the 1970s.

building permits issued each year. The use of a numerical quota was challenged legally, but it eventually survived all court tests.

To be acceptable, proposals for development in Petaluma must be consistent with the city's general plan and its "Environmental Design Plan." Proposals are reviewed by a Residential Development Evaluation Board, whose members are chosen by lottery from several different groups such as the city council and local school districts. The board checks proposals for conformity with the city's plans. It rates them numerically using criteria concerning the availability of public services, the quality of proposed projects and their "contribution to public welfare." A project is only considered further if it obtains at least 25 out of 30 points on the availability of public services criteria and 50 out of 80 points on the design quality/contribution to public welfare criteria. Proposals approved by the Residential Development Evaluation Board are forwarded to the city council. The council then issues building permits on the basis of the ratings and numerical quotas; for example, a maximum of 500 dwelling units per year might be allowed.

The approach followed by Petaluma solved some problems, but it inadvertently created others. The complex and uncertain permit application process gave developers incentives to avoid Petaluma. It was easier for them to build projects in nearby towns where the growth controls were less restrictive. Petaluma's plan succeeded in slowing the growth rate. In fact, the city's 1980 population was close to that of 1973. This was not entirely beneficial, since the city's prior investments in water and sewer facilities were predicated on a 6% annual growth rate.[11] Petaluma's strategy also had the effect of slowing the construction of low- to moderate-cost housing to levels below those desired by the city.

Local growth management schemes such as the one used by Petaluma are frequently criticized for creating an unfair burden on lower-income families. For example, by restricting the supply of new housing, Petaluma's controls increased the price of housing, and this made it more difficult for low-income families to settle there. Critics argued that Petaluma was shirking its responsibilities by not providing its fair share of the region's low-cost-housing stock. To put this criticism in perspective, observe that many schemes for regulating the use of private land have the effect of redistributing wealth. This is elaborated in Mandelker's (1981) analysis of the potential inequities associated with zoning and with state land use regulations protecting coastal resources and wetlands.

Regional planning agencies can eliminate some of the difficulties noted above by coordinating the land management plans of all cities in a region. Counties, while unable to deal with entire regions, can also develop integrated strategies. In 1976, Marin County, California, implemented a growth management plan that coordinates the land use controls used by the 11 cities within its boundaries.[12] Marin County's program is similar to that of Petaluma in that it rests on a quota

[11]The 6% figure is reported in Frieden's (1979, p. 35) assessment of the Petaluma approach. Other commentaries on the effects of the Petaluma growth management effort are given in the volume edited by Cowart (1976).

[12]Details on the Marin County approach are given by Macris (1978).

system for residential development. It also uses numerical ratings to score proposals competing for the limited number of development permits.

In addition to numerical quotas, there are many other techniques that cities can use to supplement zoning in efforts to manage growth. Cities have the power to purchase lands and to adjust their boundaries to accommodate changing conditions. They are also able to control the times at which roads, sewers, and water lines are brought into service in various areas. Many cities are using these powers to constrain increases in population and thereby minimize the environmental and social problems that often accompany rapid growth.

KEY CONCEPTS AND TERMS

LAND DEVELOPMENT AND ENVIRONMENTAL QUALITY
- Traditional land development process
- Impacts of land development

FEDERAL CONTROLS ON LAND USE
- Management of public lands
- Physical infrastructure and growth
- Environmental laws and land use
- Coastal zone management
- Flood insurance
- Flood plain zoning
- Wild and scenic rivers

LAND USE REGULATION AT THE STATE LEVEL
- Hawaii's statewide zoning
- Vermont's development permits
- Tax incentives to preserve agricultural land
- Controls on mining and power plant siting
- Management of environmentally sensitive areas
- Regional planning and permit programs

GROWTH MANAGEMENT AT THE CITY AND COUNTY LEVELS
- Zoning ordinances
- Comprehensive plans
- Subdivision regulations
- Variances
- Numerical quotas on development

DISCUSSION QUESTIONS

9–1 In your opinion, which types of physical infrastructure have the greatest influence on patterns of urban development? Describe conditions under which a major new highway would have little effect on the use of adjacent land.

9–2 Bosselman and Callies (1971) described the institutional changes in land use regulation that occurred in the late 1960s and early 1970s as a "quiet revolution in land use control." Why do you think they used the term *quiet revolution* to characterize the changes?

9–3 Many state land management programs involve the formulation of a land use plan and the use of a permit process to implement the plan. Which interests might object to this increase in government regulation? What arguments would they offer to support their opposition?

9–4 Petaluma's annual quota on new housing units was challenged in the courts. Plaintiffs argued that by curtailing its growth, Petaluma infringed on people's constitutional right to travel. Petaluma's plan was viewed as "exclusionary" of outsiders. Moreover, by increasing the price of homes, the plan allegedly excluded low-income families from the community. Petaluma's approach was also criticized for not being coordinated with the plans of neighboring communities. If you represented Petaluma in the litigation, how would you defend its growth control program? (The courts eventually upheld the legality of Petaluma's plan.)

9–5 Select a city or town that is experiencing high growth. Identify any growth management strategies that are currently being employed. Describe other possible approaches that might be useful in that particular circumstance.

REFERENCES

Bosselman, F., and D. Callies, 1971, *The Quiet Revolution in Land Use Control,* report prepared for the Council on Environmental Quality, Washington, D.C.

Council on Environmental Quality, 1976, *The Growth Shapers: The Land Use Impacts of Infrastructure Investments.* Council on Environmental Quality, Washington, D.C.

Council on Environmental Quality, 1978, Regulations for Implementing the Procedural Provisions of the National Environmental Policy Act. *Federal Register* **43**, 55978–56007.

Cowart, R. (ed.), 1976, *Land Use Planning, Politics and Policy.* University Extension Publications, University of California, Berkeley.

Delafons, J., 1969, *Land-Use Controls in the United States,* 2nd ed. MIT Press, Cambridge, Mass.

Frieden, B. J., 1979, *The Environmental Protection Hustle.* MIT Press, Cambridge, Mass.

Glickman, N. J. (ed.), 1980, *The Urban Impacts of Federal Policies.* Johns Hopkins University Press, Baltimore, Md.

Healy, R. G., and J. S. Rosenberg, 1979, *Land Use and the States,* 2nd ed. Johns Hopkins University Press for Resources for the Future, Inc., Baltimore, Md.

Kusler, J. A., 1980, *Regulating Sensitive Lands.* Ballinger, Cambridge, Mass.

Linowes, R. R., and D. T. Allensworth, 1975, *The States and Land-Use Control.* Praeger, New York.

Macris, M. W., 1978, New Growth Management Underway in Marin County, *in* F. Schnidman, J. A. Silverman, and R. C. Young, Jr. (eds.), *Management and Control of Growth,* Vol. IV, pp. 58–64. Urban Land Institute, Washington, D.C.

Mandelker, D. R., 1981, *Environment and Equity: A Regulatory Challenge.* McGraw–Hill, New York.

Moss, E. (ed.), 1977, *Land Use Controls in the United States.* Dial Press/James Wade, New York.

Popper, F. J., 1981, *The Politics of Land-Use Reform.* University of Wisconsin Press, Madison.

Schnidman, F., J. A. Silverman, and R. C. Young, Jr. (eds.), 1978, *Management and Control of Growth,* Vol. IV. Urban Land Institute, Washington, D.C.

Scott, R. W., D. J. Brower, and D. D. Miner (eds.), 1975, *Management and Control of Growth,* Vols. I–III. Urban and Land Institute, Washington, D.C.

CHAPTER 10

LAND SUITABILITY AND CARRYING CAPACITY ANALYSES

The effectiveness of new land management institutions depends heavily on planning techniques that make systematic use of information characterizing environmental quality. The importance of considering the natural environment when formulating land use plans has been emphasized for more than a century. Among the early advocates of this view are

- George Perkins Marsh, a lawyer, diplomat, and scholar, who synthesized numerous theoretical and empirical findings on how human actions affect the environment.
- Frederick Law Olmsted, often referred to as the "father of landscape architecture," who designed numerous parks in ways that demonstrated the advantages of considering natural features in land use planning.
- Sir Patrick Geddes, a Scottish biologist and planner, who made pioneering efforts to sensitize city planners to the importance of considering interactions between people and the natural environment.
- Benton MacKaye, an American forester, who used geologic and hydrologic parameters to identify land areas worth preserving on environmental grounds.[1]

By the mid-twentieth century, some of the ideas of these early observers began to be adopted in planning practice. Two professions made especially important contributions to modern procedures for including environmental factors in land

[1]Marsh and Olmsted made their principal contributions during the last half of the nineteenth century; those of Geddes and MacKaye came after the turn of the century. For historical accounts of early efforts to integrate environmental factors into land use planning, see Fabos (1979) and McAllister (1980).

use planning: biology and landscape architecture. Biologists provided a scientific rationale for considering ecosystem stability and species and habitat preservation in formulating plans. Landscape architects devised effective new ways of organizing spatial data on land characteristics, for example, slope stability and soil type.

Two ideas are central in this discussion of techniques for developing land use plans. One is that hydrologic, geologic, biologic, and other features, when viewed collectively, yield insights into the type of use "intrinsically suitable" for a particular parcel of land. A second important concept is "carrying capacity," the limits to how much growth an area can accommodate without violating environmental quality goals. Analyses of carrying capacity and the intrinsic suitability of land for certain uses provide systematic ways of utilizing environmental information to guide planning.

MAP OVERLAY TECHNIQUE

The map overlay technique is a procedure for synthesizing the spatial data used in land use planning. It involves four steps: (1) identify factors to be included in the planning exercise, for example, potential earthquake hazard and soil permeability; (2) prepare an "inventory map" for each factor showing how it varies over the study area; (3) create composite maps by overlaying two or more inventory maps; and (4) analyze the composite maps to make inferences relevant to land use planning.

A planning exercise concerning the use of Stanford University's land illustrates each of the steps in the map overlay process.[2] The study, which was conducted by the University's Planning Office, focused on 355 of the approximately 8200 acres comprising Stanford's land. The 355 acres, which were in "open space," were being considered as a possible location for either housing or research facilities. The Planning Office identified about a dozen factors that it considered important in making decisions about the area's future use. Examples of the factors are slope stability, extent of vegetation, and characteristics of the water bodies in the area.

The map overlay technique was used to identify portions of the 355 acres that were "environmentally sensitive" and therefore "less suitable for development." The analysis involved a synthesis of inventory maps for three factors: slope, slope stability, and stream zones.

Figure 10.1, which delineates slope stability in the study area, is one of the three inventory maps. The numbers in the figure are the slope stability categories described in Table 10.1. Although inventory maps are sometimes based on direct field observations, they frequently rely on published data sources. Figure 10.1 is based on a report by Nilson et al. (1979) that classified slope stability for

[2]This discussion is based on unpublished reports by the University's Planning Office and personal communication with Philip Williams, Director of Planning, Stanford University, Stanford, California, July 6, 1981.

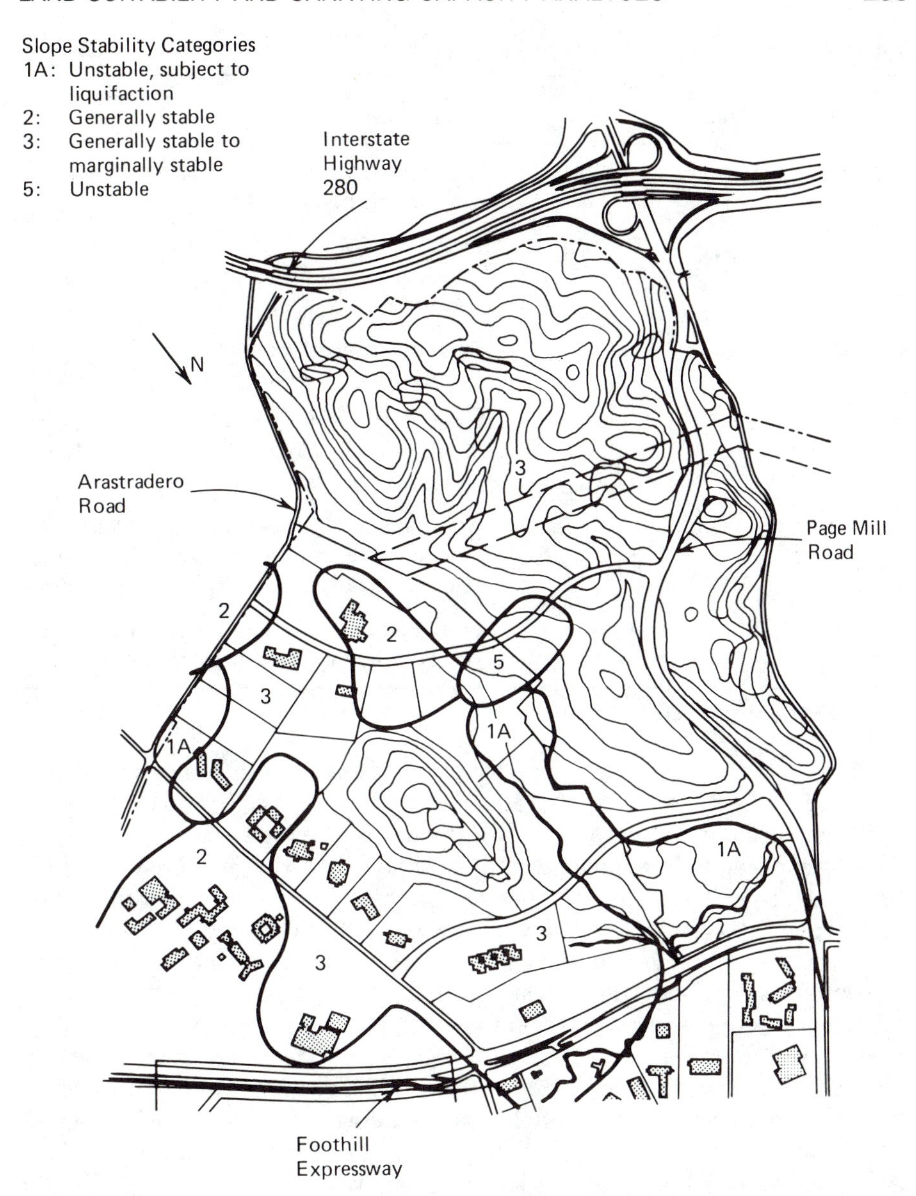

FIGURE 10.1 Inventory map for slope stability: Stanford land use study.

various parts of the study area. The expression "nominal type" refers to the labels used in categorizing factors. In this example, the slope stability factor is described using nominal types such as unstable and moderately stable.

In preparing inventory maps for the other two factors, the University's Planning Office defined the nominal types as follows. The stream zone factor was described by two types: "in stream zone" and "not in stream zone." The category in stream zone applied to land within 100 ft of either side of the centerline of

TABLE 10.1 Categories of Slope Stability[a]

	Nominal Type	Description
1	Stable	Areas of 0–5% slope that are not underlain by landslide deposits
1A	Unstable, subject to liquification	Areas of 0–5% slope that include tidelands, marshlands, and swamplands underlain by moist, unconsolidated muds
2	Generally stable	Areas of 5–15% slope that are not underlain by landslide deposits
3	Generally stable to marginally stable	Areas of greater than 15% slope that are not underlain by landslide deposits or bedrock units susceptible to landsliding
4	Moderately stable	Areas of greater than 15% slope that are underlain by bedrock units susceptible to landsliding but not underlain by landslide deposits
5	Unstable	Areas that are underlain by or immediately adjacent to landslide deposits

[a]Adapted from Nilson et al. (1979).

creeks in the study area. The slope factor was described using intervals of slope (in percent): 0–10, 10–20, 20–30, and greater than 30. Maps similar to the one in Figure 10.1 were prepared for both stream zone and slope.

In its application of the map overlay technique, the Planning Office assumed that land possessing any *one* of the following characteristics was environmentally sensitive and less suitable for development: (1) unstable slopes, (2) in stream zones, and (3) slopes greater than 30%. These conditions eliminated areas most susceptible to landslides and erosion problems. They also allowed the habitat provided by stream-side vegetation to be preserved.

The overlay process represented in Figure 10.2 identified lands possessing one or more of the above-noted characteristics. Plastic transparencies of the inventory maps were used to simplify the "reading" of all three maps together. In complex applications of the overlay technique, more elaborate procedures are often employed in combining inventory maps.[3]

The overlay technique for incorporating environmental factors into land use planning can be carried out at different levels of detail and for different purposes. In some instances, as in the Stanford example, it is used for making preliminary observations of a general nature. The overlay technique may also be applied in detailed site planning for individual facilities. Planning for the "new town"

[3]The more complex procedures are outlined by Steinitz, Parker, and Jordon (1976). They also review some of the earliest-known applications of the overlay technique.

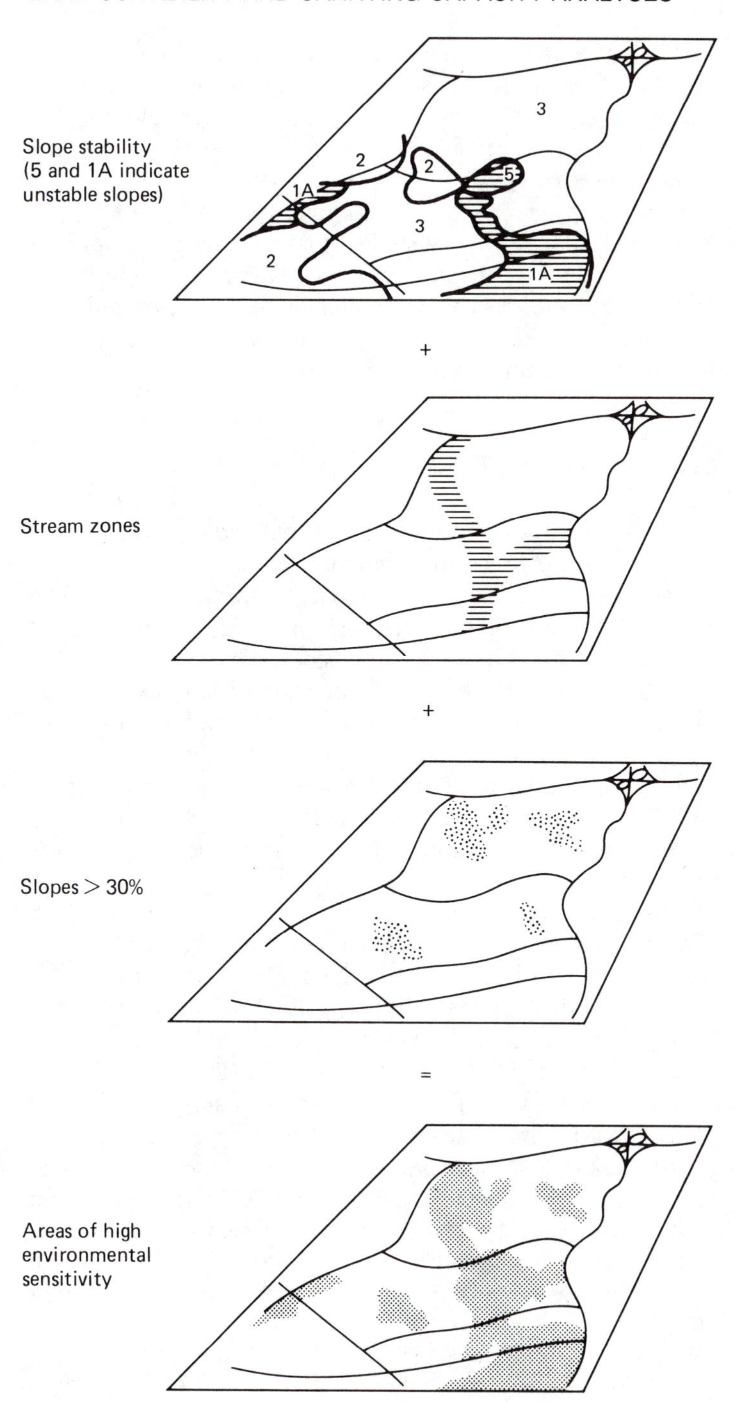

FIGURE 10.2 The map overlay technique.

of Woodlands, Texas, demonstrates the technique's versatility. Map overlays were used to lay out the town and to devise detailed guidelines for designing individual buildings. The analysis of environmental factors, especially vegetation, surface runoff, and groundwater replenishment characteristics, played a major role in the design process.[4]

The use of hand-drawn map overlays as a technique for land planning has been traced back to the early 1900s. However, the approach was not employed widely to integrate environmental factors into land use planning until the 1960s. McHarg's (1969) *Design with Nature,* which includes numerous case study applications, was instrumental in popularizing the approach. In fact, the technique is sometimes referred to as "the McHarg method."

LAND SUITABILITY ANALYSIS USING MAP OVERLAYS

An extension of the map overlay procedure is used to conduct a land suitability analysis. This is accomplished by giving ranks or numerical scores to the nominal types for each factor as part of the process of preparing inventory maps. The results from the expanded overlay analysis are maps indicating the areas most suitable for each particular land use type under consideration.

Land suitability analysis is illustrated by a study of Staten Island, New York, performed by McHarg and his associates. They wanted to identify how well suited various parts of Staten Island were for each of five land uses: conservation, passive recreation, active recreation, residential development, and commercial and industrial development. Their analysis assumed that "each area has an intrinsic suitability for certain land uses" and that "certain areas lend themselves to multiple coexisting land uses."[5]

Inventory maps that incorporate rankings of nominal types were prepared for each of 32 factors. The map preparation procedure is illustrated for the "existing habitat" factor in relation to conservation as a land use. Nominal types of habitats on Staten Island were defined as follows: intertidal, water related, field and forest, urban, and marine. Criteria for ranking the five habitat categories were based on professional judgment. McHarg and his associates decided that the best habitats for conservation were those in the shortest supply. Intertidal habitats were assigned a rank of 1 since they were the most scarce; water-related habitats were assigned a rank of 2 since they were the second most scarce, and so forth. Once the ranks had been determined for all of Staten Island, an inventory map was prepared using shades of gray to represent the ranks. The most scarce habitat (rank = 1) was given the darkest tone, the least scarce habitat (rank = 5) was shown as a blank, and other habitats were represented

[4]An account of how map overlays were used in planning Woodlands, Texas, is given by McHarg and Sutton (1975).

[5]This quotation is from McHarg (1969, p. 104); further details on the Staten Island case study are given in the same source.

using intermediate tones. Each map was photographed to obtain a transparent negative for use in preparing overlays of the individual maps.

Subsets of the 32 inventory maps were used in analyzing each of the five classes of land use included in the Staten Island study. McHarg and his associates determined the appropriate subsets of factors that would accommodate or constrain different land uses. For example, consider the commercial and industrial land use. Three factors were judged important in *accommodating* commercial and industrial development: subsurface foundation conditions, soil foundation conditions, and the existence of channels navigable by commercial vessels. Table 10.2 indicates the criteria used in ranking the three factors. The *darkest* regions on the inventory maps for these factors showed areas *most suitable* for commercial and industrial use.

Five other factors were considered significant in determining the suitability of land for commercial and industrial development. However, because of the ranking criteria employed in preparing the original inventory maps, these factors were viewed as *constraining* development (see Table 10.2). For example, slope was viewed as a constraining factor since it was ranked from highest to lowest in preparing the original inventory. Areas with slopes greater than 25% received a rank of 1 and were given the darkest tone on the original map for slope, whereas areas with slopes less than 2½% received a rank of 5 and were left blank. Since flatter slopes are more suitable for development, the tones on the original inventory map for slope were reversed in performing the overlay analysis for commercial and industrial use. After this tone reversal, the darkest regions corresponded to the areas most suitable for development. If slope had been ranked originally from lowest to highest, it would have been included among the factors accommodating commercial and industrial use. The distinction between accommodating and constraining factors depends on the ranking criteria. A high rank for any particular factor may be accommodating for one type of land use and constraining for another.

TABLE 10.2 Factors and Ranking Criteria Used in the Staten Island Land Suitability Study[a]

Important Factors* Accommodating *Commercial and Industrial Development

Subsurface foundation conditions—ranked from most to least strong
Soil foundation conditions—ranked from most to least strong
Existence of navigable channels—ranked from deepest to shallowest channels

Important Factors* Constraining *Commercial and Industrial Development

Tidal inundation—ranked from highest recorded flood to "above flood line"
Slope—ranked from highest to lowest
Erosion—ranked from most to least susceptible to erosion
Existing forest—ranked from best to poorest quality
Soil drainage—ranked from poorest to best drainage conditions

[a]Based on information in McHarg (1969, pp. 108–113). The original analysis also included factors of secondary importance; these are ignored here.

Once the factors accommodating and constraining commercial and industrial development were identified, the appropriate transparent negatives of the inventory maps were superimposed and photographed. The resulting photo was then made into a single map, with the suitability for commercial and industrial development indicated by shades of gray. The darker the area, the more suitable it was for development. This entire procedure was repeated five times, once for each of the five land use categories included in the Staten Island study. Naturally, the factors used in developing the suitability maps varied from one land use to the next. The superposition of all five suitability maps indicated that some areas were uniquely well suited for one particular land use, and other areas could readily support several uses without conflict. However, there were many areas that were reasonably well suited for uses that conflicted, such as conservation and industrial and commercial development.

Information regarding intrinsic suitability for different uses does *not* constitute a plan. To make a plan, economic and other factors influencing the demand for different land uses must also be considered. Although a land suitability study includes only part of the information needed in land use planning, it does provide a practical method for integrating environmental information into the planning process.

LAND SUITABILITY ANALYSIS USING WEIGHTED SCORES

Objections to Using Overlays in Suitability Analysis

The use of map overlays to perform suitability analyses has been criticized for leaving all the key value judgments in the hands of technical experts. Some people feel it is inappropriate to let professional planners decide which factors to consider and how the nominal types for a given factor are defined and ranked. However, this criticism is too strong, since planners who use the overlay technique may consult with elected and appointed officials in the course of their work. Thus, their choice of factors need not be made in complete isolation from community values and perspectives. In principle, at least, even the general public could be involved directly in selecting factors.

Another objection to using overlays in suitability analysis is that unless the factors are independent of each other, the same factor can be counted inadvertently several times. For example, slope and erosion are interdependent; land areas with steep slopes are frequently susceptible to erosion. By including both factors in an overlay analysis, slope may be given unintended importance. One way to avoid this is to select only factors that are not strongly related to each other. Another is to include the relative importance of factors explicitly in the analysis. This is accomplished by the weighted factor scores approach introduced below.

A more fundamental criticism concerns the theoretical legitimacy of carrying out the map overlay process. Hopkins (1977) points out that making overlays of individual maps prepared in varying shades of gray is equivalent to performing addition using a graphical procedure. The overlay technique presumes that it is proper to add factors even though they are represented by nominal types that have different physical units. For instance, overlaying the transparencies associated with slope and erosion in Table 10.2 involves adding quantities that are measured in incommensurate units.

Basis for a Weighted Scores Approach

Hopkins indicates that a solution to this theoretical dilemma is to weigh the factor scores before any addition is performed. In principle, at least, the weights can be considered to transform the scores associated with the nominal types for each factor into a common unit of measure.

Table 10.3 contains a hypothetical example demonstrating the weighted scores approach to land suitability analysis. The procedure, which in this case is for residential development, involves the following steps. First the study area is divided into grid cells. The size and shape of the cells are determined using professional judgment, and it is assumed that the areas within the cells are, in some sense, homogeneous. Next the factors relevant to assessing the suitability of land for residential development are selected. For each factor, nominal types are defined and an inventory map is prepared. The overly simple example in Table 10.3 includes only two factors, slope and quality of view, with three nominal types for each factor. Criteria must be chosen for rating the nominal types that describe each factor. For slope, the nominal types are high, medium, and low, and the corresponding ratings are 1, 3, and 5, respectively. The highest numerical score is associated with the flattest ground and indicates land with slopes most suitable for residential development. For the quality of views factor, the nominal types are also high, medium, and low. In this case, the highest ratings go to cells with the highest quality views, since these areas are judged more suitable for housing. The next step involves assigning weights indicating the relative importance of each factor in determining the suitability of the land for residences. In this instance, slope is judged to be twice as important as quality of views. The final step, illustrated at the bottom of Table 10.3, is to compute the sum of weighted factor scores for each grid cell. Higher scores indicate cells considered more suitable for residential development.

There are two practical reasons why the weighted scores approach is used more widely than the map overlay approach in major land suitability studies.[6] One reason is that in overlaying a large number of inventory maps there can be technical difficulties in distinguishing the varying shades of gray. The numerical values associated with the weighted scores approach eliminate the need

[6]It might be supposed that the increased popularity of the weighted scores approach resulted because the addition of weighted scores is permissible in theory. Assuming the factors are independent, the weighted scores can be viewed as having identical units which can then be legitimately added (Hopkins, 1977). Interestingly, many who apply the approach in real settings view this as a theoretical fine point.

TABLE 10.3 Weighted Scores Approach to Land Suitability Analysis—Hypothetical Example for Residential land Use

Division of study area into grid cells

Convention for labeling cells

Row coordinate	Column coordinate 1	Column coordinate 2
1	(1, 1)	(1, 2)
2	(2, 1)	(2, 2)

Identification of factors: slope and quality of views

Selection of nominal types for factors

Factor 1: Slope

Low	Medium
High	Low

Factor 2: Quality of Views

Medium	High
Low	Low

Determination of scores for nominal types (higher scores indicate greater suitability for development)

Factor 1: Slope

5	3
1	5

Factor 2: Quality of Views

3	5
1	1

Selection of weights

Weight for Factor 1 = 2

Weight for Factor 2 = 1

Illustrative computation of sum of weighted scores

Weighted score for grid cell (1, 1) $= 2 \times 5 + 1 \times 3 = 13$

Sums of weighted scores

Most suitable for development

Least suitable for development

13	11
3	11

for subtle distinctions in shading. A second reason is that high-speed digital computers have made it easier to perform land suitability studies. As elaborated by Fabos (1979), existing computer mapping procedures make it possible to perform an enormous range of analyses quickly and efficiently. Once the initial programming work is done, it is simple to examine the implications of changing both the criteria for scoring factors and the importance weights assigned to factors.

A Case Study Involving the Palo Alto Foothills

A study of the foothills of Palo Alto, California, a city about 35 miles south of San Francisco, illustrates how the weighted scores approach was used to integrate environmental factors into a real planning exercise. The analysis was performed for Palo Alto by Livingston and Blayney, a firm of city planning consultants.

The study area consisted of approximately 7500 acres of largely undeveloped foothills, much of which the city had annexed in the late 1950s. The area included the 1400-acre Foothills Park, owned by the city, as well as some lightly developed land. During the late 1960s, Palo Alto expected to receive several development proposals that would require variances from the one dwelling unit per acre zoning that applied to the land. In anticipation of the proposals, the city engaged the planning consultants to conduct a land suitability analysis and assess the impacts of alternative land use patterns.

The first phase of the consultants' study identified portions of the area most suitable for residential and other development. The analysis excluded the parts of the original study area that contained the 1400-acre park and the land already developed. The remaining area was divided into a rectangular grid consisting of 330 cells of 20 acres each. The consultants selected 25 factors for inclusion in the suitability analysis and developed a 5-point rating scale for each. The rating procedures were similar to those used in the above-noted Staten Island planning exercise. For example, nominal types for the "fire hazard" factor were "high, medium, and low," with factor scores of 1, 3, and 5, respectively. The highest scores represented the greatest suitability for development. Each of the 330 grid cells received a numerical score for each of the 25 factors. For 2 factors, flooding potential and suitability of soil for septic tanks, the scores were approximately equal for all grid segments. Consequently, these factors were dropped from the analysis.

The consultants then assigned weights to each of the remaining 23 factors. The weights indicated the relative importance of factors in land development. For example, fire hazard, with a weight of 8, was judged to be four times as important as the existence of air pollution. Table 10.4 lists the 23 factors and the weights assigned to each.

The final step in the suitability analysis involved computing the sum of weighted factor scores for each grid cell. Sums ranged from a maximum of 480, indicating the most suitable location for development, to a minimum of 94. The final results, shown in Figure 10.3, are presented in terms of six classes of land

TABLE 10.4 Factors and Weights Used in the Palo Alto Foothills Study[a]

	Weights
Planning and Market Factors	
Size of ownership	4
Average slope	10
Proximity to present development	1
Time-distance from freeway interchange	8
Dependence of improved access on other jurisdictions	4
Present utilities services	2
Ecological Factors	
Vegetative cover	5
Proximity to water surface features	3
Air pollution	2
Fire hazard	8
Visual and Recreation Factors	
Views	3
Visibility of study area from selected points in Palo Alto	3
Visibility of Foothills Park from study area	3
Proximity to Foothills Park and Upper Stevens Creek Park	2
Geologic and Soil Factors	
San Andreas earthquake fault zone	7
Other earthquake fault zones	3
Landslides	6
Natural slope stability	5
Cut slope stability	4
Excavation difficulty	2
Soil suitability as fill	2
Soil erosion potential	6
Soil expansion potential	3

[a]Based on information in Livingston and Blayney (1971, pp. 60–64).

ranging from the most to the least suitable for development. The consultants chose the numbers defining the classes so that approximately equal land areas were grouped in each class.

Subsequent phases of the consultants' study examined 24 scenarios representing alternative patterns for future land development. Each scenario was assessed in terms of its environmental, economic, and social impacts. Numerous development alternatives could have been formulated, corresponding to the many ways the city could divide the land into different zones. The consultants' suitability analysis helped define the particular development scenarios that received detailed consideration.

Based on their evaluation of alternative development patterns, the consultants advised the city to preserve the foothills as open space. Their recommendations

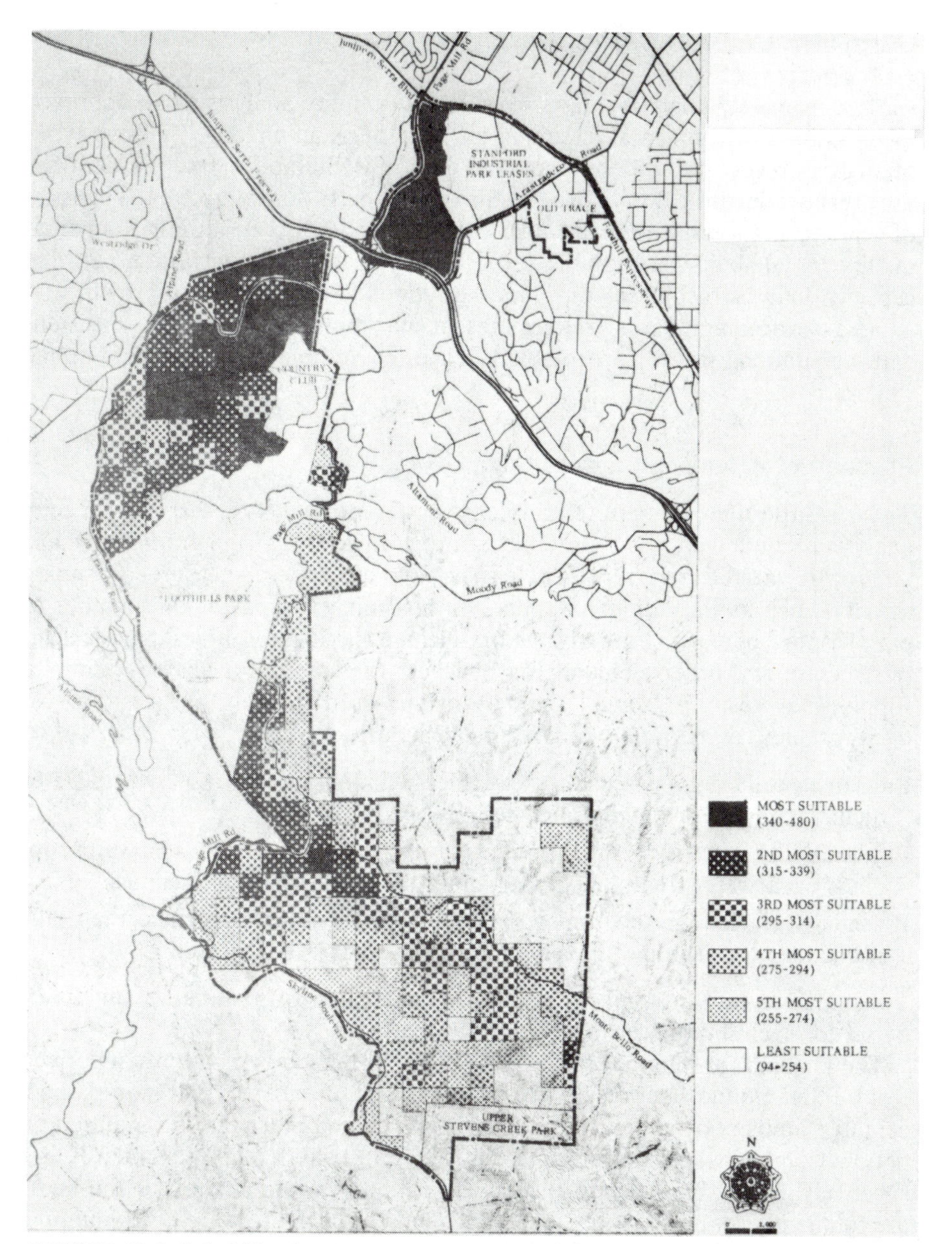

FIGURE 10.3 Suitability for development ratings: Palo Alto foothills. From Livingston and Blayney (1971).

were taken seriously. The city council subsequently preserved much of the land as open space by implementing new zoning measures that protected the foothills from intensive development.[7]

[7]For a critical assessment of the consultants' study and the city's decision making, see Frieden (1979, pp. 107–118).

CARRYING CAPACITY ANALYSIS

Carrying capacity analysis differs from land suitability analysis as an approach to including environmental factors in land use decision making. The two types of analyses respond to different land use questions. Suitability analysis identifies uses that are intrinsically well suited for various parts of a study area. The issue of whether an intensification of land use can occur without causing environmental quality to fall below acceptable levels is *not* treated. In contrast, a carrying capacity analysis recognizes that there are limits to the amount of growth that an area can accommodate. It determines the maximum level of growth consistent with maintaining socially acceptable levels of environmental quality and public welfare.[8]

Elements of a Carrying Capacity Study

There is little uniformity in terminology for describing how a carrying capacity analysis is conducted. Two useful terms are *growth variable* and *limiting factor.* A growth variable can represent either population or a measure of human activity, such as the number of new housing units per year or the number of park visitors per day. Limiting factors include natural resources, physical infrastructure and other elements that, because they are not available in infinite supply, *may* restrain growth. Limiting factors used frequently in carrying capacity studies can be grouped into three categories.

Environmental Biophysical characteristics including measures of air and water quality, ecosystem stability, and soil erosion.

Physical The capacity of infrastructure systems, including highways, water supplies, wastewater treatment plants, and solid waste disposal facilities.

Psychological Parameters concerning the way individuals perceive their surroundings: for example, the sense that an area is overcrowded.

To conduct a carrying capacity analysis, a maximum (or minimum) value must be set for each limiting factor.

The maxima (or minima) for *environmental* limiting factors are often derived from either political processes or the judgments of experts. For example, acceptable limits of air and water quality in the United States are often based on national ambient air quality standards and state water quality standards, respectively. When biological or geological parameters are employed, the limits are frequently set using professional judgment. The maxima for *physical* limiting factors are often taken as the existing capacities of the relevant infrastructure systems: for example, the dependable yield of a community's water supply. The

[8]The *carrying capacity* idea has its origins in biology. The term is used there to indicate the maximum number of individuals of a particular species that can be supported by a given area. For example, Odum (1971, p. 188) presents data showing that the island of Tasmania near Australia had a carrying capacity of 1.7 million sheep during the 1800s. Odum also explains how biologists apply the carrying capacity concept.

limits for *psychological* factors are determined either by professional judgment or by a survey of individuals in the study area. Typically, only a few limiting factors are examined since the time and effort required for analysis increase in proportion to the number of factors considered. Those conducting carrying capacity studies must make judgments about which factors are likely to place the most stringent constraints on growth.

To estimate carrying capacity, it is necessary to estimate what a maximum (or minimum) value for each limiting factor means in terms of growth. Quantitative links must be made between limiting factors and growth variables. The difficulties in establishing such connections frequently turn out to be the major impediments to conducting carrying capacity studies.

When quantitative relationships between limiting factors and growth variables can be established, they are often based on either mathematical models or expert opinion based procedures.[9] For example, suppose the dissolved oxygen (DO) of a particular stream in the study area is a limiting factor. If the minimum acceptable value of dissolved oxygen is 6 mg/l, and its current level is 7 mg/l, there is 1 mg/l available to accommodate growth. Suppose, further, that the study area is a suburban community with no major industrial or commercial effluent, and that the growth variable is population. A quantitative link between the population and the acceptable limit for DO can be found by estimating how much wastewater results from a particular increase in population. This requires assumptions about the amount of waste generated per person and the proportion of the waste removed by treatment prior to discharge. Once these assumptions are made, standard engineering formulas (similar to those introduced in Chapter 7) can be used to compute the maximum value of population that can be accommodated by the 6 mg/l limit on dissolved oxygen.

Using reasoning similar to that above, it may be possible to estimate the restriction on the growth variable imposed by each limiting factor. The bounds will be different for each factor.

Continuing the illustration, suppose the availability of DO constrains the area's population to 100,000 people and there are two other limiting factors in the analysis, traffic congestion and water supply. Table 10.5 indicates hypothetical carrying capacity analysis results for this example. It shows that water supply provides the tightest restraint on growth by restricting the area's population to 80,000. This bound is *not* fixed for all time. If new water supply

TABLE 10.5 Results from Hypothetical Carrying Capacity Analysis

Limiting Factor	Maximum Population Consistent with Limiting Factor
Stream dissolved oxygen	100,000
Water supply capacity	80,000
Traffic congestion	90,000

[9]The types of forecasting procedures reviewed in Chapter 7 can be useful in carrying capacity studies.

investments were made or if residents used the existing supply more efficiently (for example, by implementing water conservation measures), the constraining effect of water supply could be removed. If this occurred, traffic congestion would establish the carrying capacity at 90,000 people. This constraint is also changeable by making investments in new facilities and by influencing driving patterns. Furthermore, even the limits imposed by dissolved oxygen could be changed, for example, by lowering the minimum allowable DO to 5 mg/l instead of 6 mg/l, or by building more efficient wastewater treatment facilities. There is, of course, some point at which it is not practical to continue relaxing the constraints imposed by the limiting factors, and this establishes the carrying capacity for a particular setting. However, because there may be modifications in the circumstances affecting carrying capacity, it is a numerical bound that may change with time.

Procedures for applying carrying capacity concepts to land use planning are still evolving. A few examples illustrate some approaches that have been used or proposed.

Planning for the Use of Public Parks

Many carrying capacity applications have concerned the use of public parks. Consider, for example, the study by Kuss and Morgan (1980) to find the carrying capacity of picnic areas in the Patapsco Valley State Park in Maryland. They considered only a single limiting factor, soil erosion. Acceptable erosion was defined quantitatively as the maximum erosion rate that would allow the land to be used productively for an indefinite period. The growth variables in the analysis were, in principle, the type, duration, and intensity of recreational use of the picnic areas. However, Kuss and Morgan did not determine a quantitative relationship between erosion rate and picnic site use. Instead, they estimated the percentage of "ground cover" at different picnic sites required to maintain soil erosion at acceptable levels. Park managers could then restrict the number of users so that the needed ground cover could be maintained. For any particular picnic site, the percentage of ground cover consistent with the maximum allowable erosion rate was calculated using a well-established mathematical model, the "universal soil loss equation."

Less mathematically oriented carrying capacity exercises have been used for planning campsites and trail networks in many parks. Yapp and Barrow (1979) observe that the limiting factors in such studies are typically related to the ability of the trails and campsites to sustain themselves as biological systems. The growth variable is generally the number of campers per day admitted to a given system of trails and campsites. Instead of trying to estimate the relations between the limiting factors and the growth variable, a trial-and-error procedure is used to establish carrying capacity. First, a trial capacity is set. Campers who arrive after this initial value has been reached are turned away. The area is then monitored to gauge whether the trails and campsites can sustain themselves with the admitted number of campers. Adjustments to the trial value are made on the basis of the monitoring program.

Planning for Cities and Regions

Trial and error can be used to estimate the carrying capacity of park facilities because the management decisions involved are easily modified. The maximum number of campers per day can be changed by a simple administrative act, and portions of a park can be closed to restore damaged areas. Trial-and-error methods are generally not feasible in studies of the carrying capacity of cities and regions. In these cases, explicit relationships between limiting factors and growth variables must be used to estimate carrying capacity.

There have been numerous attempts to apply the carrying capacity concept in city and regional planning. An approach proposed by Nieswand and Pizor (1978) for New Jersey communities provides an illustration. They used the term *current planning capacity* in making carrying capacity determinations. The current planning capacity of an area is its "ability to accommodate growth and development within limits defined by existing infrastructure and natural resource capabilities." Their assessment of conditions likely to restrain community growth enabled Nieswand and Pizor to focus on three limiting factors—water supply, water quality, and air quality.

The example they used to illustrate the determination of current planning capacity involved a hypothetical community of 38,000 people. For each factor, the existing capacity was computed in terms of "population equivalents." For example, in the case of water supply, it was estimated that the combined yield from existing ground and surface water sources was 9 million gallons per day (gpd). On the average, the per capita consumption of water was considered to be 150 gpd. Thus, the current planning capacity in terms of water supply was

$$\frac{9 \times 10^6 \text{gpd}}{150 \text{ gpd per person}} = 60{,}000 \text{ people}$$

Analogous reasoning was used to find the number of people that could be supported while maintaining prescribed levels of water and air quality. Since the populations sustainable by the water and air quality parameters were greater than 60,000, water supply was deemed *the* limiting factor. If the community did not add to its water supply, it could only accommodate 22,000 additional people, the difference between 60,000 and its current population of 38,000.

A widely cited implementation of carrying capacity is the study for Sanibel, Florida, an island of about 11,000 acres with a seasonal peak population of 12,000. Population was the growth variable. The limiting factors included the "capacity of the island's wetlands to absorb additional pollutants and the capacity of the causeway to the mainland [to] allow for safe evacuation of the island in the event of a hurricane or severe storm."[10] The carrying capacity study provided estimates of population for each of the limiting factors. These population bounds served as a basis for restricting the number of new dwelling units on the island to 2000 over a 5-year period. The analysis focused the community's attention

[10]The quotation is from Schneider, Godschalk, and Axler (1978, p. 22); this source contains additional information on the Sanibel Island case.

on its environmental quality goals and the limits of the island to support growth that is consistent with those goals.

Carrying capacity analyses can also be used to formulate plausible alternative scenarios for how an area might develop. This is done by analyzing several different sets of maxima (or minima) for the limiting factors. Consider, for example, a hypothetical area in which only two limiting factors are examined: stream dissolved oxygen and highway capacity. Table 10.6 shows two possible sets of limits. The low-growth scenario keeps the highway capacity at its current level and maintains stream dissolved oxygen at 6 mg/l. The high-growth scenario doubles the highway capacity and allows oxygen levels to drop to 3 mg/l. A carrying capacity analysis could be undertaken for each combination of bounds. Scenarios could then be constructed around the values of the growth variable yielded by each set of limits. Jaakson, Buszynski, and Botting (1976) illustrate this use of different sets of limiting factors in their study of alternative plans for lake use and lake-side development for several Canadian lakes.

TANDEM USE OF CARRYING CAPACITY AND LAND SUITABILITY ANALYSES

Innovative applications of the carrying capacity approach involve its use in conjunction with land suitability analysis. A case study of planning in the Lake Tahoe area demonstrates the tandem use of both analytic procedures.[11] The "Tahoe region" is defined by a compact between California and Nevada. The region consists of about 500 square miles, roughly 39% of which is covered by the lake. As a result of land development during the late 1960s, the quality of Lake Tahoe was significantly degraded. The soils in the region are very susceptible to erosion. Extensive land development increased the rate at which sediments were eroded and transported to the lake. The nutrients carried by the sediments fostered the growth of aquatic plants that diminished the lake's aesthetic quality.

In the early 1970s, the U.S. Forest Service, in cooperation with the Tahoe Regional Planning Agency (TRPA), undertook a land suitability analysis based

TABLE 10.6 Using Sets of Limits in a Carrying Capacity Study

Limiting Factors	Low-Growth Scenario	High-Growth Scenario
Stream dissolved oxygen	6 mg/l	3 mg/l
Highway capacity	Maintain existing capacity	Double existing capacity

[11]This presentation is based on Schneider, Godschalk, and Axler (1978) and on two unpublished documents: The Lake Tahoe-Planning Guide Map issued by the U.S. Forest Service in 1971, and a planning report by Grove (1978). The discussion also relies on personal communication with Dennis Winslow, Executive Director, California Tahoe Regional Planning Agency, South Lake Tahoe, California, August 4, 1981.

on the region's natural characteristics. This study, referred to by the Forest Service as a "capability analysis," included the following factors: frequency of floods, landslide hazard, water table elevation, soil drainage, soil erodibility, and the fragility of flora and fauna. After considering these factors, both individually and in various combinations, the land in the region was divided into seven "capability levels." Table 10.7 indicates how the levels were defined using various landform and soil characteristics. The study results provided the basis for a TRPA ordinance that established, for each capability level, a maximum allowable percentage of the land that could be covered with buildings and other physical facilities. For example, the ordinance indicated that at most 1% of the land classified as capability level 1 could be covered.

In the mid-1970s the California Tahoe Regional Planning Agency (CTRPA) undertook a carrying capacity investigation for the California side of the Tahoe region. The CTRPA felt that the information in the Forest Service's capability studies needed to be supplemented. They reasoned that TRPA's population projections for the Tahoe region were based largely on the physical capability of land to accommodate different uses. These projections did not consider regional air and water quality goals, nor did they account for limits on water supply and highway capacity. The CTRPA study was undertaken to integrate environmental quality goals and the limits imposed by physical infrastructure into its planning and decision making.

TABLE 10.7 Land Capability Levels in the Lake Tahoe Region[a]

Capability Levels	Tolerance for Use	Slope Percentage	Relative Erosion Potential	Runoff Potential	Disturbance Hazard
7	Most	0–5	Slight	Low to moderately low	
6		0–16	Slight	Low to moderately low	Low hazard lands
5		0–16	Slight	Moderately high to high	
4		9–30	Moderate	Low to moderately low	Moderate hazard lands
3		9–30	Moderate	Moderately high to high	
2		30–50	High	Low to moderately low	
1a	Least	30+	High	Moderately high to high	High hazard lands
1b		Poor natural drainage			
1c		Fragile flora and fauna			

[a]From "Lake Tahoe Basin-Planning Guide Map," issued by U.S. Forest Service, 1971.

The CTRPA analysis focused on the following questions (Grove, 1978, p. 1):

- How much activity can the Lake Tahoe basin accommodate while maintaining desired levels of environmental quality?
- What is the ability of the region's natural and societal resources to support further activity?
- What are the interrelationships among the various natural and societal systems?
- What will it cost to increase the carrying capacity of the region to accommodate the population and development presently allowed by CTRPA's Regional Plan?
- What are alternative ways to ensure that the carrying capacity of the region is not exceeded?

Table 10.8 summarizes results from the CTRPA carrying capacity analysis for the 15 limiting factors included in the study. For each factor the table shows the peak population sustainable under conditions existing in the mid-1970s. The circa 1975 peak population of 145,000 exceeded the carrying capacity based on seven factors including air and water quality. The CTRPA also estimated how much it would cost to increase the population that could be accommodated by some of the limiting factors.

Although investments of various types can increase the Tahoe region's carrying capacity, there is a bound on how much the capacity can be augmented. This is especially true for water quality. Past decisions permitting the development of roads, houses, and commercial establishments have decreased vegetative cover, increased erosion, and transformed drainage patterns in ways that are hard to change. The increased nutrient flows to Lake Tahoe resulting from these modifications cannot be easily reversed. Moreover, leaching of nutrients from septic tank drainage fields (now no longer in use) is an additional source of nutrients to the lake that cannot be easily contained. For these reasons, the water quality goals established for Lake Tahoe provide a practical limit on population growth for the region.

The CTRPA carrying capacity study broached a subject that is only beginning to receive serious attention, namely, the connections among limiting factors. Such linkages are significant since a change in the ability of the *most* limiting factor to accommodate growth can indirectly influence the effects of other limiting factors. For example, suppose wastewater treatment plant capacity provided the most stringent restriction on growth and that relaxing this constraint by adding treatment capacity would lead to an increase in the Tahoe region's housing stock. Grove (1978) observes that the "increase in housing would also require increased water supply, generate additional sedimentation, increase travel, and generally require other infrastructure and services." He emphasizes that interdependencies can be important even when capacity is augmented for a factor that is *not* restricting growth. Such an expansion could lead to increased political pressure to enlarge the capacity of the factor that *is* the growth con-

TABLE 10.8 Carrying Capacity Estimates for the California Portion of the Lake Tahoe Region[a]

Limiting Factor	Seasonal Peak Population That Can Be Accommodated[b]
Environmental quality	
Air quality	Inadequate[c]
Land capability	175,000
Water quality	Inadequate
Noise	Unknown
Natural resources	
Water supply	223,000
Energy supply	
Electricity	Inadequate
Natural gas	185,000
Infrastructure and services	
Sewage treatment	167,000
Solid waste disposal	Inadequate
Transportation	Inadequate
Health care	185,000
Education	227,000
Police protection	Inadequate
Justice	145,000
Fire protection	Inadequate

[a]Based on information in Grove (1978, p. 4).

[b]This population includes permanent, second-home and motel/hotel occupants, campers, and day visitors; estimates are based on existing conditions. Several of the figures can be increased by making additional investments.

[c]*Inadequate* means the limiting factor cannot accommodate the circa 1975 peak population of 145,000.

straint. The interconnections among factors can be quite complicated, and they are difficult to analyze in a systematic fashion.

Although both the CTRPA's carrying capacity report and the Forest Service's land capability study had an influence on the CTRPA's 1980 land use plan, the two analyses were not synthesized in a formal way. Each study contributed to increasing people's awareness of the implications of continued growth in the Tahoe region. As more citizens recognized the adverse environmental effects of continued growth in the region, appointed and elected officials became increasingly willing to impose potentially unpopular restraints on growth. A reflection of the usefulness of the CTRPA carrying capacity study is given in the 1980 bistate compact for the Lake Tahoe area. It called upon the Tahoe Regional Planning Agency to extend the CTRPA study by analyzing carrying capacity for the entire region.

Analyses of land suitability and carrying capacity can require substantial investments of time and money. However, for highly valued environments like the Tahoe region, such analyses may be essential to preserve the very qualities which make the environments so much appreciated in the first place.

KEY CONCEPTS AND TERMS

THE MAP OVERLAY TECHNIQUE
- Nominal types
- Inventory maps
- Composite maps

LAND SUITABILITY ANALYSIS USING MAP OVERLAYS
- Intrinsic suitability
- Ranking nominal types
- Factors accommodating a use
- Factors constraining a use
- Land suitability maps
- Superposition of suitability maps

LAND SUITABILITY ANALYSIS USING WEIGHTED SCORES
- Planners' value judgments
- Factor interdependence
- Sum of weighted scores
- Computer mapping

CARRYING CAPACITY ANALYSIS
- Growth variables
- Limiting factors
- Linking factors to growth variables
- Modifying carrying capacity
- Alternative growth scenarios

TANDEM USE OF CARRYING CAPACITY AND LAND SUITABILITY ANALYSES

DISCUSSION QUESTIONS

10–1 If a factor is described in terms of categories that are not *rated* in any way, a "nominal measure" is being used. If the categories are ranked, the measure is "ordinal." Pick a factor that might be used in a land suitability study and describe it using both a nominal and an ordinal measure.

10–2 In a land suitability analysis, whether or not a factor is accommodating or constraining for a particular use depends on the criterion used in ranking nominal types. Change the ranking criteria in Table 10.2 so that all of the factors *accommodate* commercial and industrial development.

10–3 Recompute the sum of weighted scores in Table 10.3 to find the areas most suitable for preservation as open space. Assume the scoring for "quality of views" remains the same, but the scoring for "slope" is changed so that the nominal types high, medium, and low correspond to scores of 5, 3, and 1, respectively. The highest scores are associated with the highest slopes since these lands are judged most suitable for preservation as open space. Assume also that the quality of views is considered three times as important as slope in determining the suitability of lands for open-space preservation.

10–4 Suppose you owned a large parcel in the Palo Alto foothills and your land received a high weighted score on the suitability map in Figure 10.3. Criticize the consultants' recommendation that your land not be developed intensively. Focus your argument on limitations of the weighted scores approach.

10–5 Consider a region in which the carrying capacity (in terms of population) is constrained by standards limiting the annual average concentration of nitrogen oxides to 100 $\mu g/m^3$. What steps could be taken to increase the region's capacity?

10-6 Officials representing a small city ask you to determine the city's carrying capacity. The officials make it clear that your analysis should include the high noise levels in a particular neighborhood and traffic congestion in the downtown area. Prepare a short statement explaining, in general terms, how you would proceed with the analysis. Indicate how you might define growth variables and limiting factors and the problems you anticipate in linking the factors to the growth variables. Also mention the parts of your analysis that would require judgments about the community's values.

REFERENCES

Fabos, J. G., 1979, *Planning the Total Landscape: A Guide to Intelligent Land Use.* Westview Press, Boulder, Colo.

Frieden, B. J., 1979, *The Environmental Protection Hustle.* MIT Press, Cambridge, Mass.

Grove, C. F., 1978, "Carrying Capacity: Summary, Major Issues and Recommendations," report to the California Tahoe Regional Planning Agency (CTRPA) Planning Team/Technical Advisory Committee and to the CTRPA Governing Board, CTRPA. South Lake Tahoe, Calif. (September), unpublished.

Hopkins, L. D., 1977, Methods for Generating Land Suitability Maps: A Comparative Evaluation. *Journal of the American Institute of Planners* **43** (4) 386–400.

Jaakson, R., M. D. Buszynski, and D. Botting, 1976, Carrying Capacity and Lake Recreation Planning: A Case Study for North-Central Saskatchewan, Canada. *Town Planning Review* **47** (4), 359–373.

Kuss, F. R., and J. M. Morgan III, 1980, Estimating the Physical Carrying Capacity of Recreation Areas: A Rationale for the Application of the Universal Soil Loss Equation. *Journal of Soil and Water Conservation* **35** (2), 87–89.

Livingston and Blayney, 1971, "Open Space vs. Development: Foothills Environmental Design," Final Report to the City of Palo Alto prepared by Livingston and Blayney, City and Regional Planners. San Francisco, Calif., unpublished.

McAllister, D. M., 1980, *Evaluation in Environmental Planning.* MIT Press, Cambridge, Mass.

McHarg, I. L., 1969, *Design with Nature.* Natural History Press, New York.

McHarg, I. L., and J. Sutton, 1975, Ecological Plumbing for the Texas Coastal Plain. *Landscape Architecture* **65** (1), 78–89.

Nieswand, G. H., and P. I. Pizor, 1978, How to Apply Carrying Capacity Analysis, *in* F. Schnidman, J. A. Silverman, and R. C. Young, Jr. (eds.), *Management and Control of Growth,* Vol. IV. Urban Land Institute, Washington, D.C. (originally published in the December 1977 issue of *Environmental Comment*).

Nilson, T. H., R. H. Wright, T. C. Vlasic, and W. E. Spangle, 1979, "Relative Slope Stability and Land Use Planning in the San Francisco Bay Region, California," Geological Survey Professional Paper No. 944, U.S. Geological Survey, Washington, D.C.

Odum, E. P., 1971, *Fundamentals of Ecology,* 3rd ed. Saunders, Philadelphia, Pa.

Schneider, D. M., D. R. Godschalk, and N. Axler, 1978, *The Carrying Capacity Concept as a Planning Tool,* Report No. 338. Planning Advisory Service, American Planning Association, Chicago, Ill.

Steinitz, C., P. Parker, and L. Jordon, 1976, Hand Drawn Overlays: Their History and Prospective Uses. *Landscape Architecture* **66** (9), 444–455.

Yapp, G. A., and G. C. Barrow, 1979, Zonation and Carrying Capacity Estimates in Canadian Park Planning. *Biological Conservation* **15,** 191–206.

CHAPTER 11

PHYSICAL INFRASTRUCTURE AND ENVIRONMENTAL QUALITY

This chapter focuses on a particular type of land use, namely, physical infrastructure. The term *infrastructure* includes the large-scale physical facilities associated with the development of cities and regions. Examples are water supply reservoirs, transportation networks, and electric power plants.

EFFECTS OF INFRASTRUCTURE ON ENVIRONMENTAL QUALITY

One reason for examining facilities like water works and highways together is that they often cause similar types of environmental impacts. Some of these impacts, such as the noise from new highway traffic, are fairly direct. Others are quite indirect: for example, the influence of expanding wastewater facilities on air pollution in a region. As illustrated below, air quality can diminish as a result of the increased residential development accommodated by the new wastewater facilities.

Typical Environmental Effects Caused by Infrastructure

To examine some direct environmental impacts associated with infrastructure investments, consider four project types: airports, dams, electric power plants, and highways. Checklists indicating the categories of environmental impacts associated with such projects invariably include items related to air and water quality, noise, biological systems, and scenic resources. For each of these categories, Table 11.1 contains a sampling of *direct* effects associated with the four types of facilities. For example, the first column shows some typical water quality impacts. For highways and airports, water quality changes are due to residuals washed off road surfaces and runways when it rains. Electric power facilities affect water quality via the discharge of "waste heat" from power plant cooling water systems. Finally, retarding the flow in a river by building a dam is sufficient to influence numerous water quality indicators. Although Table 11.1 is *not* a comprehensive summary of infrastructure-related impacts, it illustrates that there are many similarities in the environmental effects caused by different facilities.

The direct effects indicated in Table 11.1 have been studied for many years, and they commonly receive detailed treatment in EISs for airports, power plants, and other infrastructure investments.[1] There are, however, numerous *indirect* impacts of physical infrastructure that have only recently received serious attention in environmental assessments. These are effects that result when infrastructure influences the location, density, and timing of new land development.

The presence of infrastructure rarely causes new regional growth. Individuals and firms do not locate in a region simply because it has unused capacity in its road networks, sewer systems, and so forth. Other important factors, such as opportunities for employment, determine whether a region increases or decreases in size. However, infrastructure can play a pivotal role in determining where, *within a region,* new growth takes place. This is especially the case when the absence of a major facility such as a highway constrains development in a part of a region otherwise ready to yield to growth pressures.

Influence of Infrastructure on Land Development

The expansion of a wastewater facility in the Livermore-Amador Valley in California demonstrates how infrastructure may cause indirect environmental impacts.[2] The valley is located about 35 miles east of San Francisco and contains three major communities: Livermore, Pleasanton, and Dublin-San Ramon (see Figure 11.1). Many residents of Pleasanton and Dublin-San Ramon commute by car to work in either Oakland or San Francisco. Long distances are involved in these trips. The average one-way commute for valley residents has been estimated by Little and Reid (1974) to be 27.5 miles.

Emissions from the motor vehicles of valley commuters have not been accommodated well because of local meteorological and topographic conditions.

[1]The procedures used in analyzing the direct impacts shown in Table 11.1 are surveyed in Chapters 12–16.

[2]The discussion here relies on unpublished analyses of the Livermore-Amador wastewater facilities prepared by Violetta Cavalli-Sforza and Francis Sweeney while they were students in the Department of Civil Engineering at Stanford University. For a more dramatic demonstration of the influence of infrastructure on land use, see Kahrl's (1982) account of the role of water supply on the growth of Los Angeles.

TABLE 11.1 Direct Impacts Commonly Associated with Infrastructure Projects

Project Type	Water Quality	Air Quality	Noise	Biological Effects	Visual Effects
Airports	Residuals in runoff from roads and runways	Emissions from aircraft, airport buildings, and motor vehicles	Emissions from aircraft and motor vehicles	Habitat destruction	Views of aircraft, airport structures, and roadways
Dams	Chemical, physical and biological effects of reservoir development	Effects are indirect (e.g., results of induced land development)	Effects are indirect (e.g., results of induced land development)	Change from running water to standing water ecosystem	Views of dam and reservoir
Electric power plants	Waste heat discharges affecting water temperature	Emissions from smokestacks	Emissions from electric power transformers	Habitat destruction	Views of smoke-stacks and stack emissions
Highways	Residuals in runoff from road surfaces	Emissions from motor vehicles	Emissions from motor vehicles	Habitat destruction and creation of barriers to animal migration	Views of highways and related facilities

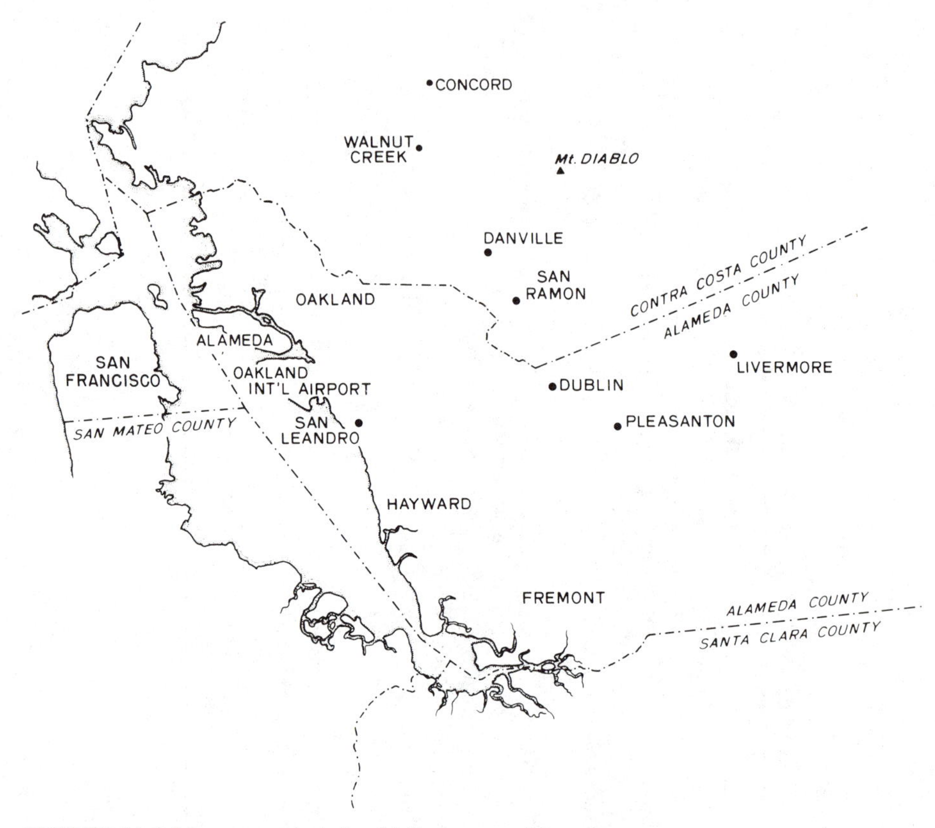

FIGURE 11.1 Livermore-Amador Valley: general location map.

Essentially, the Livermore-Amador Valley is a flat inland plain almost completely surrounded by mountains ranging in height between 1000 and 2000 ft. Circumstances leading to photochemical smog occur frequently. The valley's violations of the national ambient air quality standards for photochemical oxidants indicate the seriousness of the smog problem. For each year during the period from 1969 to 1974, the standards were exceeded between 35 and 180 days.

In addition to increased air pollution, the valley experienced water quality problems following the rapid population growth that occurred between 1950 and 1970. Although Livermore's high growth period started somewhat earlier, the population of the remaining valley communities increased by almost a factor of 10 during the 1960s. The combined population for Pleasanton and Dublin-San Ramon jumped from about 5000 in 1960 to over 40,000 in 1970. During the early 1970s, however, the population growth rates of the valley communities declined abruptly from figures ranging between 7 and 22% for the years 1970 to 1972 to between 1 and 6% for 1973. The main reason for this decline was that existing wastewater collection and treatment facilities could not accept additional growth and meet the water quality standards established by the California Water Quality Control Board.

During the 1970s, consultants to the communities and agencies within the Livermore-Amador Valley conducted several investigations of how much additional wastewater collection and treatment capacity should be added to serve the valley. These studies recognized that, as of the early 1970s, limitations in sewerage capacity were constraining population growth, and new growth could make the existing air quality problems even worse than they already were.

Table 11.2 indicates the steps followed by one set of consultants in a study of indirect environmental impacts of wastewater facility expansion. Their analysis assumed that once the constraint imposed by sewerage capacity limitations was removed, the number of valley residents would increase. New residents would depend heavily on commuting by auto to jobs in San Francisco and Oakland. The new trips would cause an increase in motor vehicle emissions of hydrocarbons and nitrogen oxides, which, in turn, would increase the concentrations of photochemical oxidants in the ambient air. Mathematical models, based on several assumptions, were used to estimate how much a particular enlargement in wastewater facilities would increase the photochemical smog problem.[3] The analysis outlined in Table 11.2 helped public officials decide how much wastewater collection and treatment capacity to add.

INSTITUTIONAL SETTING OF INFRASTRUCTURE DEVELOPMENT

In examining relationships between infrastructure and environmental quality, it is useful to consider the institutional setting in which planning is carried out. This setting evolves around the agencies or other organizations doing the plan-

TABLE 11.2 Estimating Indirect Effects of Wastewater Facility Expansion in Livermore-Amador Valley, California[a]

Select, for analytic purposes, a size for the new wastewater management facilities
Estimate the increase in residential population accommodated by the new facilities
Compute the additional "vehicle miles traveled" (VMT) by the new valley residents
Determine the increased emissions of air-borne residuals associated with the added VMT
Compute the ambient concentrations of photochemical oxidants that reflect the increased emissions

[a]Steps shown in the table were followed by the URS Company and John Carollo Engineers (1975).

[3]Details regarding the mathematical modeling are given by the URS Company and John Carollo Engineers (1975). Both mathematical models and expert opinion procedures such as the Delphi method have been used to forecast the effect of infrastructure on land use. Wendt (1976) illustrates some approaches that have been tried.

ning and the various client groups they serve. Among the clients are individual citizens, interest groups, and government agencies that have an interest in how infrastructure is developed. Other institutional elements are the laws, regulations, and policies that determine some of the evaluative factors considered in planning. The institutional context greatly influences how environmental concerns are integrated into the process of formulating and evaluating alternative projects.

The basis for planning infrastructure depends on circumstances specific to the organization doing the planning. There are many different infrastructure planning organizations, even for a single type of facility like a highway. In the United States, highway planning is carried out at all levels of government. The National Park Service develops roads in national parks and monuments. State transportation agencies implement elements of the interstate highway system using funds administered by the Federal Highway Administration. Numerous county and city departments of public works develop roads within their individual jurisdictions. Each of these governmental organizations uses different criteria as the basis for planning and evaluating highways. In some cases, most of the attention is given to technologic, financial, and economic considerations, whereas in others, environmental issues receive substantial weight.

Although the details of planning depend on the particular setting, there are similarities in the processes used to plan different types of infrastructure projects. Facilities as diverse as sanitary landfills, electric power plants, and airports are planned typically by either a public agency or a private entity (such as an electric utility) that is highly regulated by a public body. In principle, many infrastructure projects are intended to serve the same overall end, "the public interest." Because these facilities typically affect numerous individuals and groups, it is necessary to consider different views on what it means to serve the public interest. Account must be taken of dissimilar opinions about criteria that should be used in formulating and evaluating alternative proposals.

Another common feature of infrastructure development is that the organization doing the planning coordinates extensively with agencies and public officials at all levels of government. In addition, individual citizens and groups are sometimes involved directly in planning. High levels of interagency coordination and citizen involvement became common in the United States following the passage of the National Environmental Policy Act of 1969. The many infrastructure projects in the United States that are either federally funded or require permits from federal agencies are all subject to NEPA's environmental impact reporting requirements.

Prior to the 1960s, infrastructure planning organizations often relied heavily on economic criteria in assessing alternative proposals. A common goal was to provide an adequate level of service, such as delivering *X* million gallons per day of drinking water, at the minimum cost. In some cases, for example federal water resources projects in the United States, it was deemed appropriate to use a benefit–cost analysis to evaluate alternatives. A project's size and other features were selected to maximize the difference between economic benefits and costs. Invariably, the fund raising needed to implement plans also had a significant influence on which of several alternative projects would be selected.

Public concern over environmental quality issues, as reflected in new laws such as NEPA, augmented the criteria used in planning infrastructure in the United States. An example of the mandates for expanded criteria is Section 102(2)(B) of NEPA; it requires federal agencies to develop procedures ensuring "that presently unquantified environmental amenities and values may be given appropriate consideration in decision making along with economic and technical considerations." In addition to reflecting a broader range of evaluative criteria, infrastructure planning after the 1960s was much more "open." NEPA and other environmental quality laws have given citizens and agencies concerned with the environment new opportunities to influence the way plans are formulated and decisions are made.

INFRASTRUCTURE PLANNING IN PRACTICE: THE SAN PEDRO CREEK STUDY

The San Pedro Creek study demonstrates how infrastructure planning is carried out and how environmental concerns can influence decision making. San Pedro Creek runs through Pacifica, a small coastal community several miles south of San Francisco. A flood control investigation was conducted during the 1970s by the San Francisco District Office of the U.S. Army Corps of Engineers ("the district") The District Engineer assigned the study to the district's "Planning Branch." A member of that branch, referred to as the "study manager," was made responsible for orchestrating all aspects of the investigation. This included coordination with technical specialists within the district office and with citizens, local officials, and agencies having an interest in the study.[4]

The initial phase of the San Pedro Creek study focused on clarifying the nature of the flooding problem and the criteria to be used in formulating and evaluating alternative approaches to dealing with the problem. Part of this "problem identification" task included organizing a program to involve local officials and citizens in various aspects of the study. (The San Pedro Creek public involvement program is discussed in Chapter 8.)

Based on a "reconnaissance investigation," the study manager acquired initial impressions regarding possible ways of reducing flood damages. Early efforts to involve citizens in the San Pedro Creek study led to some refinements in the study manager's thinking. The following factors were found to be especially important to the local community: the maintenance of San Pedro Creek as a steelhead fishery, the visual character of the creek, and the costs to Pacifica of

[4]The San Pedro Creek flood study discussion summarizes a presentation by Wagner and Ortolano (1976).

any flood control projects. Preliminary initiatives to include other agencies in the study were directed toward the U.S. Fish and Wildlife Service and the California Fish and Game Department. The views of these agencies were important since the creek was used for sport fishing. Any structural measures proposed to reduce flooding had to be sensitive to the creek's value as a steelhead fishery. In addition to the evaluative factors that were important to local citizens and other agencies, the Corps of Engineers' planning regulations had to be considered. For example, the corps' regulations specified two planning objectives: one concerned the contributions of a project to net national income, and a second concerned the ability of a project to "enhance environmental quality." The regulations also included cost-sharing requirements indicating the funds that had to be provided by nonfederal entities on a corps flood control project. All of these factors were considered in developing alternative proposals to deal with flooding on the creek.

Another part of the problem identification task involved the study manager and technical specialists within the San Francisco District Office. The San Pedro Creek study was performed by specialists from the Planning, Design, and Environmental Branches within the district's Engineering Division. The Planning Branch participants included members of the Hydraulics and Hydrology Section and the Economics Section. A partial organization chart for the district is given in Figure 11.2. One representative from each of the above-noted specialty units was assigned to work with the study manager on a San Pedro Creek "planning team." During the problem identification phase of the study, the team contributed (1) a hydrologic analysis of the extent of flooding and (2) a preliminary description of several structural and nonstructural actions that could decrease the damages due to flooding.

The next phase of the study produced a more detailed conception of alternative plans to reduce flood damages and a preliminary analysis of the impacts of the different schemes. There is no simple way to characterize the activities during this phase because the interactions between the planning team, other agencies, and Pacifica citizens were so numerous. Moreover, decisions regarding which proposals to study in detail were modified continually as a result of the information exchanged during these interactions. Individuals, citizens' groups, and the fish and game agencies made important contributions to the process of delineating and assessing alternatives. Members of the planning team provided "sketch-level" descriptions of alternative plans and assessments of their economic and environmental impacts.

Consider now the role of environmental factors in these preliminary planning efforts. The Environmental Branch, which included biologists, archeologists, and landscape architects, was the main organizational unit concerned with environmental issues. The branch prepared a *working paper,* consisting of a preliminary environmental appraisal of the alternative proposals under consideration. In deciding on which issues to address in their assessment, the branch's specialists relied on their own professional judgments as well as the concerns expressed

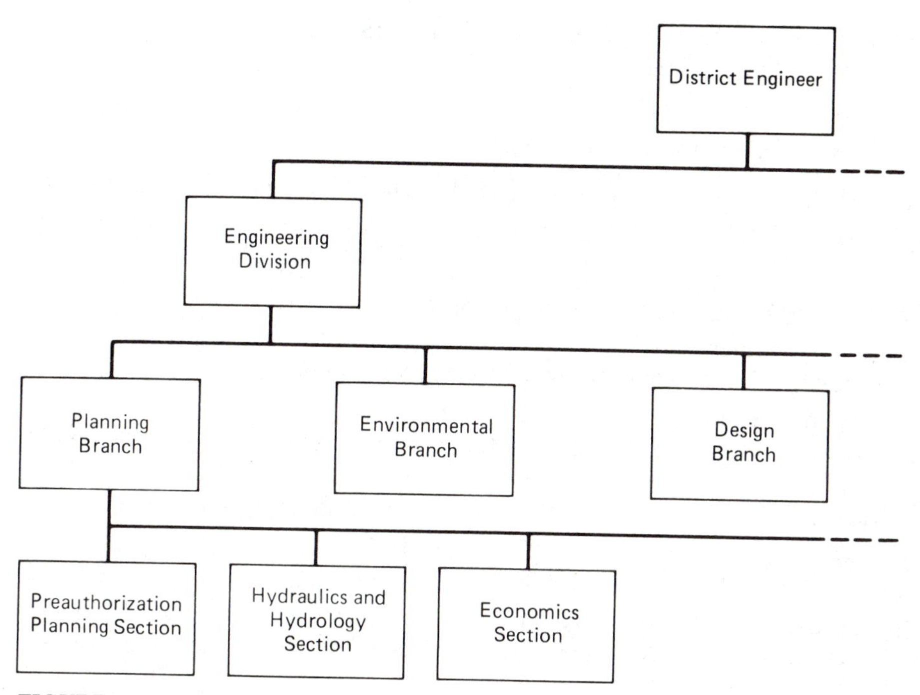

FIGURE 11.2 Organizational units represented on San Pedro Creek planning team (U.S. Army Corps of Engineers, San Francisco District Office, circa 1974).

by various citizens and agencies. The Environmental Branch also delineated an *environmental quality plan.* This plan maximized contributions to the enhancement of environmental quality objective, while addressing the flooding problems that were at the center of the district's study. The corps' planning regulations required that an environmental quality plan be formulated and seriously considered, regardless of whether it made a positive contribution to the net national income objective.

The final phase of the San Pedro Creek study emphasized the *evaluation* of alternative plans. Several appraisals were made. Individual citizens ranked the possible actions as part of various public involvement activities. The fish and game agencies evaluated the proposals in the context of the district's interagency coordination efforts. The study manager, representing the position of the planning team, also made an appraisal. In addition, the Pacifica City Council ranked the alternatives. The district engineer was responsible for assimilating all the evaluations and choosing what he considered to be the most suitable plan for reducing flood damages. The selected plan was sensitive to the environmental concerns raised throughout the study and it was endorsed by citizens and other agencies. However, the city of Pacifica was unable to provide its share of the total project cost, and the chosen plan was not implemented.

INFLUENCE OF ENVIRONMENTAL SPECIALISTS ON PLANNING OUTCOMES

Alternative Roles of Environmental Specialists

The San Pedro Creek study illustrates only some of the roles that environmental professionals can play in integrating environmental factors into infrastructure planning. More generally, six different roles may be distinguished:[5]

1. Planning—working on a planning team.
2. Report preparation—preparing statements that disclose and assess the environmental impacts of proposed plans and alternatives.
3. Internal report review—evaluating environmental impact documents prepared by other staff members within the environmental specialist's organization.
4. Design—participating in detailed design of a proposed plan.
5. System monitoring—checking operations of a completed project to ensure that desired levels of environmental quality are maintained.
6. External report review—commenting on environmental impact documents for projects proposed by entities outside of the environmental specialist's organization.

Of the six roles, the one that received the greatest attention in the United States during the early 1970s was report preparation. When NEPA was enacted, there was a multitude of federal projects in various stages of planning, and many of them required environmental impact statements before they could be carried out. Environmental professionals in federal agencies were under great pressure to produce EISs complete enough to withstand legal challenges. The demand for reports kept many environmental experts from becoming actively involved in project planning or design.

In several infrastructure agencies, this narrow perception of the role of environmental professionals has persisted. Planning guidelines of some agencies describe the functions of environmental specialists almost entirely in terms of preparing EISs. Not surprisingly, many study managers view the role of environmental experts in this way.

When the main function of environmental specialists is to prepare impact documents, their influence in integrating environmental factors into infrastructure planning is limited. This view is supported by research on federal water resources agencies in the United States during the late 1970s.[6] It found that environmental factors had a notable influence on the formulation of alternative water plans in studies where environmental professionals were given a broad range of duties. This strong influence was not evident for water planning studies

[5]This discussion of alternative roles of environmental specialists relies on Jenkins and Ortolano (1978).

[6]The research examined the influence of environmental information on decision making in two water agencies and one highway agency (Ortolano et al., 1979).

in which the roles of environmental experts were confined largely to EIS preparation. The research on U.S. water resources agencies also suggested that environmental specialists were more likely to influence planning outcomes if they functioned as planners, and not just as "staff advisors" responding to specific requests made by study managers.

Organizational Location of Environmental Specialists

Consideration of the various roles played by environmental specialists raises several questions: Where should environmental professionals be placed in an infrastructure planning organization to most effectively carry out their different functions? Should they all be in one organizational unit such as the Environmental Branch in the San Francisco District Office of the Corps of Engineers? Alternatively, would it be better to disperse the specialists in different units such as the Planning Branch and the Design Branch of the San Francisco District Office? Although research on these questions has not provided definitive answers, it has established that the questions are important and worth pondering.[7]

The optimal organizational locations for environmental specialists may depend on whether an infrastructure planning study is in its early or late stages. The early stages are illustrated by the San Pedro Creek flood study. Planning investigations in their late stages provide design details for a particular proposal.

For studies in their *early stages,* the planning, report preparation, and internal report review functions of environmental specialists are relevant. For performing the planning role, there are several advantages to locating environmental experts in the same organizational unit as engineers and planners ("the planning unit"). It allows environmental specialists to become familiar with the jargon and viewpoints of other professionals engaged in planning. This organizational arrangement also recognizes the interdependencies among planning tasks and the need for frequent coordination among professionals in the early stages of planning.

Although environmental experts may effectively contribute to planning if they are part of the planning unit, this is *not* the best location for carrying out report preparation and internal report review functions. Report preparation and review require the ability to make impartial environmental assessments that may be critical of proposals developed by the planning unit. Environmental specialists *outside* the planning unit can provide a more objective perspective. Moreover, if they differ with engineers and planners, environmental experts outside the planning unit at least have the option of referring disputes to a higher level in the office hierarchy. For example, in the San Francisco District Office of the Corps, the head of the Environmental Branch can raise objections directly with the head of the Engineering Division (see Figure 11.2). Although this mechanism for influencing the Planning Branch may not be exercised frequently, it is a useful option. For the reasons above, when planning investigations are in their early stages, no single organizational arrangement increases the effectiveness of all the roles played by environmental specialists.

[7]The following discussion is based on research by Jenkins and Ortolano (1978) and Ortolano et al. (1979).

During the *late stages* of planning there is much less emphasis on formulating new plans than there is on preparing reports and coordinating with outside agencies and citizens' groups. Report preparation and external coordination activities are performed advantageously by environmental experts in a unit separate from the planning unit. In addition to having an ability to make referrals to higher offices and increased impartiality in assessing plans, a separate unit that retains its objectivity can be effective in coordinating with outside agencies and groups.

Perhaps the best overall arrangement is to place environmental specialists in more than one organizational unit. For example, the environmental experts engaged in defining planning objectives, formulating alternatives, and performing other planning activities can be part of the planning unit. Other environmental professionals can be located outside that unit to review proposals and reports and to provide an interface between the planning unit and various environmental agencies and groups. This makes it possible to obtain both good coordination and integration of environmental perspectives on routine planning tasks *and* critical in-house reviews of the planning effort.

ENVIRONMENTAL FACTORS AND THE USE OF PLANNING TEAMS

Planning teams have the potential for enhancing the consideration given to environmental factors in decision making. There are wide variations in the extent to which teams are used in planning infrastructure. Some organizations do not use them at all. Instead, a study manager works with various technical specialists on a one-to-one basis. Some organizations form a planning team, but it exists in name only. The "team" never meets as a group. Planning teams that exist in more than a nominal sense meet on one or more occasions during a study.

Planning teams can be either multidisciplinary or interdisciplinary. Members of multidisciplinary groups contribute to a study only within their particular areas of expertise. For example, economists on a multidisciplinary water resources planning team estimate monetary costs and benefits of alternative proposals, but they probably would not assist in formulating alternatives. In contrast, members of an interdisciplinary team participate in all aspects of a study. The interdisciplinary concept envisions different specialists working together and contributing to the team's conception of study objectives, the formulation of alternative plans, and the assessment and ranking of plans. The judgments of several individuals are synthesized to yield an integrated perspective that is broader and more informed than that of any one team member.

Research by Ortolano et al. (1979) suggests that when teams are used in infrastructure planning they are frequently of the multidisciplinary type. The research examined 68 planning studies conducted in the United States by two federal water resources agencies and one state transportation agency. Although the research results were not definitive, they suggest that using teams in the *early stages* of a study can increase the attention given to environmental factors in formulating alternative proposals. This was especially true when environmental experts were included as team members and when the teams met frequently.

Of the three organizations included in the research, the one that made the most sophisticated use of teams was the Bureau of Reclamation, an agency responsible for federal irrigation projects in the western United States. The bureau included representatives of other agencies and environmental interest groups as full-fledged team members and used the team as an important mechanism for coordinating with those agencies and groups. The bureau also employed subteams and task forces to assist a main team on particular issues, such as determining ways of preserving important wildlife habitats. The specialized subteams increased the overall team effectiveness by allowing members with specific interests to concentrate on their areas of expertise without being distracted by other issues. Representatives from subteams participated in the deliberations of a main team to allow for a synthesis of all viewpoints.

Although teams have the potential for increasing the attention given to environmental factors, they are not always effective. Difficulties have resulted from team members being confused about their roles and feeling that they would not be rewarded for exceptional individual contributions to the group.[8] Actions to reduce these impediments typically require a substantial organizational commitment to using the team planning concept.

GENERATING ENVIRONMENTAL INFORMATION VIA EXTERNAL COORDINATION

Coordination with citizens, groups, and agencies plays a key role in integrating environmental factors into infrastructure planning and decision making. In the early stages of planning, coordination can enhance an infrastructure organization's ability to identify potential environmental impacts that need to be investigated. As a study progresses, coordination provides additional information about alternative actions, impacts, and values. Extensive coordination can increase both the range of environmental factors considered and the weight given to such factors in evaluating alternative proposals.

The previously cited research on three infrastructure agencies confirms the importance of coordination. For these agencies, the influence of environmental information on decision making was enhanced when a substantive effort was made to involve citizens and other agencies in planning. The research results suggest that the role of the public is especially significant in early stages of planning, whereas information from other agencies is more important in later stages. The research on Corps of Engineers' water planning demonstrates this finding. In the early stages of the corps planning studies, other agencies often

[8]For a discussion of problems in using planning teams, see Wagner and Ortolano (1976). Difficulties in performing an interdisciplinary environmental impact assessment are elaborated by Burdge and Opryszek (1983).

seemed hesitant to commit their limited resources toward making contributions. However, individual citizens and citizens' groups were active in making the corps aware of environmental issues. The opposite was often true for corps studies in their late stages. The research results indicated that as the corps studies drew to a close, other agencies had more opportunities than citizens' groups to influence the study outcomes. This does not mean that citizens' concerns were disregarded in the late stages. In fact, conservation agencies such as the U.S. Fish and Wildlife Service sometimes became strong advocates of the positions of citizens' groups, which they viewed as their constituents.

For infrastructure proposals subject to NEPA's requirements, there are several EIS-related coordination exercises that can generate information on environmental issues. For example, many agencies use preliminary environmental assessment documents that are circulated to other agencies and the public *prior* to the preparation of a draft EIS. In an extensive survey of Corps of Engineers and Soil Conservation Service planners, well over half the respondents reported that the most useful comments on environmental documents resulted from reviews of "pre-EIS" reports circulated well before planning was completed.[9] Even when such documents are modest in scope and level of detail, they seem to stimulate useful communications with agencies and the public in the early stages of a planning study. The importance of a timely exchange of information should not be underestimated. When environmental issues are clarified early, they have a greater likelihood of being considered seriously in planning and decision making.

KEY CONCEPTS AND TERMS

EFFECTS OF INFRASTRUCTURE ON ENVIRONMENTAL QUALITY
- Direct versus indirect impacts
- Location and timing of growth
- Infrastructure as a growth constraint
- Residential location decisions
- Vehicle miles traveled

INSTITUTIONAL SETTING OF INFRASTRUCTURE DEVELOPMENT
- "The public interest"
- Interagency coordination
- Public involvement in planning
- Multicriteria evaluation

INFRASTRUCTURE PLANNING IN PRACTICE: THE SAN PEDRO CREEK STUDY
- Problem identification
- Formulating alternatives
- Forecasting impacts
- Evaluating alternatives
- Environmental specialists
- Planning teams
- Environmental quality as a planning objective

INFLUENCE OF ENVIRONMENTAL SPECIALISTS ON PLANNING OUTCOMES
- Roles of environmental professionals
- Planners versus staff advisors
- Alternative locations for environmental experts

[9] A description of the survey results relating to interagency coordination is given by Hill and Ortolano (1976).

ENVIRONMENTAL FACTORS AND THE USE OF PLANNING TEAMS
- Multidisciplinary versus interdisciplinary teams
- Representing interest groups on teams
- Subteams and task forces
- Factors influencing team effectiveness

GENERATING ENVIRONMENTAL INFORMATION VIA EXTERNAL COORDINATION
- Early identification of environmental issues
- Information on actions, impacts, and values
- Role of citizen involvement early in planning
- Role of other agencies late in planning

DISCUSSION QUESTIONS

11–1 Review the steps in Table 11.2 for computing the indirect air quality impacts due to wastewater facility expansion in the Livermore-Amador Valley. Speculate on the types of assumptions the analysts had to make to obtain estimates of new residents in the valley, new increments of VMT, and increased concentrations of photochemical oxidants.

11–2 Select an infrastructure planning exercise that was completed recently in a nearby community. Characterize the institutional setting for the exercise in terms of applicable laws, regulations, and policies, and affected citizens, groups, and agencies. What aspects of this setting dictated that environmental factors be considered in decision making? How did the organization conducting the study attempt to define the public interest in evaluating alternative projects?

11–3 Identify the main participants and affected interests in a recently completed infrastructure planning study. Did the planning organization utilize a study manager or a planning team, or both? What mechanisms were used to encourage communications among the study participants and affected interests both inside and outside of the planning organization? Did the coordination strategies allow the planners to learn about key environmental issues early in the study? What additional coordination techniques could have been applied to enhance the consideration given to environmental factors?

11–4 Obtain the organization chart for an infrastructure planning agency office. How many environmental professionals are in the office and where are they located? Speculate on the advantages and disadvantages of the organizational arrangement in terms of integrating environmental factors into planning. Distinguish between the early and the late stages of planning in preparing your response. If possible, consult with engineers and environmental specialists in the office to see if they agree with your speculations.

11–5 Consider the partial organization chart in Figure 11.2. Why might the branch chiefs oppose the use of teams set up to undertake specific planning studies? What arguments might persuade the branch chiefs to support the use of planning teams?

REFERENCES

Burdge, R. J., and P. Opryszek, 1983, On Mixing Apples and Oranges: The Sociologist Does Impact Assessment with Biologists and Economists, *in* F. A. Rossini and A. L. Porter (eds.), *Integrated Impact Assessment,* pp. 107–117. Westview Press, Boulder, Colo.

Hill, W. W., and L. Ortolano, 1976, Effects of NEPA's Review and Comment Process on Water Resources Planning: Results of a Survey of Planners in the Corps of Engineers and Soil Conservation Service. *Water Resources Research* **12** (6), 1093–1100.

Jenkins, B. R., and L. Ortolano, 1978, Environmental Specialists in Water Agencies. *Proceedings of the American Society of Civil Engineers, Journal of the Water Resources Planning and Management Division.* **104** (WR1), 61–74.

Kahrl, W. L., 1982, *Water and Power, the Conflict over Los Angeles' Water Supply in the Owens Valley.* University of California Press, Berkeley.

Little, Arthur D., Inc., and Thomas Reid Associates, 1974, *Environmental Impact Report Supplemental Analysis: Population Growth and Air Quality in the Livermore-Amador Valley,* report prepared for the Valley Community Services District Stage III Wastewater Treatment Plant Enlargement, Document No. C-77102.

Ortolano, L., C. M. Brendecke, J. E. Price, and J. J. Meersman, 1979, *Environmental Considerations in Three Infrastructure Planning Agencies: An Overview of Research Findings,* Report IPM-6, Department of Civil Engineering, Stanford University, Stanford, Calif.

URS Research Company and John Carollo Engineers, 1975, *Draft Environmental Impact Statement: Proposed Wastewater Management Program, Livermore-Amador Valley, Alameda County, California,* Document No. C-06-1031-010, prepared for the U.S. Environmental Protection Agency, Region IX, San Francisco, Calif.

Wagner, T. P., and L. Ortolano, 1976, *Testing an Iterative Open Process for*

Water Resources Planning, Report 76-2, U.S. Army Engineer Institute for Water Resources, Ft. Belvoir, Va.

Wendt, P. F. (ed.), 1976, *Forecasting Transportation Impacts upon Land Use.* Mortinus Nijhoff Social Sciences Division, Leiden, The Netherlands.

PART FIVE

TECHNIQUES FOR ASSESSING IMPACTS

CHAPTER 12

BIOLOGICAL CONSIDERATIONS IN PLANNING

Although many government planners are now concerned about the effects of their decisions on biological systems, this is a recent development. In the United States it was not until the 1930s that assessments of the impacts of physical projects began to include the perspectives of biologists. An impetus for this was the Coordination Act of 1934, which required federal water resources development agencies to coordinate their plans with the U.S. Fish and Wildlife Service. The latter commented on how proposed water projects would affect fish and wildlife as biological *resources.* These early biological assessments did not emphasize the integrity and stability of biological systems. Instead, an anthropocentric orientation was used. The focus was on particular species of fish and wildlife that were valuable commercially or for recreational hunting and fishing.

A more modern perspective, which views *all* species as being potentially significant to the proper functioning of biological systems, was expressed in Leopold's (1949) *A Sand County Almanac.* Leopold argued that human actions should be tempered by an understanding of how they might affect the "biotic community." For Leopold, man was just another member of this community, not an outside element divorced from other species. His views on the need for careful stewardship of biotic communities and on the shortsightedness reflected in economic (as opposed to biological) evaluations of proposed actions were embraced by many during the 1970s. However, they were not widely held before that time.

Some efforts to consider biological factors in planning have been resource oriented in that they focused on individual species directly valuable to humans. Others have been systems oriented and have emphasized the structure, function, and long-term stability of entire ecosystems. Government policies affecting infrastructure and land use often insist that both perspectives be considered.[1]

BIOLOGICAL CONCEPTS USEFUL IN PLANNING

The discussion below introduces vocabulary and concepts widely used in integrating biological concerns into planning, especially planning for physical facilities. By its nature, such an introduction must be highly selective.[2]

The basic unit employed in many efforts to include biological factors in planning is the *ecosystem,* defined as a biotic community and its abiotic (nonliving) environment. Any change in an area will affect the ecosystem of that area. Although an ecosystem may be resilient and capable of withstanding disturbances, human or natural forces can destroy enough of an ecosystem so that its integrity is not maintained and it collapses completely.

A *community* is any assemblage of living populations coexisting in a particular area. Communities are often classified by a characteristic type of vegetation; one speaks of a Redwood forest community or a California chaparral community. However, the term *community* includes all organisms associated with the vegetative type used to name it. For example, the blue-gray gnatcatcher is a member of the chaparral community. Although many species, such as starlings or dandelions, are not associated with any one type of community, one would not expect to see an elk in a desert, or a cactus in a swamp. They are not characteristic of those communities. The term *habitat* refers to the physical location where an organism lives. Each species requires a habitat capable of providing the space, food, cover, and other requirements for its survival. For example, one would say that a pine forest is a suitable habitat for elk, but a desert is not.

Each organism occupies a position in a community as either prey or predator or both. The organization of feeding patterns is represented by a *food chain.* Organisms occupying the lowest level in a food chain derive nourishment from the abiotic components of an ecosystem. A simple example is given in Figure 12.1.[3] Solar energy is transformed into chemical energy (new grasses) by a process known as photosynthesis. In addition to sunlight, this process uses water, carbon dioxide, and nutrients, as well as the chlorophyll in grasses. The grasses produced photosynthetically are eaten by herbivores such as mice, which are

[1]In the United States, policies emphasizing both single species as resources and ecosystem-level considerations are contained in NEPA and various environmental laws passed in the 1970s. Camougis (1981, pp. 6–16) summarizes U.S. laws mandating a consideration of biological factors in planning.

[2]Daniel Belik, a graduate student in biology at Stanford University during the 1979–1980 academic year, prepared an early draft of the material in this section. A more complete treatment is provided by the emerging literature on the application of biological sciences in planning and management contexts. See, for example, Edington and Edington (1978), Ward (1978), and Holling (1978).

[3]Food chains are not typically as simple as the one in this example. Often several chains interconnect in complex configurations referred to as "food webs."

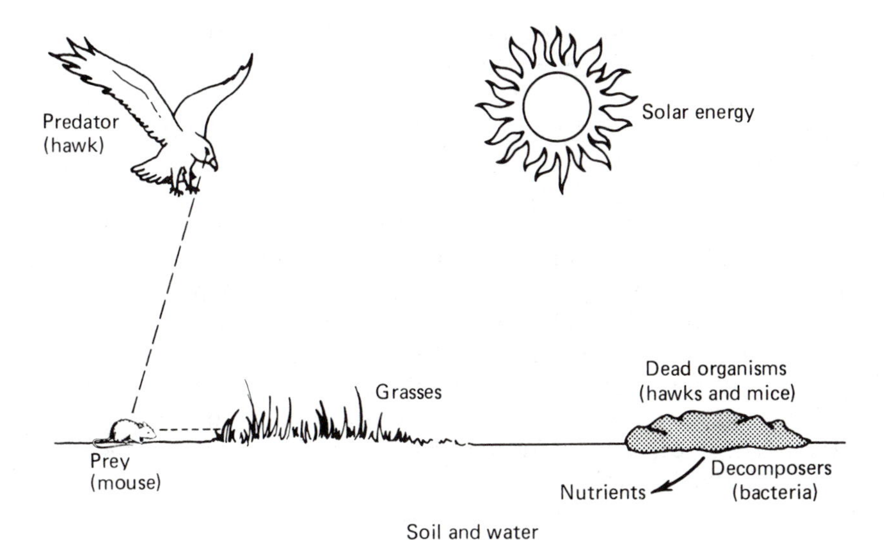

FIGURE 12.1 Illustrative food chain relationships.

eaten in turn by carnivores like hawks. The hawks are said to be at the top of the food chain. Dead hawks are a food source for organisms like bacteria whose activities as decomposers of organic matter refortify the physical environment. Disruption of a food chain affects both the predator and the prey. If the predator is deprived of food it will starve. The prey, with its source of population control gone, will increase in numbers. If the prey population becomes so large that it consumes its entire food source, the prey will starve in turn.

The neglect of food chain interactions by planners sometimes leads to unintended, adverse effects. Consider an example involving field mice in Israel several decades ago. Every 10 years the mice greatly increased in numbers, for just 1 year, and consumed large quantities of grain harvested in agricultural areas. During these periods, an increase was noted in birds that preyed on the mice, such as eagles and falcons. In response to a 1949–1950 increase in mouse population, the Israeli Plant Protection Department of the Ministry of Agriculture implemented a program to poison the mice using thallium coated wheat. Thallium takes several days to kill field mice and does so through slow paralysis. Partially paralyzed mice were easy prey for the birds, and before long the birds were dying from thallium poisoning. The results from the government program were not those intended: the birds of prey decreased in numbers and the mouse population was not brought under control.[4]

Another example demonstrating the need for planners to consider food chain interactions concerns the *biological magnification* of substances, such as DDT, that do not decompose readily. During the 1950s, many toxicological studies of individual species suggested that the use of DDT for pest control would not

[4]This example is from Mendelssohn (1972).

cause significant problems. Subsequent studies revealing how DDT was passed through various food chains refuted the earlier findings. In general, much of the living matter ("biomass") passed from lower to higher positions in a food chain is excreted or used up in respiratory activities, whereas only a small amount stays in the consuming animal. DDT, however, is very soluble in fats and is not easily broken down to simpler compounds. Consequently, as biomass is transferred up through the food chain, very little DDT is "lost" through each transfer. The result is that DDT concentrations often increase dramatically from one food chain level to the next (see Figure 12.2). For example, studies of marshes on Long Island, New York, showed the following magnification in DDT concentrations (measured in parts per million): 0.04 in plankton, 1 in minnows, and 75 in the tissues of carnivorous scavenging birds.[5] High DDT concentrations due to biological magnification have been a cause of mortality for some species of birds and interference with reproduction for others.[6] This example shows the importance of ecosystem-level analyses of biological impacts.

Productivity is another aspect of a biological community that merits consideration by planners. In biology, *productivity* refers to the rate at which an ecosystem can take abiotic components and use them to produce living matter. The arctic tundra, with sparse vegetation, is a relatively unproductive community, whereas a tropical rain forest is highly productive. Components of the abiotic environment that determine an ecosystem's productivity are water, nutrients, and energy. The energy input is generally sunlight, which is photosynthesized into chemical energy. In various forms, energy is cycled through an ecosystem via its food chains. Human actions reducing the abiotic elements used in forming organic matter can decrease an area's productivity. For example, in some parts of the western United States, groundwater is being withdrawn more rapidly than it is being replenished by natural sources. As a result, the surrounding land is becoming increasingly arid, surface water is not being held in the topsoil, and biological productivity is being decreased. The reduction in productivity affects both the natural biological communities and the agricultural use of the land. If enough vegetation is lost over time, the soil will not be held in place and erosion problems and dustbowls will appear.

The change in a community's characteristics over time is termed *ecological succession.* Natural population fluctuations occur continually in a community, but succession refers to changes in the type of species and thus in the community itself. A typical example of an area's natural succession begins with the colonization of bare sand by bacteria and fungi, and then mosses and lichens. These simple organisms break down the sand into soil that is then colonized by annual grasses and herbs. The "annuals" are often crowded out by perennial grasses and herbs. The grasses change the characteristics of the soil so that the area is eventually colonized by shrubs. The shrubs might eventually be replaced by trees, which would be part of the final or "climax" community to occupy that

[5]These figures, and a more complete discussion of biological magnification, are given by Woodwell (1967).

[6]Further details on the adverse effects of DDT are given by Ward (1978).

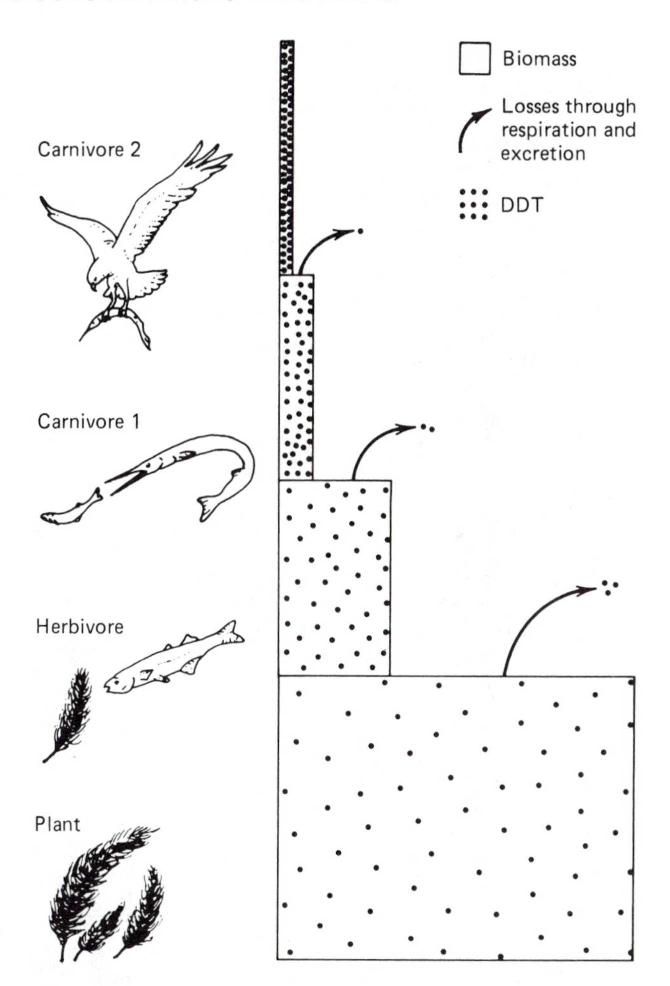

CONCENTRATION of DDT residues being passed along a simple food chain is indicated schematically in this diagram. As "biomass," or living material, is transferred from one link to another along such a chain, usually more than half of it is consumed in respiration or is excreted (*arrows*); the remainder forms new biomass. The losses of DDT residues along the chain, on the other hand, are small in proportion to the amount that is transferred from one link to the next. For this reason high concentrations occur in the carnivores.

FIGURE 12.2 Passage of DDT up a simple food chain. From "Toxic Substances and Ecological Cycles" by G. M. Woodwell. Copyright © 1967 by Scientific American Inc. All rights reserved.

site.[7] Figure 12.3 shows a pattern of ecological succession observed on abandoned upland agricultural fields in the southeastern United States. Over the first 100 years the bare field progressed through several stages to become a pine

[7]This illustrative sequence is from Weisz (1973, p. 375). He indicates that the development from sand to forest may take about 1000 years. If soil is present at the outset, the succession to forest may occur in about 200 years.

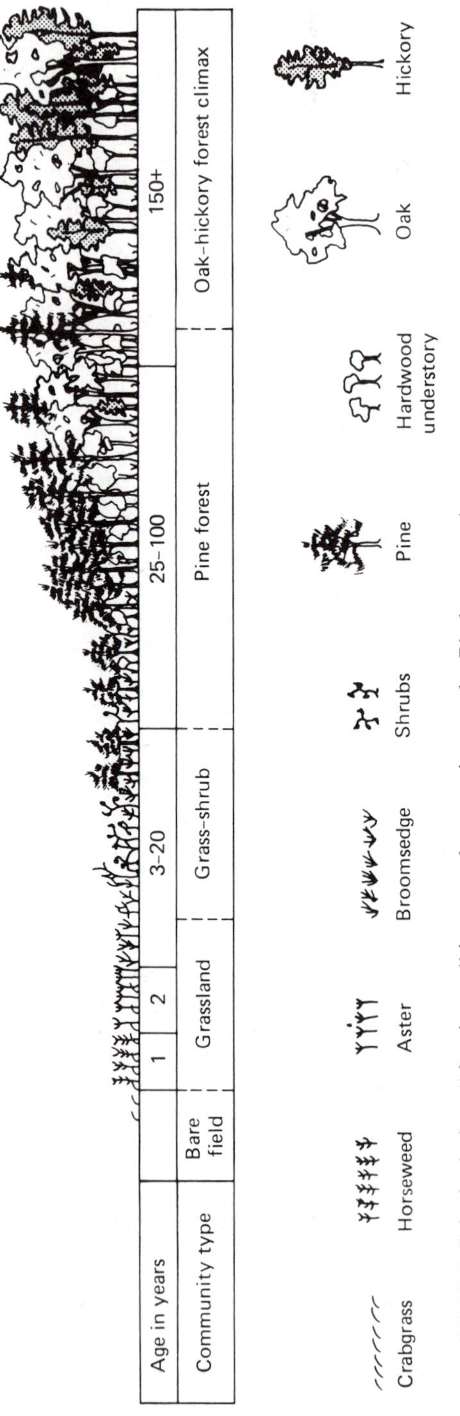

FIGURE 12.3 Principal plant "dominants" in natural succession on the Piedmont region of the southeastern United States. From *Fundamentals of Ecology*, third edition, by Eugene P. Odum. Copyright © 1971 by W. B. Saunders Co. Reprinted by permission of Holt, Rinehart and Winston, CBS College Publishing, New York.

forest. The climax community, which occurred after 150 years or so, is an oak–hickory forest. In any situation, the particular climax community that develops is determined largely by climate and other physical factors.[8]

Ecological succession is an important concept for planners since it can have an impact on land development decisions. For example, as a chaparral community matures, it relies on natural fires to burn out old individuals for replacement by new ones. Development plans for chaparral communities should consider both the risk of periodic fires and the effect that controlling fires will have on succession.

Actions taken by governments sometimes influence succession in undesirable ways. Consider, for example, the case of Lake Washington in Seattle. Wastewaters were released into the lake during the early 1900s. The discharges were stopped in the late 1930s, but started again in the early 1940s due to increased population around the lake. The phosphorus and nitrogen in the effluents spurred algal growth in the lake and the lake's succession toward becoming a marshland was proceeding much faster than would be expected in an undisturbed lake. Fortunately for the residents near Lake Washington, this rapid succession has been reduced by the diversion of wastewater to another location.[9]

CRITERIA USED IN ECOLOGICAL EVALUATIONS

The application of biological concepts in formulating physical development proposals requires that biologists articulate their science in a form useful to planners. For a planner, whether or not the biological productivity of an area increases as a result of a proposed action is meaningless. The planner needs to know whether higher values of productivity, a purely technical measure, are "good or bad." This places biologists in a difficult position. On the one hand, if a biologist insists on maintaining a purely scientific attitude and reporting only objective measures like biological productivity, then planners and decision makers may not know how to interpret biological findings, and the biologist's contributions to decision making will be diminished. On the other hand, if a biologist makes statements that planners can comprehend (for example, decreased biological productivity is an adverse consequence from a societal perspective), then he or she is open to criticism by peers who argue that such observations are value judgments that have no scientific basis.[10]

Increasing numbers of biologists are working on improving the interface between biological science and physical planning. The emerging literature on this

[8]A climax community is often viewed as being self-perpetuating and in equilibrium with its physical environment. However, the climax concept is not as simple as it appears. Biologists have developed numerous concepts to explain the differences in climax communities observed in neighboring locations. For an introduction to this complex subject, see Odum (1971, pp. 264–267).

[9]The Lake Washington example is from Krebs (1972, pp. 548–551).

[10]For an expanded discussion of this point, see van der Ploeg and Vlijm (1978).

subject recognizes that value judgments must be made to appraise the biological changes associated with proposed actions. There is no consensus on evaluative criteria in this literature. However, the following topics pertinent to evaluating biological impacts appear frequently: (1) species diversity and its hypothesized linkage with ecosystem stability, (2) species and habitat rarity, (3) criteria for identifying lands suitable for inclusion in preservation programs, and (4) criteria for judging the ability of habitats to support particular fish and wildlife species.

Species Diversity and Ecosystem Stability

Although most biologists are not prepared to argue that communities characterized by high species diversity are necessarily "good," the concept of diversity has been used extensively in evaluating ecosystems.[11] The meaning of species diversity is explained with an example. Consider three unrelated communities occupying different areas. Community A has only two species, whereas communities B and C each have four. As indicated in Table 12.1, the total number of "individuals" is the same for each community. The diversity of species in community A is clearly less than that in either of the others.

One of the many indexes of species diversity in the literature is defined as

$$D = \frac{N(N-1)}{n_1(n_1-1) + n_2(n_2-1) + \cdots + n_s(n_s-1)}$$

where

D = Simpson's index of diversity
s = number of species
n_i = number of individuals of species i ($i = 1, \ldots, s$)
N = total number of individuals of all species

TABLE 12.1 Number of Individuals in Three Hypothetical Communities

	Community		
	A	B	C
Species 1	80	60	25
2	20	30	25
3	0	5	25
4	0	5	25
Number of individuals (N)	100	100	100
Simpson's diversity index (D)	1.48	2.22	4.13

[11] Van der Ploeg and Vlijm (1978) provide numerous case examples demonstrating how diversity is used in ecosystem evaluation.

Simpson's measure of diversity is sensitive to both the number of species and their relative abundance. The lowest possible value, $D = 1$, occurs when there is only one species in a community. If each individual in a community belongs to a different species, D takes on an infinite value.[12]

Table 12.1 demonstrates how Simpson's measure is influenced by both the number of species and their distribution. Community B has more species than community A, and this is reflected in the index values. Communities B and C have the same number of species, but the distribution of species within C is more uniform. Using Simpson's index, the diversity of community C is greater than that of B.

Biologists often consider species diversity in evaluating ecosystems because it has been statistically correlated with *stability,* the ability of an ecosystem to return to equilibrium after stress. A possible explanation for this correlation is that in ecosystems with high diversity the interrelationships among species are complex and there are many alternate mechanisms for adjusting to stress. Despite the substantial data showing correlations between stability and diversity, there is much debate as to whether or not a cause–effect relation exists: Does the existence of high species diversity bring about stability? Even though relationships between diversity and stability must be viewed as tentative hypotheses, species diversity receives considerable attention in biological impact assessments.[13]

Rare Species and Habitats

"Rarity" is related to species diversity. However, rarity itself has come to occupy a singularly important position as a criterion for evaluating biological impacts. One reason to preserve rare species and habitats concerns the ability of ecosystems to respond successfully to the changes in climate and other physical characteristics that inevitably occur over long periods. The more "genetic diversity" there is among species and the more genetic patterns ("genotypes") there are within species, the greater the likelihood that new species able to withstand the rigors of a changing environment will evolve. Another reason for preserving rare species is that scientists cannot tell which of the existing millions of plant and animal species will be of practical value in the future. There are numerous studies documenting advances in medicine, agriculture, and industry

[12]Simpson's index can be clarified by considering an experiment in which two individuals are drawn at random from a community. The value of D represents the number of randomly selected pairs that "must be drawn from the population to have at least a fifty percent chance of obtaining a pair with both individuals of the same species" (Collier et al., 1973, p. 343). An alternate form of Simpson's index consists of the probability that a random selection of two individuals will be from different species. Pielou (1969, p. 223) presents this variation as well as the widely used Shannon-Wiener diversity index derived from information theory.

[13]For a discussion of the hypothesized relationships between stability and diversity and their applications in planning, see Dearden (1978). A related concept, *ecosystem resilience,* is explained by Holling (1978, pp. 30–33).

that occurred only because a wide variety of genetic information was available.[14] A case involving the melon industry in the United States is illustrative. As reported by Timothy (1972), threats to the industry caused by mildew led to a worldwide search for species of melons resistant to mildew. After extensive crossbreeding, the mildew problem was considered solved. Shortly thereafter a virus attack threatened the melon industry. A worldwide search was conducted again, and a program of crossbreeding to obtain virus-resistant strains of melons was successfully implemented. The existence of many species of melon, each with a different genetic composition, made it possible to make exchanges of genetic information that maintained the industry.

Rare habitats are also valued for recreational purposes. (Wilderness areas provide an example.) In addition, their scientific and educational worth is well established. Among the less direct, but no less important, reasons for preserving rare species and habitats is the role they sometimes play in maintaining the integrity of ecosystems. Although eliminating one particular species may not have serious impacts, it is often impossible to predict what the consequences will be. The more species are eliminated, the greater the chance that the functioning of ecosystems will be impaired significantly. To put this point in context, note that of all the extinctions known to have taken place during the last 2000 years, more than half have occurred during the past 60 years.[15] Moreover, it has been estimated that "between half a million and 2 million species—15 to 20% of all species on earth—could be extinguished by 2000, mainly because of loss of wild habitat, but also in part because of pollution."[16]

Although there are numerous practical arguments for protecting rare species and habitats, many people feel that species and habitats should be preserved simply because they exist. As elaborated in Chapter 1, the ethical view that rights should be granted to plants and nonhuman animals is one that has a rich intellectual heritage.

In the United States, the goal of preserving rare species has been translated into legislation. The Endangered Species Act of 1973 established a program protecting habitats of "endangered" or "threatened" species. Because of this act, all environmental assessments under NEPA must determine whether areas to be affected by a proposed action contain habitats of species declared endangered or threatened.[17] In addition to the Endangered Species Act, there are a number of federal and state programs that either protect individual species or preserve particular ecosystems.[18]

[14]Ehrlich and Ehrlich (1981) provide extensive documentation of the practical value of preserving rare species and habitats.

[15]This observation is from Turk (1980, p. 69).

[16]The quotation is from the *Global 2000 Report to the President of the U.S.*, edited by Barney (1980, p. 37).

[17]Lists of such species are published periodically in the *Federal Register;* in addition, the U.S. Fish and Wildlife Service identifies specific areas known to contain "critical habitats." Much pertinent information is summarized by Golden et al. (1979).

[18]Several of these programs are outlined by Camougis (1981, pp. 8–14).

Ranking Natural Areas for Preservation

The usefulness of preserving natural areas for scientific and educational purposes is widely recognized. A growing volume of literature relates the experiences of biologists from many countries in ranking natural areas in terms of their value as preserved lands.[19] Typically, the criteria used for ranking include the size of the area involved, its availability for purchase, its utility for educational purposes, and the extent to which the area is threatened by pressures for development. However, the criteria of concern here are the *biological* characteristics used in deciding whether to preserve an area. These factors provide further insights into how biologists evaluate ecosystems.

A survey of techniques for ranking natural areas reveals several biological characteristics that are frequently employed. Tans' (1974) description of a ranking scheme used by Wisconsin's Scientific Area Preservation Council is illustrative. The Wisconsin approach includes biological criteria for "quality, commonness and community diversity." An area receives a numerical score for each criterion, and the sum of the scores represents a measure of its "natural area value." It is used with several other measures (which are *not* combined into a single index) in ranking different areas for preservation purposes.

In the Wisconsin scheme, *quality* refers to the "excellence of an area's main features," and is measured in terms of

> *(1) Diversity of native plant or animal species present . . . ; (2) plant community structure and integrity; (3) the extent of significant human interference (disturbance) to the community. . . . [and](4) the extent to which a community corresponds with . . . [the Scientific Area Preservation Council's] concept of the identified natural community as it existed before settlement.*[20]

The highest quality rating (10 on a scale from 1 to 10) is given to areas which contain no visible disturbances and satisfy the above-noted conception of excellence.

The second criterion, "commonness," is rated on a scale from 1 to 6. To receive the highest score an area must contain two or more "rare or endangered species" or be the only known location of a nonbotanical natural feature. An area receives the lowest score for commonness if it contains communities that occur frequently in the Wisconsin landscape.

"Community diversity" is the third criterion for determining natural area value. Using a 5-point scale, the highest diversity scores go to areas containing four or more different plant communities or other natural features. The lowest scores are for areas with only a single type of plant community.

There are many schemes for ranking natural areas, and they employ a variety of terms and criteria. Rarity and diversity are among the most widely used.

[19]The British journal, *Biological Conservation,* contains many contributions to this literature.

[20]This quotation is from Tans (1974, p. 35).

Numerous indexes are available to characterize diversity. Rarity is measured either by the existence of rare or endangered species or by the presence of a rare or unique habitat. Van der Ploeg and Vlijm (1978) review the several rarity and diversity measures used in 10 ecological evaluations in the Netherlands.

Procedures for Habitat Evaluation

Additional insights into evaluative criteria are provided by examining the Habitat Evaluation Procedure (HEP) developed by the U.S. Fish and Wildlife Service (1980) for assessing water projects.[21] The procedure assumes that the suitability of a habitat for a particular animal species can be determined by analyzing an area's vegetative features along with various physical and chemical characteristics. Habitat suitability is linked to the biologists' conception of carrying capacity, the maximum number of individuals of a given species that can be supported by an area.[22] HEP evaluates how changes in habitat conditions influence an area's *potential* to support a species. Providing a suitable habitat does not, in itself, guarantee that the species will develop at maximum potential density levels.

HEP begins by defining the study area and dividing it into homogeneous "cover types." These are either terrestrial communities such as grassland and evergreen shrubland or aquatic zones with similar chemical and physical properties. "Evaluative species" are then selected based on economic and ecological considerations. For example, a species may be chosen because it is highly valued for recreational fishing or hunting or because it plays a key role in the functioning of an ecosystem. Assessments may include several different species. Consideration is given to "baseline conditions"—habitat characteristics in a given area *prior* to any change in land or water use, and future conditions—the most likely future habitat expected to exist under a particular proposal. In addition to alternative development projects, HEP also considers the no-action plan.

In the remaining HEP steps, the baseline and projected habitats are described in terms of *habitat units* (HUs), calculated as a land area times a habitat suitability index (HSI). The computation of an HSI is based on a delineation of habitat conditions (demonstrated below) and the concept of an *optimum habitat.* The latter is defined as the habitat that supports the maximum density of an evaluative species. Several HSI values are calculated. For any particular species, the HSI for the baseline case equals the baseline habitat condition divided by the optimum condition. A similar calculation determines HSI values for various future habitat conditions. An HSI of 1 indicates the habitat is ideal for the species, whereas a value of 0 means it is totally unsuitable.

A Fish and Wildlife Service (1981) analysis demonstrates how to compute habitat suitability indexes and habitat units using the red-tailed hawk as an

[21]Virginia Rath assisted in preparing early versions of this discussion of habitat evaluation methods in 1980 while she was a biology student at Stanford University.

[22]The application of carrying capacity in land planning (as discussed in Chapter 10) is an adaptation of its much earlier use in biology.

evaluative species. Although the agency's study included two cover types, only the grasslands portion of its analysis is considered here. Using available scientific information, the Fish and Wildlife Service concluded that three variables were of central importance in determining the suitability of grasslands as a habitat for the red-tailed hawk:

V_1 = percentage herbaceous canopy cover

V_2 = percentage herbaceous canopy 3 to 18 in. tall

V_3 = number of trees greater than or equal to 10 in. dbh/acre[23]

For the baseline case, values of V_1, V_2, and V_3 in the grasslands portion of the study area were determined from field data. Professional judgment was used to estimate these values under various future conditions.

A suitability index was calculated for each of the variables using relationships derived by the Fish and Wildlife Service and shown in Figure 12.4. For example, if $V_1 = 25\%$, the curve in Figure 12.4*a* indicates a suitability index of 0.39. The computation of the *overall* HSI for the red-tailed hawk in the study area employed mathematical formulas involving suitability indexes for V_1, V_2, and V_3 and for variables describing the second cover type. These formulas were based on scientific facts and professional judgments and are described by the Fish and Wildlife Service (1981). Once the overall HSI was computed, it was multiplied by the study area to yield the number of habitat units. The entire computational procedure is repeated for each of the several scenarios describing future habitat conditions.

Habitat units are used to assess alternative water plans. A proposed project's impact on a particular evaluative species is adverse if the estimated HUs with the project are less than the HUs without the project. Differences in habitat units with and without a project are also used in evaluating "mitigation features" to offset negative impacts. For example, suppose a proposed Bureau of Reclamation reservoir was expected to produce a loss of 100 HUs for the red-tailed hawk. The bureau might try to compensate for this loss by managing a different land area containing a similar cover type. Its goal would be to provide 100 new HUs (in terms of habitat for the red-tailed hawk) by actions such as selective burning and timber cutting.

Weaknesses of the Habitat Evaluation Procedure are its narrow species orientation and its failure to examine species diversity and ecosystem structure and function. In addition, many mathematical relationships used in HEP give the impression of being more scientifically rigorous than they are. For example, index curves such as those in Figure 12-4 suggest that habitat suitability relationships are understood precisely. In reality, suitability index curves often represent qualitative judgments by fish and wildlife specialists. Despite these limitations, HEP provides a way of organizing scientific information concerning habitat suitability and has been used widely in federal water resources planning.

[23]*dbh* is defined as "diameter at breast height" (that is, 132 cm above the ground).

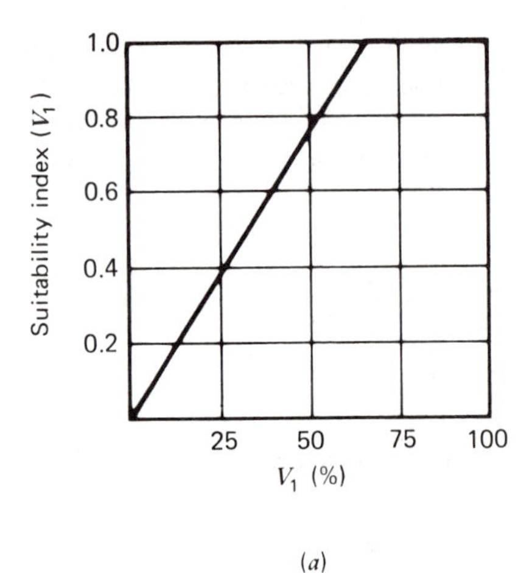

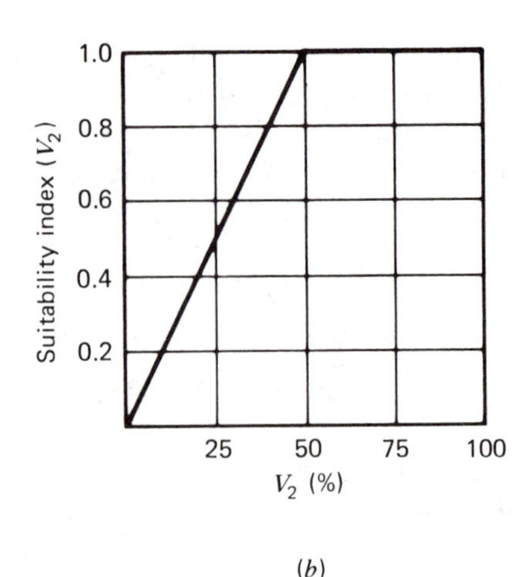

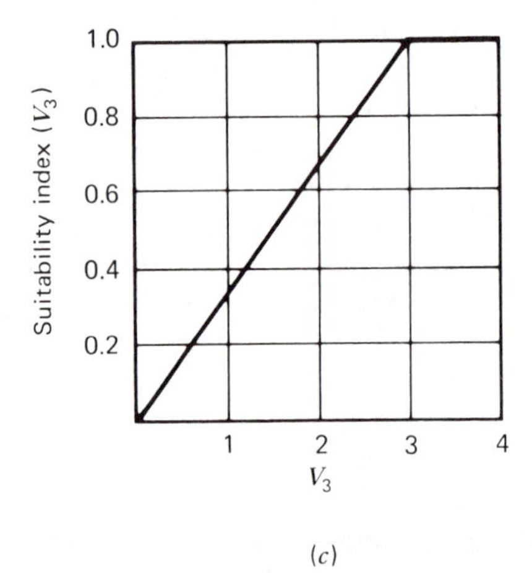

FIGURE 12.4 Suitability index curves for red-tailed hawk example. From U.S. Fish and Wildlife Service (1981). (*a*) Percentage herbaceous canopy cover. (*b*) Percentage herbaceous canopy 3 to 18 in. tall. (*c*) Number of trees $\geq$ 10 in. dbh/acre.

ACTIVITIES UNDERTAKEN IN BIOLOGICAL ASSESSMENTS

Biological impact assessment procedures vary greatly depending on both the preferences of individual biologists and the context in which the work is carried out. One reason there is no standardized process is that biological systems are incompletely understood. In addition, many of the available methods have been developed only recently.[24] In the United States, for example, biologists have been conducting impact assessments routinely only since the passage of the National Environmental Policy Act of 1969.

[24]Holling (1978) reviews widely used approaches to biological impact assessment. He also introduces an adaptive strategy that views environmental assessment as a process that continues while a project is being designed, implemented, and used.

Although detailed procedures vary, a general approach to assessing biological impacts of proposed actions can be identified. It consists of six activities that are often carried out in an iterative, as opposed to sequential, fashion: (1) describe the proposed action and alternatives, (2) identify key biological concerns, (3) inventory the study area's ecosystem, (4) forecast biological impacts, (5) evaluate projected effects, and (6) influence the decision process.

Identifying Details of Proposed Actions

Obtaining an adequate description of a proposed action is not always simple. Information of central importance from a biological perspective may be viewed by others as minor details to be settled in the late stages of project design. For example, suppose a proposed dam requires that an existing roadway be relocated. To the engineers planning the dam, the key issues are the location of the dam and the storage capacity of the reservoir. The road relocation may not be finalized until late in the planning process. To a biologist, however, the impacts of the new roadway (especially if the road is routed through undeveloped land) may be as important as the effects of the dam and reservoir.

In general, biologists are often as concerned about ancillary facilities and construction methods as they are about major project features. For example, in constructing an electrical transmission line, the greatest biological impacts are commonly associated with access roads and not the lines themselves. In building a highway, procedures for making road cuts and conducting grading can significantly influence biological communities. The details associated with ancillary facilities and construction operations are often overlooked in preliminary project proposals, and thus they are of special concern.

Identifying Issues of Biological Concern

Because of stringent time and budget limits, biologists participating in an impact assessment must focus on a limited number of key issues. As explained in Chapter 8, important issues are usually identified on the basis of three sources: institutions, community interactions, and technical and scientific judgments.

Information from institutional sources is contained in laws, plans, and policy statements indicating the importance of specific biological resources. Examples are the lists of endangered and threatened species issued by federal agencies in response to the Endangered Species Act of 1973. Other illustrations are the federal and state laws protecting special types of ecosystems such as wetlands.

Community interaction often includes meetings with local officials, conservation group representatives, and fish and wildlife agency personnel to identify significant impact assessment issues. These meetings may also reveal the attitudes of citizens toward aspects of the local environment. For example, a particular fish species may be specially valued by local anglers.

The scientific judgments of biologists play a key role in determining issues to be examined. For example, biologists may deem it important to assess impacts on species that are neither protected by law nor valued by the local populace.

This is illustrated in a biological assessment of the influence of mosquito control chemicals on salt marsh ecosystems in New Jersey. Ward (1978) indicated that two species of grasses were among the organisms the biologists chose to study carefully. In their judgment, adverse effects on these particular grasses would interfere significantly with the functioning of the entire salt marsh ecosystem.

Preparing Inventories to Characterize Existing Biological Systems

Biologists often rely heavily on published information to characterize the area affected by a proposed action. Time and budget constraints frequently do not permit the detailed field surveys that many biologists would like to conduct. De Santo, a biologist with considerable experience in impact assessment, provides the following "rule of caution" for inventory preparation:

> *Do not give free reign to data collection. Anyone responsible for the collection and/or interpretation of data, regardless of its nature, must understand why the data is needed and what level of comprehensiveness and detail is required. In the absence of careful control over this function, great quantities of time, let alone budget, can be needlessly spent with no useful conclusion resulting from the work.*[25]

Standard manuals that list species found in different communities are typically consulted in describing existing conditions.[26] Publications documenting past biological investigations of the study area are also useful. These include reports by local fish and game agencies, native plant societies, and high school biology classes. Inventory information can also be obtained by interviewing local experts, such as bird watchers and biology teachers who have a long-term familiarity with the area.

Many agencies routinely collect data useful for inventory preparation. For example, the National Cartographic Information Center of the U.S. Geological Survey annually publishes a series of "land use and land cover maps" based on photographs from orbiting satellites. In addition, the U.S. Department of Agriculture has aerial photos useful for determining vegetation, soil types, and other land features.[27]

Field trips to the site of a proposed project are another source of inventory information. For small, routine projects, a few days are typically spent in the study area observing significant habitats and natural processes. If data from the literature and local experts are inadequate, a more comprehensive field investigation may be required. This might involve systematically identifying com-

[25]This quotation is from De Santo (1978, p. 134).

[26]For an introduction to literature useful in preparing inventories, see the references cited by Camougis (1981).

[27]The U.S. Department of Agriculture's Aerial Photography Field Office in Salt Lake City, Utah, provides details on the availability of aerial photos. Information on the many data gathering activities of the U.S. Department of Interior is available from its Office of Library and Information Services in Washington, D.C.

munity types, endangered species, and fish and wildlife habitats, and studying structural and functional characteristics of the local ecosystem. Field surveys may also be needed to interpret data obtained using remote sensing techniques such as infrared aerial photography.[28]

Forecasting Biological Effects

Biologists, like other environmental specialists, often rely on professional judgment in forecasting impacts of a proposed action. In these instances, projections are based on the expert's prior experience, including formal education and knowledge of pertinent scientific literature. Often biologists collaborate and make forecasts based on group judgments.

Ward (1978) has argued that biologists can greatly improve their forecasting abilities by going beyond intuitions and judgments and relying on systematic observations of ecosystems that have been manipulated. This position is illustrated by the alternative methods that Ward and her colleagues considered in predicting how a mosquito insecticide (tempephos) would influence salt marsh ecosystems in New Jersey. The following forecasting strategies were considered:

Comparative Analysis Find salt marshes treated with insecticides in the past, and see how they differ from untreated marshes.

Monitoring Approach Study salt marshes that the New Jersey Mosquito Control Commission had already decided to treat with insecticides, and compare them to untreated marshes.

Limited Experiment Apply experimental insecticide doses to a small section of an existing salt marsh, and contrast it with the larger untreated section of that marsh.

In this case, the limited experiment strategy was used to make predictions.

Mathematical models of ecosystems provide another approach to forecasting. However, these models are not yet developed to where they can be used in routine planning studies. One reason is that many ecosystems are not sufficiently well understood to be adequately represented by equations. In addition, numerous biologists performing impact assessments have neither experience nor training in the use of ecosystem models. Another difficulty is that enormous amounts of field data are often needed to estimate model coefficients and properly test the models. In many instances, ecosystem model forecasts have not been compared systematically with observed outcomes. This type of model testing can extend over decades. In the future, as biologists gain more experience with modeling, mathematical representations of ecosystems may play a greater role in impact assessment.[29]

[28]Standard field methods used in biological impact assessments are reviewed by De Santo (1978) and Ward (1978).

[29]Although mathematical models were used in ecology during the 1920s, widespread interest in ecosystem modeling occurred only in the 1960s. Hall and Day (1977) demonstrate how mathematical models have been used in forecasting biological impacts of government decisions.

Evaluating Biological Effects

Evaluative factors are defined in Chapter 8 as the goals criteria and constraints that decision makers and segments of the public consider important in ranking alternative proposals. They are determined using the above-mentioned sources of evaluative information: institutions, community interactions, and technical and scientific judgments.

Although scientific judgments are a source of evaluative factors, many biologists feel uncomfortable in making *value* judgments about the societal significance of changes in ecosystem characteristics. For example, a biologist's training may highlight the adverse impact of overgrazing on the long-term biological productivity of livestock pastures. However, biological science does not address whether a decrease in productivity is good or bad from a societal perspective.

Although many biologists have chosen not to argue that a change in a biological measure like productivity is good or bad, some have taken stands on social issues. As previously noted, these positions often support the preservation of rare species and unique habitats, and the maintenance of high levels of species and habitat diversity.

Influencing the Decision Process

Biological impact assessments can influence project planning in several ways. Results from assessments made in the early stages of planning can suggest completely new alternatives. If an assessment is conducted after much of the planning budget has been spent, the influence of the biological analyses may be limited to designing new project features that reduce the adverse effects of the original proposal ("mitigations"). Consider, for example, a proposed reservoir that would ruin sport fisheries and wildlife habitats. Partial mitigations may be attained by providing fish hatcheries and purchasing and preserving habitats similar to the ones to be destroyed. Some mitigation features involve construction procedures, such as measures to control erosion during the building process and to revegetate impacted areas after a project is erected. Efforts to reduce adverse impacts may also be based on project operations such as special handling practices at oil terminals that reduce the likelihood of oil spills.

BIOLOGICAL ASSESSMENTS IN PRACTICE: AN ENERGY FACILITY PLANNING STUDY

The framework for impact assessment outlined above is clarified using an energy planning example.[30] In 1973, a consortium of utilities called the Western LNG Terminal Association proposed to construct a liquified natural gas (LNG) ter-

[30]This discussion relies on an unpublished study by Christopher Slaboszewicz while he was a Stanford University student during the 1979–1980 academic year. The information was developed largely from Mr. Slaboszewicz's interviews of Thomas Reid and Karen Weissman of Thomas Reid Associates, Palo Alto, California.

minal at Point Conception along the California coast near Santa Barbara (see Figure 12.5). The utilities engaged an engineering firm, Dames and Moore, to perform biological studies and prepare an impact assessment. Dames and Moore conducted extensive field investigations, including fish and plankton sampling, and prepared a species inventory and a report analyzing the field data.

Before the LNG terminal could be constructed, the approval of the California Public Utilities Commission (PUC) had to be obtained. The PUC engaged a firm of biological consultants, Thomas Reid Associates, to provide an independent assessment. After reviewing the data assembled by Dames and Moore, the PUC's consultants concluded that the majority of significant biological impacts would involve marine organisms.

Thomas Reid Associates decided that the impacts associated with the proposed LNG project could be examined in terms of several project components including a small boat harbor and the associated dredging to maintain the harbor, a pier and trestle for tankers, and a "kelp exclusion zone" around the LNG terminal. The latter consisted of an area where the kelp harvesters would not be permitted. The consultants were also concerned about possible spills of LNG, boat fuel, and other chemicals. A seawater intake, designed to draw in 160,000 gal of water/min, was singled out as the component likely to cause the most

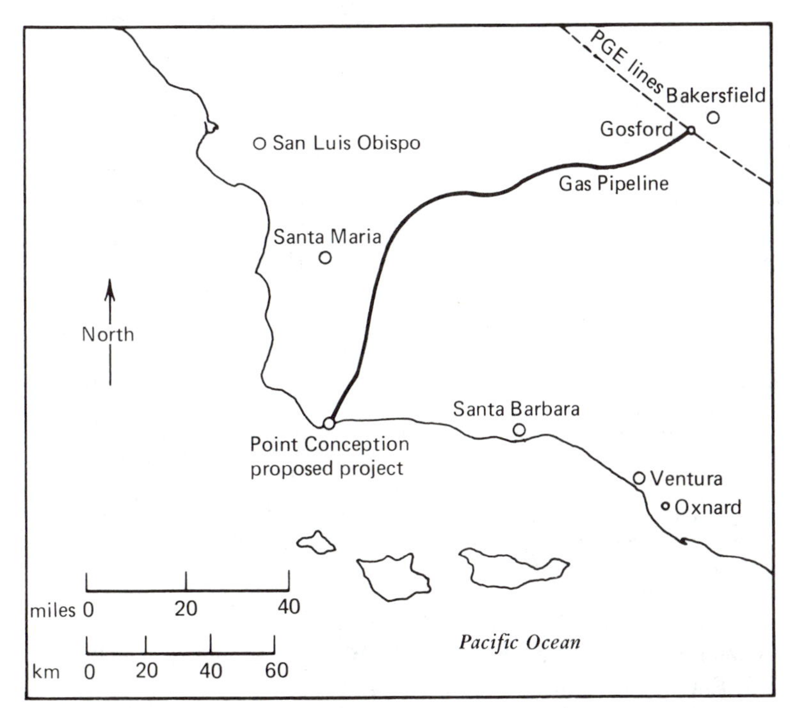

FIGURE 12.5 Location of proposed LNG terminal at Point Conception, California. From Thomas Reid Associates, 1978, *LNG Vaporizor Seawater System*, Technical Report No. 26 in support of the Point Conception LNG Facility DEIR, report to the California Public Utilities Commission, San Francisco, Calif.

significant biological impacts (see Figure 12.6). The seawater was to provide heat for vaporizing the liquified natural gas. Officials of the California Department of Fish and Game had expressed particular concern about the possible entrainment of plankton and fish in the seawater intake. In addition, the fish and kelp that would be affected were considered to be economically important resources.

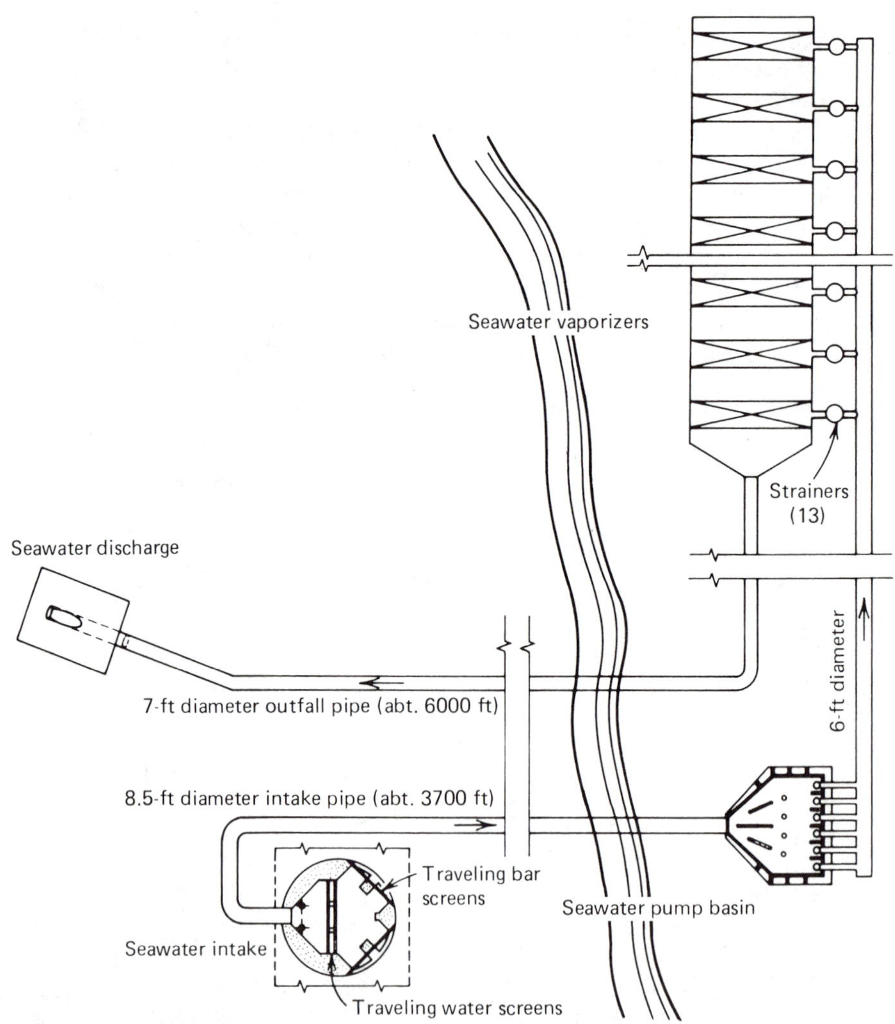

FIGURE 12.6 Seawater intake and discharge system for proposed LNG terminal. From Singh, D., T. S. Reid, and K. G. Weissman, 1979, *An Evaluation of Western LNG Terminal Associates' Proposed Seawater System submitted in Compliance with Condition No. 4 of the California Public Utilities Commission Decision No. 89177 for an LNG Facility at Point Conception,* report prepared for the California Public Utilities Commission LNG Task Force, San Francisco, Calif.

The consultants examined both the construction and the operation of each of the project components. During the construction phase, blasting, pile driving, and kelp removal were each felt to have potentially significant effects on biological resources. Impacts related to the LNG terminal's operations centered on the seawater intake and discharge system. Other impacts associated with operations concerned the ecosystem of the kelp exclusion area. The PUC's consultants acquired detailed information on various project components, especially the seawater intake and discharge system. For example, they determined the velocity of the seawater at the ocean end of the intake pipe and the proposed method of dealing with fish inadvertently drawn into the pipe.

The next stage of the biological impact analysis was inventory preparation. The earlier studies by Dames and Moore, which included surveys of fish and plankton populations, provided much of the data needed to characterize the local ecosystem. Experts at the Scripps Institution of Oceanography and the National Marine Fisheries Service provided supplementary information regarding the plankton community. This included details on the life cycles of rare and endangered species in the area and the reproductive capacities of various planktonic species.

The impact forecasting process was guided by an analysis of biological effects of similar seawater intakes at other facilities. Much pertinent data were available because EPA had required several nearby power plants with seawater intakes to collect data on the amount and species composition of fish drawn into the intakes. Using these data, Thomas Reid Associates predicted that 40,000 lb of fish would be entrained annually. They also gave a species breakdown of the fish. The consortium of utilities had proposed the following fish return system: the fish that were sucked in through the seawater intake pipe were to be isolated in a collection tank, transported up an elevator to another pipe, and then propelled back out to sea after an 8400-ft round trip. The PUC's consultants concluded that this system would probably wind up killing many of the entrained fish.

In addition to finding the fish return system unacceptable, Thomas Reid Associates felt that the discharge of chlorine, used to sterilize the seawater to prevent fouling of the vaporizors (see Figure 12.6), might have significant impacts on ocean plant and animal communities. The consultants also used a mathematical model to assess the cooling effect of the discharge water, which would be as much as 12°F colder than the ambient ocean temperature. Based on the model, and the fact that the ocean temperature fluctuated up to 9°F as a result of natural currents, no significant impact on ambient temperature was predicted. The total biomass of the plankton that would be killed by entrainment in the seawater intake system was also estimated. This was done using data from the California Cooperative Fish Investigation Studies performed by several universities and government agencies. The overall effect on plankton productivity was then predicted. The plankton loss was not considered significant since the amount killed would be a very small fraction of the total local plankton population and because planktonic species are able to reproduce rapidly. The

PUC's consultants predicted no impact on rare and endangered species, since the rare and endangered species in the area did not have a planktonic stage in their life cycles. They also foresaw no major construction-related impacts if construction procedures were modified to use techniques other than blasting.

Another part of the impact assessment focused on measures to mitigate the negative effects of the seawater intake and discharge scheme and the discharge of chlorine. A system of three offshore caissons, eventually modified to one, was proposed to cover the ocean end of the intake pipe. It was designed to reduce annual fish entrainment from 40,000 to 200–300 lb. At the request of the PUC, the consultants also examined alternatives to chlorination, including the use of ozone, manual cleaning, and a newly invented antifoul rubber coating to protect the vaporizers from being damaged by impurities in the seawater. Eventually, chlorination was recommended, because the alternatives were either too costly or unreliable. Because of concerns expressed by the California Fish and Game Department regarding plankton entrainment in the seawater intake system, additional mitigation measures were examined. The PUC's consultants looked at artificial filter beds, fine-mesh screens, and other techniques to prevent plankton entrainment. They felt that the impact on saving the plankton would be only minimal and that these techniques were neither effective nor worth the cost.

The LNG example demonstrates each of the steps mentioned in the previous discussion of approaches to biological impact assessment. In this case, much project planning had been completed before the PUC's consultants did their study. Despite this, the PUC gave its consultants a mandate to consider a wide range of alternative seawater systems. As with other environmental experts, biologists should be involved at the earliest stages of project planning. As they learn more about the context in which facility planning studies are carried out, biologists will become increasingly effective in contributing to the initial formulation of alternative plans.

KEY CONCEPTS AND TERMS

BIOLOGICAL CONCEPTS USEFUL IN PLANNING
- Biotic communities and habitats
- Abiotic elements
- Predator–prey relations
- Food chains
- Biological productivity
- Biological magnification
- Ecological succession
- Climax communities

CRITERIA USED IN ECOLOGICAL EVALUATIONS
- Species diversity
- Ecosystem stability
- Species and habitat rarity
- Endangered or threatened species
- Genetic diversity
- Carrying capacity
- Habitat evaluation procedure
- Habitat suitability index

ACTIVITIES UNDERTAKEN IN BIOLOGICAL ASSESSMENTS
- Principal versus ancillary project features
- Controlled ecosystem experiments
- Ecosystem modeling
- Biological inventory preparation
- Impacts of analogous projects
- Sources of evaluative criteria
- Mitigation measures

DISCUSSION QUESTIONS

12-1 Pesticide chemicals have been found in nearly all kinds of foods. Explain why they are found in foods such as dairy products which are not subject to direct applications of pesticides.

12-2 In 1973, a University of Tennessee biologist surveyed the fish populations of the Little Tennessee River and discovered a previously unknown member of the perch family, the Snail Darter. The habitat of this rare, 3-in. fish was just above the site of the Tennessee Valley Authority's proposed Tellico Dam. As explained by Ehrlich and Ehrlich (1981), the dam was about 80% complete by the time the courts began to consider whether construction should be halted to preserve the Snail Darter's habitat. What arguments do you think were raised by those trying to save the Snail Darter? After a series of court challenges the dam was eventually completed. Speculate as to why the proponents of the dam were able to implement their plans.

12-3 Suppose the U.S. Fish and Wildlife Service's Habitat Evaluation Procedure is used to assess the impacts of a proposed reservoir on habitat for deer. Discuss three potentially significant biological impacts of the reservoir that would *not* be considered if HEP were the only assessment technique employed.

12-4 Ward (1978, p. 55) offers the following observations regarding the use of ecosystem modeling as a forecasting procedure:

[M]odels often fail to predict the measured system responses. In such cases, the model is frequently useful in pointing out errors in the concepts used to develop the model. New or altered models can then be constructed. . . . A computerized mathematical model can be used to investigate the possible consequences of many options rapidly. . . . Thus mathematical models have many advantages that should be considered

regardless of the possibility of predictive failure (inaccuracy). In environmental impact analysis, precision of systems description and exploration of options, as well as predictive power, are very important.

Discuss the extent to which you agree with Ward's position. Does it encourage you to use ecosystem models in forecasting impacts?

12-5 Imagine you are a planner for a state transportation agency. Suppose you are in charge of selecting a consultant to perform a biological impact assessment for a proposed highway, part of which traverses an undisturbed forest. List three questions you would ask of biological consultants seeking to carry out the assessment. What types of answers would you be looking for?

REFERENCES

Barney, G. O., 1980, *The Global 2000 Report to the President of the U.S.* Pergamon, New York.

Camougis, G., 1981, *Environmental Biology for Engineers.* McGraw–Hill, New York.

Collier, B. D., G. W. Cox, A. W. Johnson, and P. C. Miller, 1973, *Dynamic Ecology.* Prentice–Hall, Englewood Cliffs, N.J.

Dearden, P., 1978, The Ecological Component in Land Use Planning: A Conceptual Framework. *Biological Conservation* **14** (3), 167–179.

De Santo, R., 1978, *Concepts of Applied Ecology.* Springer-Verlag, New York.

Edington, J. M., and M. A. Edington, 1978, *Ecology and Environmental Planning.* A Halsted Book, Wiley, New York.

Ehrlich, P. R., and A. H. Ehrlich, 1981, *Extinction, the Causes and Consequences of the Disappearance of Species.* Random House, New York.

Golden, J., R. P. Ouellette, S. Saari and P. N. Cheremisinoff, 1979, *Environmental Impact Data Book.* Ann Arbor Science Publishers, Ann Arbor, Mich.

Hall, C. A. S. and Day, J. W., Jr., 1977, *Ecosystem Modeling in Theory and Practice: An Introduction with Case Histories.* Wiley, New York.

Holling, C. S. (ed.), 1978, *Adaptive Environmental Assessment and Management.* Wiley, Chichester, England.

Krebs, C. J., 1972, *Ecology: The Experimental Analysis of Distribution and Abundance.* Harper & Row, New York.

Leopold, A., 1949, *A Sand County Almanac.* Oxford University Press, New York (reissued in 1970 as a Sierra Club/Ballentine Book by Ballentine Books, New York).

Mendelssohn, H., 1972, Ecological Effects of Chemical Control of Rodents and Jackals in Israel, *in* M. T. Farvar and J. P. Milton (eds.), *The Careless Technology, Ecology and International Development.* pp. 527–544. Natural History Press, Garden City, N.Y.

Odum, E. P., 1971, *Fundamentals of Ecology,* 3rd ed. Saunders, Philadelphia.

Pielou, E. C., 1969, *An Introduction to Mathematical Ecology.* Wiley, New York.

Tans, W., 1974, Priority Ranking of Biotic Natural Areas. *The Michigan Botanist* **13,** 31–39.

Timothy, D. H., 1972, Plant Germ—Plasm Resources and Utilization, *in* M. T. Farvar and J. P. Milton (eds.), *The Careless Technology, Ecology and International Development,* pp. 631–656. Natural History Press, Garden City, N.Y.

Turk, J., 1980, *Introduction to Environmental Studies.* Saunders, Philadelphia, Pa.

U.S. Fish and Wildlife Service, 1980, *Habitat Evaluation Procedure,* Report 102 ESM, Department of the Interior, Washington, D.C.

U.S. Fish and Wildlife Service, 1981, *Standards for the Development of Habitat Suitability Index Models,* Report 103 ESM, Department of the Interior, Washington, D.C.

van der Ploeg, S. W. F., and L. Vlijm, 1978, Ecological Evaluation, Nature Conservation and Land Use Planning with Particular Reference to Methods Used in the Netherlands. *Biological Conservation* **14,** 197–221.

Ward, D. V., 1978, *Biological Environmental Impact Studies, Theory and Methods.* Academic Press, New York.

Weisz, P. B., 1973, *The Science of Zoology.* McGraw–Hill, New York.

Woodwell, G. M., 1967, Toxic Substances and Ecological Cycles. *Scientific American* **216** (3), 24–31.

CHAPTER 13

SIMULATING AND EVALUATING VISUAL QUALITIES OF THE ENVIRONMENT

Since the 1960s, a number of landscape architects and psychologists have developed procedures to assess the visual characteristics of landscapes and cityscapes. Some of the impetus for this activity took the form of regulations and laws, such as the Highway Beautification Act of 1965, calling for the consideration of "scenic values" in government decision making. The National Environmental Policy Act of 1969 also focused attention on visual aspects of the environment. The act requires the development of techniques that appropriately weigh aesthetic values in federal agency decision making.[1]

A visual impact assessment for a proposed project requires judgments about visual attributes of both the existing site area (preproject conditions) and the same area after the project is constructed (postproject conditions). Although visual assessments are sometimes made after a project has actually been built, this is *not* considered here. Instead, the concern is with visual impact assessments made during the planning stages, before any proposal is implemented. These assessments provide information to help people decide which, if any, project to build. Even though the term *postproject* is used here, it refers to the situation that would result *if* a proposed project were carried out.

Visual impact studies are often performed using perspective drawings, artist's impressions, three-dimensional models, and photography to simulate views un-

[1]Two specialists in graphic design provided many suggestions for improving early versions of this chapter: Stephen R. J. Sheppard and Patti J. Walters.

der different conditions.[2] For example, Figure 13.1 represents Westminister Cathedral in London using a sequence of perspectives drawn from several viewing locations. Other "visual simulation" techniques are illustrated later in this chapter.[3]

A simulation of postproject conditions is essentially a forecast, and the term *forecast* is generally defined to be *free* of judgments about whether the future situation is better or worse than the preproject conditions. The term *evaluation* is used when value judgments are made about which circumstances are preferred. For example, imagine two perspective sketches of a flat, open field: one sketch with and one without a proposed shopping center. If the view of the undeveloped field is judged more aesthetically pleasing than the view showing the shopping center, that judgment constitutes an evaluation.

Visual impact specialists distinguish between two bases for evaluation: evaluative appraisals and preferential judgments. Craik (1972) describes these as follows: *evaluative appraisals* are made when the quality of a specific landscape is judged against some explicit (or implicit) standard of comparison. In contrast, *preferential judgments* are those that reflect a wholly personal, subjective appreciation of (or repugnance for) specific landscapes. For example, suppose a landscape architect used "aesthetic criteria" concerning "unity" and "variety" to judge the views of the above-mentioned field with and without the shopping center. That constitutes an evaluative appraisal of the two conditions. In contrast, a local resident who feels the landscape would be more beautiful without the shopping center is making a preferential judgment. Both evaluative appraisals offered by design experts and preferential judgments of those likely to be viewing a completed project have been used in visual impact assessments.

ISSUES EXPLORED USING LANDSCAPE PREFERENCE RESEARCH

Psychologists and others have investigated how people perceive landscapes and cityscapes. "Landscape preference research" attempts to understand which factors influence a person's preference for one landscape (or cityscape) over another. If landscape attributes that positively or negatively influence an individual's preference could be identified, visual impact assessments might be made more objective.

Predicting Landscape Preferences

One approach to landscape preference research is illustrated by Shafer, Hamilton, and Schmidt's (1969) analysis of the "scenic perceptions" of Adirondack

[2]Architects and engineers frequently rely on plan, elevation, and section drawings to represent physical facilities. These drawings are more useful in providing design information than in realistically portraying the appearance of a facility, and they are not considered further.

[3]The term *simulation* is also used in describing mathematical representations of reality, such as "computer simulation models" of pollutant dispersion in the atmosphere. Since this chapter is concerned exclusively with visual impacts, the existence of multiple uses for the term *simulation* should not cause confusion.

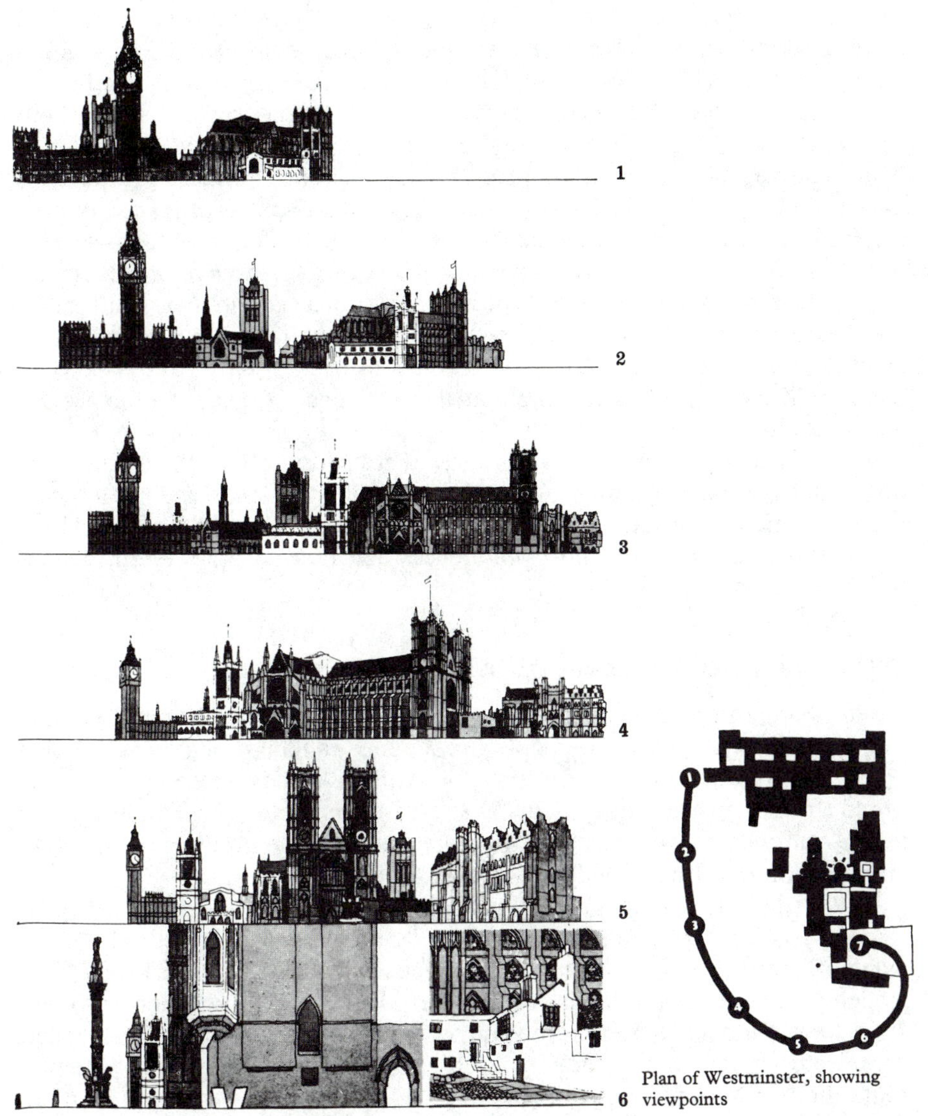

FIGURE 13.1 Sequence of perspectives of Westminister Cathedral in London. From *The Concise Townscape* by Gordon Cullen, published by Architectural Press, London. Hardback 1961; revised paperback edition 1971.

campers. A random sample of 250 campers participated in the study. Each was asked to rank his or her preferences for one hundred 8 × 10-in. black-and-white photographs of landscapes. The rank orderings of all 250 people were used to develop a "preference score" for each photo. Shafer and his colleagues tried to find a mathematical relationship between how much a photo was preferred, as indicated by the preference scores, and various landscape attributes. For each photo, they measured characteristics such as *perimeter of immediate vegetation,* defined as the perimeter of that part of the photo where tree barks and individual leaves could be easily distinguished; and *area of water,* defined as the area within the photo that included water. Shafer and his associates then performed a statistical analysis to discover relations between the preference score for each photo and combinations of the measured photo characteristics, such as perimeter of immediate vegetation and area of water. The resulting statistical relationships have been used to forecast preference scores for photographs of landscapes not included in the original 100 photos. The analysis by Shafer and his co-workers (which is one of many possible landscape preference research approaches) has been criticized for the "total lack of theoretical or even intuitive justification for the variables."[4] In other words, even though variables like perimeter of immediate vegetation may give good predictions from a statistical perspective, there is little basis for postulating that they influence people's preference for one landscape over another.

Many visual assessment specialists agree that factors such as the existence of water and natural vegetation are useful predictors of landscape preference. However, there is dispute over how these variables should be measured. There is also disagreement about how preferences are affected by different combinations of factors.

Response Equivalence between Simulations and Reality

An interesting issue examined by some researchers concerns "response equivalence," the extent to which people's preference rankings for real landscapes are the same as their rankings for simulations of those landscapes.[5] For example, would a camper participating in Shafer's research on preferences for landscapes in the Adirondacks rank the real landscapes in the same order as the 8 × 10-in. photographs of those landscapes?

To illustrate the kinds of investigations that have explored the response equivalence issue, consider a study by Appleyard and Craik (1979) at the Environmental Simulation Laboratory of the University of California, Berkeley. (The modeling facilities at this laboratory are introduced in the discussion of Figure 7.1.) Two randomly chosen groups of residents of Marin County, California, participated in the study. One group took a 9-mile tour by auto of a particular route through Marin County. A second group viewed a 16-mm color film of an equivalent tour through a three-dimensional model of the research area. The film was made by moving a remotely guided optical probe (equipped with a movie camera) through the model. Both groups of participants then recorded their impressions, and the responses of the two groups were compared. Based on this comparison, Appleyard and Craik observed that "the character of the region conveyed by the direct and simulated presentations and captured by this simulation technique is essentially identical." They also noted that, owing to

[4]The quoted criticism is from Kaplan (1975, p. 124). For a further discussion of the approach used by Shafer and his associates, see Brush and Shafer (1975).

[5]The term *response equivalance* is elaborated by Sheppard (1982, p. 14). In general, it applies to many different types of visual responses, not just preference rankings.

the lack of human and vehicular motion, the model was viewed as being somewhat cold and barren compared to the real environment. This less positive impression is probably caused, in part, by limitations of the model in replicating the texture of vegetation and road and sidewalk surfaces.[6] Overall, however, there was much agreement between those viewing the actual area and those viewing the color film.

Sheppard's (1982) analysis indicates that the high response equivalence in Appleyard and Craik's model studies of Marin County cannot be expected when less sophisticated simulation media are used. He argues that for any one simulation technique, the subject matter (or scene content), field of view and "image elements" can affect responses to a simulation. Image elements commonly felt to influence the realism of visual simulation include color, detail, and texture. Sheppard's empirical research on how image elements affect response equivalence supports his opinion that few generalizations can be made regarding this complex topic. Although many landscape preference research investigations have been conducted, very few of them have systematically analyzed the effects of varying either subject matter or image elements.

Preferences of Lay Persons versus Design Experts

Another issue explored with landscape preference research is whether rankings of various landscapes by professionals using aesthetic criteria would turn out the same as preference rankings of those landscapes by lay persons. This issue was investigated by Zube, Pitt, and Anderson (1975) using data originally obtained by Zube (1973). The research involved 185 participants, each of whom viewed twenty-seven 35-mm color slides of the "everyday rural landscape" of the northeastern United States. Each person used the same procedure to describe and evaluate the 27 slides. For example, each slide was evaluated using scales to measure beauty versus ugliness, urban versus rural, and so forth. The resulting data were divided into seven groups corresponding to the occupations of the participants. The "expert group" consisted of individuals classified as "professional environmental designers." There were six groups of nonexperts corresponding to different categories of students and workers. For each group, the data were converted into scores representing group-wide summary statistics for the various descriptions and evaluations made by each participant.

Scores were analyzed to determine the extent to which groups agreed with each other. To do this, coefficients measuring the strength of correlation between the scores for each *pair* of groups were computed. (A value of 1.0 for a "correlation coefficient" indicates the scores for two groups are perfectly correlated.) Twenty-one different pairs can be formed from seven groups. Results from this analysis are summarized by Zube, Pitt, and Anderson (1975, p. 155):

> *The correlation between the seven subgroups on scenic evaluation of all 27 landscapes ranged from 0.43 (environmental design students and secretaries)*

[6]For a discussion of this point, and other aspects of the Marin County simulation, see Appleyard and Craik (1979).

to 0.91 (resource managers and teachers). The mean correlation for the 21 between-group associations was 0.77.

Based on their analyses, Zube, Pitt, and Anderson (1975, p. 156) concluded that "overall, there was generally high agreement between the seven subgroups on both landscape evaluation and description." These findings are for existing rural landscapes only. Whether nonexperts would agree with experts if the landscapes were modified by physical developments remains to be tested.

A subsequent, more elaborate study by Zube, Pitt, and Anderson involved 307 participants divided into 13 groups. The landscapes included in this research were from a portion of the Connecticut River Basin in Massachusetts and northern Connecticut. All participants in the study were from this region. With one notable exception, congruence in the patterns of response among different groups was similar to that observed in the earlier study by Zube (1973). The exception concerned responses of a group of black, center-city residents from Hartford, Connecticut. The correlations between the Hartford group's responses and those of each of the other groups were consistently low. Of a total of 66 between-group correlations that were below 0.8, 58 involved the Hartford residents. These results led Zube, Pitt, and Anderson to suggest that cultural and socioeconomic factors may affect landscape perception.

Other research reinforces the important influence of cultural and socioeconomic variables on visual perception. Brush (1976) reviews several investigations addressing this issue and concludes that the backgrounds and prior experience of observers can have a substantial effect on how landscapes are evaluated. This is consistent with findings in the more general literature on aesthetics and visual perception.[7]

Truth in Simulation

Landscape preference research has provided much information relevant to visual impact assessment. However, it has not thoroughly examined questions of bias and misrepresentation or, to use the phrase of Appleyard et al. (1973), "truth in simulation." There are several issues involved here, and they overlap partially with the previously introduced concept of response equivalence.[8]

Consider the complexities in preparing an unbiased simulation of a proposed project. The many choices required in simulating a project's appearance make it difficult to define what constitutes an unbiased representation. At the outset, viewing stations and simulation media must be selected. Then numerous additional decisions must be made about color, scale, field of view, and so forth. Many of the choices may be affected inadvertently by the simulation expert's

[7]See, for example, Arnheim's (1974) work on visual perception.

[8]This discussion of truth in simulation relies on Mellander (1974) and on personal communication with Stephen S. J. Sheppard, Department of Landscape Architecture, University of California, Berkeley, October 27, 1981. Sheppard's (1982) research is one of the few systematic investigations of bias and misrepresentation in visual impact assessment.

educational background. For example, an illustrator's training often emphasizes the preparation of aesthetically pleasing renderings, not the attainment of absolute realism. In addition, many technical judgments may unintentionally bias a simulation. For instance, when photography is used in the simulation process, the choice of film and lighting can significantly affect the outcome. There are many subtleties that influence the extent to which a simulation looks better or worse than the constructed project.

Another truth in simulation issue concerns deliberate misrepresentations of proposed projects. Examples include the addition of flashy automobiles and well-dressed people to provide a prestigious appearance, and the use of sunlight and blue skies to project visions of cleanliness and comfort.[9] Sometimes false impressions result when a simulation omits things such as traffic congestion and billboards. Perspective drawings are often criticized for being misleading, especially when building projects are involved. Appleyard (1977, p. 74) raised this objection:

> *The elaborate and costly perspectives used to describe many projects are usually taken from the viewpoint that shows off the building to best advantage and minimizes its impact on the surrounding environment. Frequently, these views are not ones from which the building will be most often seen. . . . Deception, too, is relatively easy. A draftsman has only to use dramatic shadows, beautiful textures, and fine trees for most people to like any scheme he portrays.*

Blatant misrepresentations are also possible by the choice of viewing location. Appleyard (1977) cites a case involving two simulations of a proposed high-rise building on the waterfront in downtown San Francisco. Project proponents showed the building from a near-overhead location several thousand feet in the air. It was hardly discernible in the cluster of downtown office buildings. Opponents of the project used an unflattering vantage point, a low "birds-eye" view over San Francisco Bay that showed the profile of the downtown skyline. The proposed building was painted in black on an old picture of downtown San Francisco that omitted many high-rise buildings that had already been built.

The question of what constitutes an unbiased simulation is complex because a high degree of realism is not necessarily associated with a good response equivalence (that is, correspondence between reactions to a simulation and to the completed project).[10] A viewer's past experience with different simulation media has an important effect on response equivalence. For example, architects can gain a realistic and accurate impression of a proposed project by viewing plans and elevations that are meaningless abstractions to most people. Another example involves artist's renderings that reflect "artistic license" and show proposed projects in a favorable light. Sheppard (1982) speculates that, based on

[9]These examples are from Appleyard (1977, p. 60).

[10]Sheppard (1982, p. 224) provides empirical support for this statement.

past experience, professional planners view such renderings with skepticism and may thereby avoid being misled by them.

Still another complication is that a simulation that looks realistic is not necessarily unbiased. Consider an authentic looking simulation that is inaccurate, for example, because it omits buildings adjacent to the proposed project. It can be misleading precisely because it appears so credible.

PREPARATION OF VISUAL INVENTORIES

Having introduced several important visual simulation issues, the chapter now focuses on techniques for visual impact assessment. The first stage in an assessment involves developing a "visual inventory," a record of visual features likely to be affected by the proposed project. Topographic maps and aerial photos are widely used in preparing inventories. In the United States, these maps and photos are available from the U.S. Geological Survey. Black and white panchromatic stereo pairs of aerial photos are especially useful, since they can be viewed through special stereo glasses to provide three-dimensional images. Information from topographic maps and aerial photos is often supplemented by field observations. In fact, field surveys are generally required in interpreting aerial photos. Survey data typically include sketches, notes, and ground level photos clarifying characteristics of the study area.

Litton (1972) elaborates on two basic inventory types: route inventories and area inventories. A route inventory is appropriate when the main visual impacts involve views from a specific hiking trail, city street, highway, or waterway. An area inventory is appropriate if the visual effects are extensive (for example, the influence of new timber harvesting practices) or if the views from several routes are affected by a proposal.

Figure 13.2 is part of a route inventory for a part of Highway 89 near Lake Tahoe in the Sierra Nevada mountains. It includes lines of sight from selected viewing stations to significant visual features. The outside boundary of the inventory is defined by the limits of the landscapes visible from the route. The selection of observation points is an important part of inventory preparation. Criteria for choosing these locations often depend on both the number of likely viewers and the quality of the scene itself. Sketches and photos of views from important viewing points are a part of the inventory.[11] The level of detail in preparing an inventory is influenced by the likely travel speed of viewers along routes. For example, a motorist driving at 40 mph sees much less detail than a pedestrian traveling along the same route.

Inventory preparation often includes mapping "viewsheds," that is, areas that can be seen from a particular location. This can be done in the field. Alternatively, viewsheds can be delineated with existing topographic maps. Computer

[11]The choice of views to record involves more than just observer position. Other important factors include season and lighting conditions. A systematic discussion of these and other factors which influence a view is given by Litton (1972).

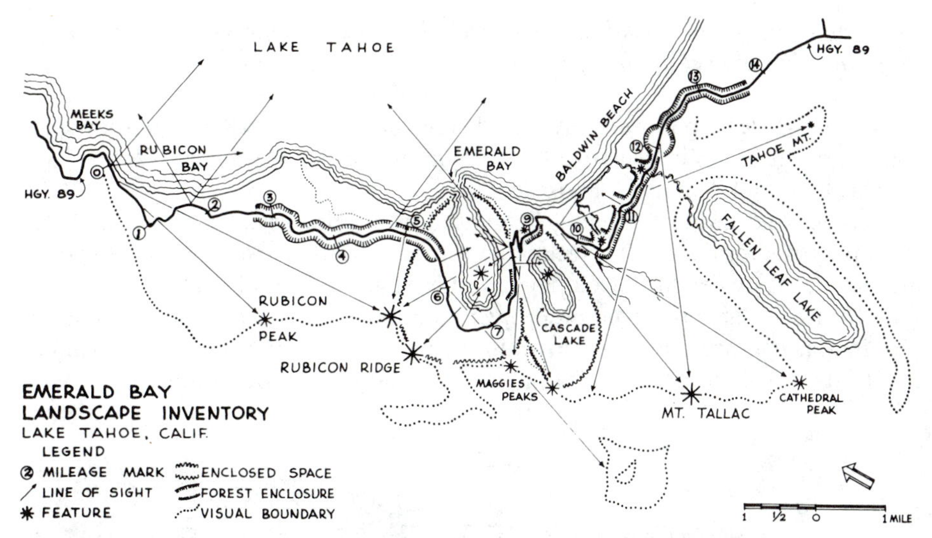

FIGURE 13.2 Portions of a visual inventory (route-type) of Highway 89 near Lake Tahoe. From R. B. Litton, Jr. (1972), in *Natural Environments: Studies in Theoretical and Applied Analysis,* published for Resources for the Future, Inc. by the Johns Hopkins University Press, Baltimore, Md.

programs for determining all locations viewable from a particular spot are available.[12]

TECHNIQUES FOR SIMULATING POSTPROJECT CONDITIONS

Postproject conditions are often represented using only verbal descriptions. Sometimes the language used conveys a visual impression by evoking stereotypical images, for example, "the countryside will be filled with billboards." There are numerous opportunities for bias as illustrated by the following: "The proposed highway will obliterate the soft, rolling hills which are filled with tiny poppies." Sometimes verbal descriptions employ specialized terminology that may be confusing to lay persons. For example, consider the testimony by Gussow (1977), an expert in landscape evaluation, in hearings on a proposed electrical transmission line through Riverhead, New York. Gussow was asked to describe the visual impact of a corridor of transmission towers along a particular route. His response included the following:

> *The effect would be visually severe. The towers would overwhelm all other elements, including the tallest trees in the area. Their verticality would impinge upon the horizontal plane of the farmlands. Because the landscape*

[12]Detailed procedures for plotting viewsheds using topographic maps are given by the U.S. Department of Transportation (1981). Travis et al. (1975) describe a widely used computer program for delineating viewsheds.

is confined, a wide swath of such overwhelming structures would have a galvanizing impact upon the viewer.

This type of description conveys much information to landscape design specialists, but it is less informative to those unfamiliar with specialized terms like "verticality" and "confined landscape."

Perspective drawings have long been used to describe visual changes resulting from proposed projects. This technique was employed in assessments of several alternative strategies for developing a terminal at the Port of Seattle, Washington. Figure 13.3*a* represents preproject conditions. A scheme for a container terminal using yard cranes is shown in Figure 13.3*b*. Both sketches are from a point on a scenic route that is popular with local residents and tourists.[13]

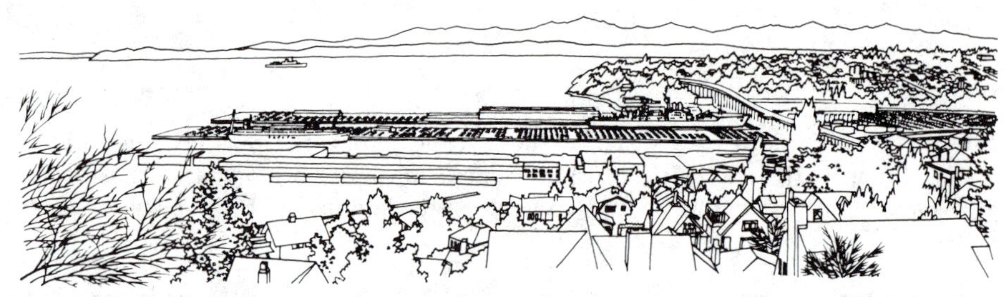

EXISTING: AUTOMOBILES / LARGE VESSEL MOORAGE / CHEMPRO

(*a*)

CONTAINERS / YARD CRANE

(*b*)

FIGURE 13.3 Perspective sketches used in appraising a proposed port development in Seattle, Washington. (*a*) Preproject conditions. (Courtesy of Jones & Jones, Inc., Seattle, Wash.) (*b*) Proposed container terminal using yard cranes. From Port of Seattle, Washington, 1981, Final Environmental Impact Statement on Alternative Uses for Terminal 91 (Pier 90/91), Seattle, Wash.

[13]William Blair of Jones & Jones Inc. in Seattle helped in obtaining the sketches in Figure 13.3 and the photos in Figure 13.5. Lawrence Isaacson of the Federal Highway Administration also assisted in this effort.

A sequence of perspective sketches can enhance the accuracy of visual simulation. As demonstrated in Figure 13.1, sequenced sketches can simulate a journey through an environment.

Digital computers have been programmed to yield perspective drawings from different viewing angles and locations. Figure 13.4 shows computer generated perspectives depicting pre- and postproject conditions for a highway expansion proposal in Oregon. Using a computer, many different perspectives can be developed rapidly. The cost of computer generated drawings can be quite low once the computer-plotter equipment is in place and the terrain and project development data are put in the required form.[14]

Another approach to simulation involves the manipulation of photographs. Two approaches are common: photomontage and photoretouching. The effects obtainable with these approaches are demonstrated in a study of a proposed highway in Colorado. Figure 13.5*a* shows a preproject view from an existing road. The postproject condition is simulated in Figure 13.5*b*. It was produced using a special "photomontage program" to obtain a computer generated perspective sketch of the proposed highway from the same vantage point used in the photo of existing conditions. An artist added color and tone to the perspective sketch and then superimposed it on the original photo.

Three-dimensional models are also used in the photomontage process. This is demonstrated by the visual analysis of pre- and postproject conditions for the Kentucky Center for the Arts in Louisville. The postproject condition was portrayed by superimposing a photo of a model of the proposed arts center on a photo of downtown Louisville. The results are shown in Figure 13.6.[15]

Sometimes postproject conditions are simulated by viewing a model directly. This is the case in Stockholm where a three-dimensional model of the city is used to evaluate proposals for new buildings. Appropriately scaled models of proposed structures are placed in the model representing existing conditions in Stockholm. Planners and others can then assess the impacts of proposed developments.[16]

Models viewed from above provide a birds eye or aerial perspective, which is not representative of the way most people would see a new project. Model scopes with cameras attached make it possible to take photos from an eye level perspective within a model. Figure 13.7 shows photos made using a scope in a three-dimensional model to assess the visual impacts of a proposed bridge across a lake in Louisiana. The photographs depict pre- and postproject conditions from the same lakeside location. Compare these to Figure 13.8, photos of the real lake from the same viewing station. The postproject condition in Figure

[14]Procedures for producing computer generated sketches such as those in Figure 13.4 are given by the U.S. Department of Transportation (1978).

[15]James J. Walters and his associates at Humana, Inc. in Louisville provided the photos in Figure 13.6.

[16]For additional information on models of Stockholm and other cities, see Appleyard (1977, pp. 77–79).

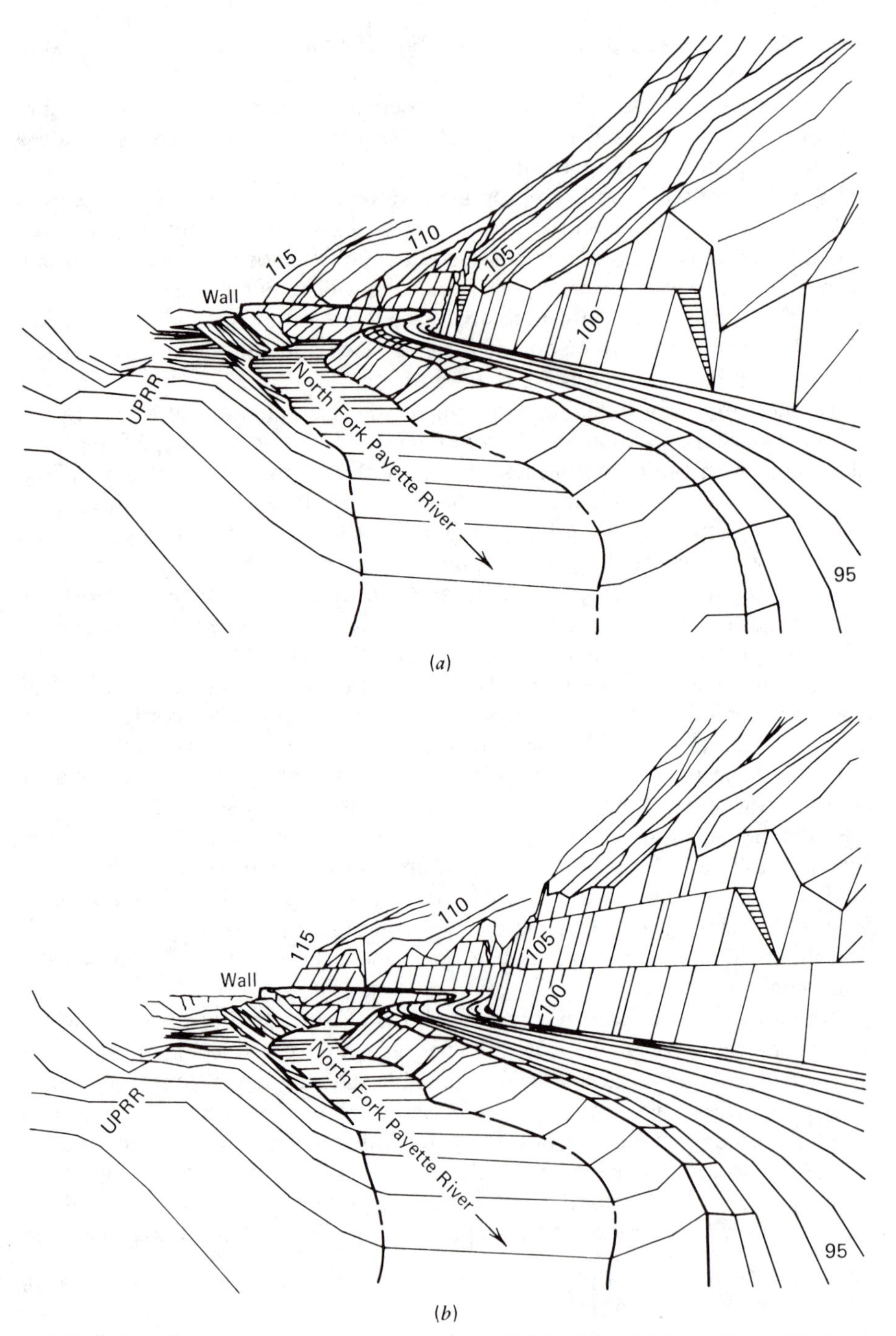

FIGURE 13.4 Computer-generated perspectives of Idaho forest highway Route 23. From U.S. Department of Transportation, 1978, Idaho Forest Highway Route 23, North Fork Payette River Highway, Final Environmental Impact Statement, Federal Highway Administration, Vancouver, Wash. (*a*) Preproject conditions. (*b*) Four-lane roadway alternative.

FIGURE 13.5 Photographic simulations used to assess a proposed highway in Colorado. (*a*) Preproject conditions. (Courtesy of the Colorado Department of Transportation.) (*b*) Photomontage showing postproject conditions based on a computer generated perspective. From U.S. Department of Transportation (1981).

13.8*b* was prepared by first photographing a scale model of the bridge. The picture of the model was then superimposed on a photo representing preproject conditions.[17]

More complex three-dimensional modeling procedures are used to simulate motion. This requires use of extensive structural supports allowing for (1) the attachment of movie or television cameras to the model scope and (2) the ability

[17]Figures 13.7 and 13.8 were provided through the efforts of Kevin Gilson and Peter Bosselmann of the Environmental Simulation Laboratory at the University of California, Berkeley.

FIGURE 13.6 Photos used to portray the Kentucky Center for the Arts in Louisville. (Courtesy of Humana, Inc., Louisville, Ky.) (*a*) Preproject conditions. (*b*) Photomontage showing postproject conditions based on a model of the proposed arts center.

to move the scope in the model area to simulate walking, driving, or flying. Modeling laboratories capable of simulating motion exist in several countries. In the United States, a widely known facility is at the University of California, Berkeley (see Figure 7.1). Remarkable degrees of realism can be obtained when sophisticated movie-making techniques are used with three-dimensional models.[18]

[18]An example is the film made for the U.S. Congress to represent Lawrence Halprin's proposal for a monument to Franklin D. Roosevelt. Color photos made from the film are presented by Aidala (1979).

(b)

EVALUATING VISUAL EFFECTS AND LANDSCAPE QUALITY

As mentioned earlier, visual impacts can be evaluated using both preferential judgments of potential project users and aesthetic criteria applied by design specialists. Landscape evaluation experts do not agree fully among themselves as to whether aesthetic criteria or user preferences should play the dominant role. However, the context in which an evaluation is conducted often dictates the approach to be followed. When well-defined, controversial projects are being appraised, the preferential judgments of users are frequently the dominant consideration. Evaluations made exclusively by design specialists are often appropriate when large landscape areas are assessed and potential users are not

FIGURE 13.7 Model simulation of a proposed bridge across a lake in Louisiana. (Photos by Kevin Gilson. Courtesy of the Environmental Simulation Laboratory, University of California, Berkeley.) (*a*) Scale model of preproject conditions. (*b*) Scale model including a proposed bridge.

FIGURE 13.8 Photomontage used to portray a proposed bridge across a lake in Louisiana. (Courtesy of the Environmental Simulation Laboratory, University of California, Berkeley.) (*a*) Preproject conditions. (*b*) Photomontage showing postproject conditions based on a model of the proposed bridge. (Photomontage prepared by Kevin Gilson.)

easily identified. The discussion below demonstrates two of the several approaches applied in evaluating visual effects.[19]

The Berkeley Environmental Simulation Lab

The Environmental Simulation Laboratory at the University of California, Berkeley, has been used to involve citizens in visual impact studies.[20] One exercise assessed the appearance of a waterfront development proposal in Richmond, California. The proposed project included 2000 new housing units, a commercial development, and a large marina. A three-dimensional model of the project was constructed at a scale of 1 in. = 30 ft. Then a film was produced showing aerial views of the development and views that would be seen during a drive into its center. The film had the imaginary driver get out from the car and take a stroll along a proposed shopping promenade. The driver also took a sailing trip through the marina to show how the project would look from a boater's perspective. Both the film-making process and the film itself led to much discussion of design details and debate on various aspects of the project. A citizens' committee made evaluations that led to modifications in the original proposal.

Another study undertaken at the Berkeley laboratory involved portraying views from a proposed road intended to replace San Francisco's "Great Highway" along the Pacific Ocean. One criterion for selecting the new route was that it provide "an exciting and varied sequence of views of the ocean to visitors."[21] Because drawings alone were not felt to adequately represent the views from the new route, a model of a prototypical ¼-mile road segment was developed. Slides taken from inside the model simulated a sequence of views from the proposed route as well as views from adjacent homes. The proposed project was very controversial and the level of public interest was high. The simulations prepared by the Berkeley laboratory allowed both citizens and design experts to evaluate the project's visual impacts.

Litton's Application of Aesthetic Criteria

In contrast to the above-noted exercises at the Berkeley laboratory, Litton's approach to landscape evaluation relies on professional judgments of specialists trained in the application of aesthetic criteria. His concepts have been employed by several resource management agencies in the United States. The discussion below is from work by Litton et al. (1974) on the visual effects of water resources developments.

[19]Julie Lane assisted in the preparation of an early draft of this section while she was a student at Stanford University in 1980. Cerny (1974) summarizes several landscape evaluation procedures.

[20]The two evaluation studies described below are from Appleyard et al. (1979).

[21]This criterion is quoted in Appleyard et al. (1979, p. 510).

Litton's water project evaluation procedure uses a classification system composed of landscape units, setting units, and waterscape units. A landscape unit is regional in nature, covering areas characterized by a dominant topographic pattern or vegetative cover. An example is California's Central Valley, a relatively flat area extending for hundreds of miles down the length of California west of the Sierra Nevada Mountains. A setting unit exists within a landscape unit and is a visual corridor enclosed by a group of land forms. A lake surrounded by hills on all sides provides an example; the setting unit includes both the lake and the hills. A waterscape unit is a topographic entity visually dominated by water, such as a lake and its immediate shore as seen by a person near the lake's edge.

In evaluating a proposed water project, two visual inventories are prepared: one for existing conditions and one for postproject conditions. Inventories for each of the three different units are based on "typologies." For example, the typology for waterscape units includes the "spatial expression" of the shore, "vertical edge definition" of the shore, and so forth. The terms used in the inventory typologies are defined by Litton et al. (1974).

Pre- and postproject conditions are appraised by applying aesthetic criteria for "unity, vividness and variety" to each element of the inventory typologies. Use of the criteria is consistent with the opinion that "any composition . . . having aesthetic merit must represent some combination of" unity, vividness, and variety.[22] Litton and his associates define these criteria and provide guidelines for distinguishing between high- and low-quality unity, vividness, and variety.

Unity exists when various parts of a landscape are joined into a coherent and harmonious visual entity. For example, a large mountain can give unity to the streams and trees that lie upon it. Variety is the presence of richness and diversity of objects and relationships in a landscape. However, high-quality variety is not characterized by a large number of objects per se; there "needs to be order and control over numerous and diverse parts." Vividness in a landscape is "that quality which gives distinction or creates a strong visual impression."[23] Such distinction is created primarily through contrast and is illustrated by an effervescent waterfall plunging into a still pool.

The three aesthetic criteria cannot be considered in isolation since they overlap to some extent. In discussing who should apply the criteria in evaluating landscapes, Litton et al. (1974, p. 113) observe

> *aesthetic criteria are not whimsical nor are they spur of the moment ideas. They are not determined by popularity contests. They represent a body of knowledge and need to be applied by those who are competent in their application.*

[22]This quotation is from Litton et al. (1974, p. 104).

[23]The quotations defining variety and vividness are from Litton (1972, p. 286) and Litton et al. (1974, p. 111), respectively.

In relying on specialists competent in the use of aesthetic criteria, Litton's approach to landscape appraisal differs fundamentally from the citizen-oriented evaluation exercises conducted at the Berkeley Environmental Simulation Laboratory. The two approaches represent polar cases, with commonly used methods falling somewhere in between. It is often considered appropriate to have design experts conduct visual appraisals using criteria that reflect the preferences of those affected by proposed projects. This position is embodied in an excerpt from a "field guide" on visual impact assessments for highway projects:

> *It may be impractical to obtain a random and completely representative sample of the public to rate the visual effects of highway alternatives. Expert judgment may be a valid and reliable substitute if it is based on criteria derived from research about public perceptions. Its validity can be further strengthened by direct but limited public response in project community involvement programs.*[24]

Because assessments of the visual effects of physical facilities have only been carried out routinely since 1970, it is not surprising that experts disagree on how they should be conducted. The excerpt above from the field guide encourages an appropriate balance of perspectives. It recognizes the contributions that can be made by design specialists, and, at the same time, is sensitive to the values of citizens likely to be affected by a proposed project.

KEY CONCEPTS AND TERMS

SIMULATION VERSUS EVALUATION
- Pre- and postproject conditions
- Simulations as forecasts
- Evaluative appraisals
- Preferential judgments

ISSUES EXPLORED USING LANDSCAPE PREFERENCE RESEARCH
- Response equivalence
- Expert versus nonexpert preferences
- Factors affecting landscape preference
- Bias due to aesthetic training
- Misrepresentation in simulation
- Realistic versus accurate simulations

PREPARATION OF VISUAL INVENTORIES
- Topographic maps
- Aerial photos
- Field surveys
- Route inventories
- Area inventories
- Viewsheds

TECHNIQUES FOR SIMULATING POSTPROJECT CONDITIONS
- Perspective sketches and sequences
- Photomontage
- Retouched photos
- Computer generated perspectives
- Three-dimensional models
- Simulating motion with models

EVALUATING VISUAL EFFECTS AND LANDSCAPE QUALITY
- Citizen involvement in evaluation
- Aesthetic criteria
- Unity, variety, and vividness

[24]These remarks are from the U.S. Department of Transportation (1981, p. 13).

DISCUSSION QUESTIONS

13–1 Characterize the types of judgments involved in preparing visual simulations of a proposed project. Do these judgments obscure the distinctions between simulation and evaluation?

13–2 Sheppard (1982) studied response equivalence by having one group of design specialists respond to visual portrayals of proposed projects and another group react to photographs of the same projects after they were built. The response equivalence in the two sets of evaluations was low for three quarters of the 30 projects included in the research. Speculate on the factors that might explain this research outcome.

13–3 Identify a large-scale development project that is proposed for an area you are familiar with. How would you go about preparing an inventory of existing visual features that might be affected by the proposal? In discussing your approach, consider variables such as viewer location, time of day, and season of the year.

13–4 Alternative simulation media include photomontage, perspective drawings, and so forth. List and define five criteria you would use to characterize "good" media for portraying visual impacts of physical development projects.

13–5 Suppose graphic design specialists have prepared models, sketches and photos to portray a proposed electric power plant in a rural location. You are asked to use these simulations in designing a process for evaluating the visual impacts of the proposed facility. How would aesthetic criteria and preferential judgments fit into your evaluation scheme?

REFERENCES

Aidala, T., 1979, The FDR Memorial—Halprin Redefines the Monumental Landscape. *Landscape Architecture* **69** (1), 42–52.

Appleyard, D., 1977, Understanding Professional Media: Issues, Theory and a Research Agenda, *in* I. Altman, and J. F. Wohlwill (eds.), *Human Behavior and the Environment: Advances in Theory and Research,* Vol. 2, pp. 43–88. Plenum, New York.

Appleyard, D., and K. H. Craik, 1979, "Visual Simulation in Environmental Planning and Design," Working Paper No. 314, Institute of Urban and Regional Development, University of California, Berkeley.

Appleyard, D., K. H. Craik, M. Klapp, and A. Kreimer, 1973, "The Berkeley Environmental Simulation Laboratory: Its Use in Environmental Impact Assessment," Working Paper No. 206, Institute of Urban and Regional Development, University of California, Berkeley.

Appleyard, D., P. Bosselmann, R. Klock, and A. Schmidt, 1979, Periscoping Future Scenes, How to Use an Environmental Simulation Lab. *Landscape Architecture* **69** (5), 487–510.

Arnheim, R., 1974, *Art and Visual Perception.* University of California Press, Berkeley.

Brush, R. O., 1976, Perceived Quality of Scenic and Recreational Environments: Some Methodological Issues, *in* K. H. Craik, and E. H. Zube (eds.), *Perceiving Environmental Quality: Research and Applications,* pp. 47–58. Plenum, New York.

Brush, R. O., and E. L. Shafer, 1975, Application of A Landscape Preference Model to Land Management, *in* E. H. Zube, R. O. Brush, and J. G. Fabos (eds.), *Landscape Assessment: Values, Perceptions and Resources,* pp. 168–182. Dowden, Hutchinson & Ross, Stroudsburg, Pa.

Cerny, J. W., 1974, Scenic Analysis and Assessment, *in* C. P. Straub (ed.), *CRC Critical Reviews in Environmental Control* **4** (2), 221–250.

Craik, K. H., 1972, Appraising the Objectivity of Landscape Dimensions, *in*

J. V. Krutilla (ed.), *Natural Environments: Studies in Theoretical and Applied Analysis,* pp. 292–346. Johns Hopkins University Press for Resources for the Future, Inc., Baltimore, Md.

Gussow, A., 1977, In the Matter of Scenic Beauty. *Landscape* **21** (3), 26–35.

Kaplan, R., 1975, Some Methods and Strategies in the Prediction of Preference, *in* E. H. Zube, R. O. Brush, and J. G. Fabos (eds.), *Landscape Assessment: Values, Perceptions and Resources,* pp. 118–129. Dowden, Hutchinson and Ross, Stroudsburg, Pa.

Litton, R. B., Jr., 1972, Aesthetic Dimensions of the Landscape, *in* J. V. Krutilla, (ed.), *Natural Environments: Studies in Theoretical and Applied Analysis,* pp. 262–291. Johns Hopkins University Press for Resources for the Future Inc., Baltimore, Md.

Litton, R. B., Jr., R. J. Tetlow, J. Sorensen, and R. A. Beatty, 1974, *Water and Landscape, An Aesthetic Overview of the Role of Water in the Landscape.* Water Information Center, Inc., Port Washington, N.Y.

Mellander, K., 1974, Environmental Planning: Can Scale Models Help? *American Engineering Model Society—Seminar '74,* (October), pp. 61–67. Available as Reprint No. 125, Institute of Urban and Regional Development, University of California, Berkeley.

Shafer, E. L., J. F. Hamilton, Jr., and E. A. Schmidt, 1969, Natural Landscape Preferences: A Predictive Model. *Journal of Leisure Research* **1,** 1–19.

Sheppard, S. R. J., 1982, *Landscape Portrayals: Their Use, Accuracy and Validity in Simulating Proposed Landscape Changes.* Ph.D. dissertation. University of California, Berkeley.

Travis, M. R., G. H. Elsner, W. D. Iverson, and C. G. Johnson, 1975, "VIEWIT: Computation of Seen Areas, Slope and Aspect for Land Use Planning," General Technical Report PSW-11, USDA Forest Service, Pacific Southwest Forest and Range Experiment Station, Berkeley, Calif.

U.S. Department of Transportation, 1978, *Highway Photomontage Manual,* Report No. FHWA-DP-40-1, Federal Highway Administration, Arlington, Va.

U.S. Department of Transportation, 1981, *Visual Impact Assessment for Highway Projects.* Federal Highway Administration, Office of Environmental Policy, Washington, D.C.

Zube, E. H., 1973, Rating Everyday Rural Landscapes of the Northeastern U.S. *Landscape Architecture* **63,** 370–375.

Zube, E. H., D. G. Pitt, and T. W. Anderson, 1975, Perception and Prediction and Scenic Resource Values of the Northeast, *in* E. H. Zube, R. O. Brush, and J. G. Fabos (eds.), *Landscape Assessment: Values, Perceptions and Resources,* pp. 151–167. Dowden, Hutchinson & Ross, Stroudsburg, Pa.

CHAPTER 14

ELEMENTS OF NOISE IMPACT ASSESSMENT

Noise problems have become increasingly severe, especially in urban areas. The Council on Environmental Quality (1979) indicated that in the late 1970s over 40% of the U.S. population was regularly subjected to noise that interfered with activities or caused annoyance. Often these high noise levels were associated with construction activities and with transportation facilities, especially highways and airports.

Noise is often defined as "unwanted sound." The well-developed science of sound provides a basis for defining noise indicators and making forecasts of noise. Although acoustics is a complex and specialized topic, it must be introduced to pursue even a rudimentary discussion of techniques to predict noise impacts of proposed projects.

SOUND AND ITS MEASUREMENT

Most people are familiar with the concentric circles of waves that result from dropping a stone into a still pond. Analogous waves propagate through the air when a metal tuning fork is struck with a hard object. The vibrations of the tuning fork push some air molecules close together and allow others to move farther apart. The resulting fluctuations in air pressure, relative to undisturbed air, are sensed by the human ear as sound.

Pure Tones and Sound Pressure Levels

Suppose that immediately after the tuning fork is struck, measurements of air pressure are made at a nearby point. A plot of the air pressure levels at this point might look like the graph in Figure 14.1. When air pressures associated with a sound can be described by the simple, regular shape in Figure 14.1, the sound is termed a *pure tone*. The regular pattern in the graph is called a *simple harmonic oscillation.*[1] Although most commonly heard sounds are *not* pure tones, they can be represented as combinations of pure tones.

Figure 14.1 introduces terms used to describe sound. The time it takes for the pressure to go through one full cycle of the pattern in the figure is called the *period of oscillation*. Its inverse is termed the *frequency of oscillation* and is generally expressed in Hertz (Hz), a shorthand way of saying "cycles per second." Frequency is related to what people perceive subjectively as "pitch"; the higher the frequency, the higher the pitch.

Because sound pressures are only a minute fraction of atmospheric pressure, they are generally reported in terms of their differences from atmospheric pressure [see $P(t)$ in Figure 14.1]. The maximum sound pressure, P_{max}, is called the *amplitude of the oscillation*. Amplitude is associated with the subjective perception of "loudness."[2] It is tempting to characterize the sound pressure in Figure 14.1 in terms of average pressure. However, the regular pattern in the figure makes the average pressure over a period equal to zero. Instead of a simple average, the *root-mean-square* (rms) pressure is frequently used. It is computed by first determining the square of all values of $P(t)$ in a period of oscillation and then computing the average of the squared values. Finally, the square root of this average is taken and it is called the *root-mean-square pressure.* For pure tones like the one in Figure 14.1, it can be shown that

$$p = \frac{P_{max}}{\sqrt{2}} \quad \textbf{(14-1)}$$

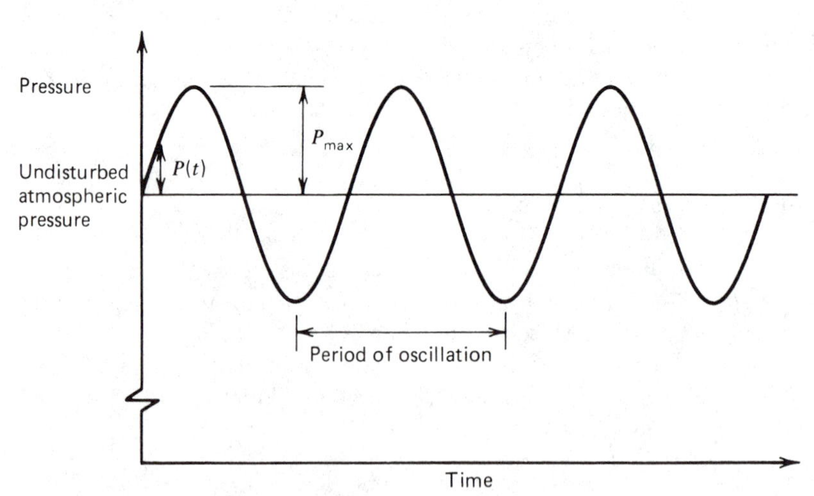

FIGURE 14.1 Representation of a pure tone.

[1]Harmonic oscillations can be described using trigonometric functions. Figure 14.1 has the shape of a sine curve.

[2]The links between loudness and amplitude have been determined empirically. For a discussion of these relationships, see Cunniff (1977, pp. 89–91).

where p is the rms pressure and is generally measured in Pascals (Pa).[3] Sound meters are used to record rms pressure amplitudes.

The range of audible sound pressures is enormous. At a frequency of 1000 Hz, barely audible sound has an rms pressure of 2×10^{-5} Pa and the maximum rms pressure that will not damage the human hearing mechanism is 20 Pa. Because audible sound spans a range of 10^6 Pa, a logarithmic scale is used to describe pressure.[4] For any rms pressure, p, the *sound pressure level, L,* expressed in units of decibels (dB), is defined as

$$L = 10 \log \left(\frac{p^2}{p_0^2} \right) \tag{14-2}$$

where $p_0 = 2 \times 10^{-5}$ Pa, the threshold of audible sound, and the logarithm is to the base 10.

Sound pressure level is defined using the *square* of the rms pressure because the amount of power contained in sound is proportional to the pressure squared. Since equation 14-2 employs the logarithm of the pressure squared, a 10-fold increase in sound pressure corresponds to a rise of only 20 dB of sound pressure level.[5]

In doing noise impact assessments, it is often necessary to estimate the sound pressure level resulting from several sources. As shown by Cunniff (1977), a relationship for combining the sound pressure levels of n different sources is

$$L = 10 \log (10^{L_1/10} + 10^{L_2/10} + \cdots + 10^{L_n/10}) \tag{14-3}$$

where L_i is the sound pressure level of source i ($i = 1, \ldots, n$), and L is the sound pressure level of the n sources combined. For example, the overall effect of two 80-dB sources ($L_1 = L_2 = 80$) is a sound pressure level of

$$L = 10 \log (10^{80/10} + 10^{80/10}) = 83 \text{ dB}$$

The logarithmic additions in equation 14-3 are performed conveniently using special graphs and tables such as those presented by Magrab (1975).

Sound Spectra and the *A*-Weighted Decibel

Discussions of environmental noise usually do not focus on pure tones or even simple combinations of pure tones. Commonly heard sounds have complex

[3]A Pascal is defined as the pressure exerted by a force of 1 newton (N) over an area of 1 m^2.

[4]A logarithm to the base 10 is defined as follows: $10^a = b$ is equivalent to saying "the logarithm of b to the base 10 equals a," or $\log_{10} b = a$. Cunniff (1977, pp. 31–39) provides an introduction to how logarithms are used in noise measurements.

[5]A property of logarithms is that $\log a^c = c \log a$. Therefore, $L = 20 \log (p/p_0)$. In addition, $\log 10^n = n$, and thus $\log 10 = 1$, $\log 100 = 2$, and so forth. Hence, if $p = 10\, p_0$, the decibel level is 20, whereas if $p = 100\, p_0$, the decibel level is 40.

frequency and pressure characteristics that can be represented on sound spectrum plots like the ones in Figure 14.2. Frequency is plotted on the abscissa and sound pressure level is on the ordinate. Figure 14.2a, which contains two pure tones, represents a note from a flute. Figure 14.2b approximates the continuous sound spectrum at a distance of 5 ft from an air compressor.

The human ear does not perceive sounds at different frequencies in the same way. Numerous experiments have shown that humans are far more sensitive to sounds in the 2000- to 5000-Hz frequency range than they are to either very low or very high frequencies. For example, the sound pressure level required to

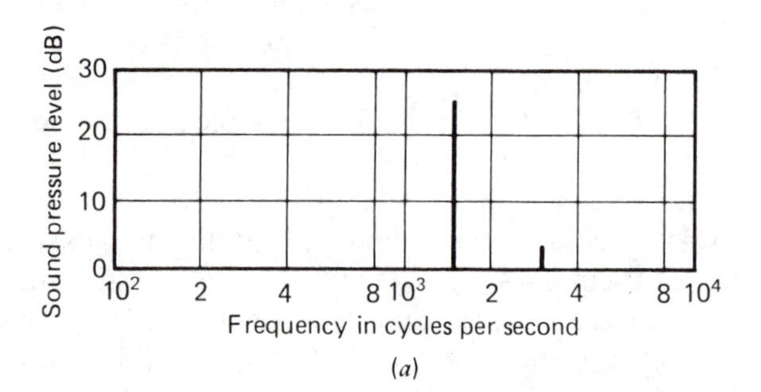

(*a*)

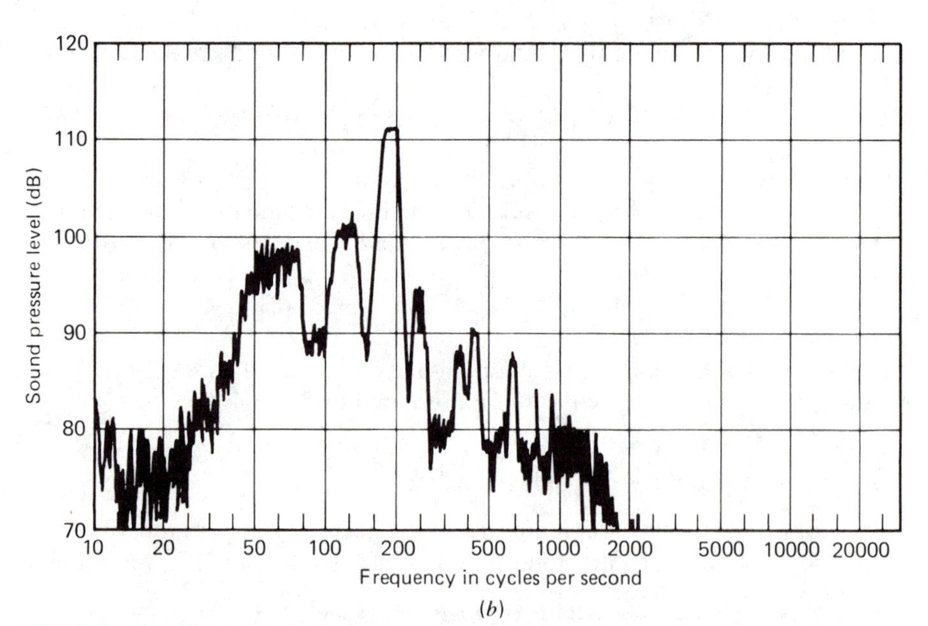

(*b*)

FIGURE 14.2 Illustrations of sound spectra. (*a*) A note from a flute (lowest frequency, termed *fundamental harmonic* is at 1568 Hz). From *Music, Physics and Engineering,* by H. G. Olson, 1967, Dover, New York. (*b*) Sound from an air compressor at a distance of 5 ft. (Courtesy of C. E. Hickman, Southern Company Services, Inc., Birmingham, Ala.).

barely hear a sound at 200 Hz is substantially higher than that needed to just hear one at 2000 Hz. The subjective perception of a sound's loudness depends very much on the frequency of the sound.

Sound measurement equipment has been designed to account for the sensitivity of human hearing to different frequencies. Correction factors (in dB) for adjusting actual sound pressure levels to correspond with human hearing have been determined experimentally. Several different correction factors are used. However, for measuring noise in ordinary environments ("community noise"), *A-weighted correction factors* are generally employed.[6] The term *weighting* is used because sound level meters make the corrections electronically using a weighting network.

Table 14.1 illustrates the computation of *A*-weighted sound pressure levels, recorded in units of dB*A* (the *A* signifies *A* weighting). The table contains measurements from a spectrum analyzer, an electronic device that separates or "filters" the incoming sound into frequency bands and records the sound pressure level within each band. "Octave bands" are widely used in practice and they are employed in this example. By definition, the upper frequency limit of each octave band is exactly two times the lower limit. Beginning with 11 Hz, the band limits (in Hz) are 11 to 22, 22 to 44, 44 to 88, and so forth. The sound pressure level recorded for each octave band is associated with the band's "center frequency."[7] Although the human ear can hear sounds with frequencies between 16 and 20,000 Hz, the sound represented in Table 14.1 consists only of frequencies in the range from 44 to 355 Hz. A continuous sound has been filtered into three octave bands and the table indicates the pressure level of the sound in each band.

TABLE 14.1 Spectral Analysis and *A*-weighted Sound Pressure Levels

Octave Band Limits (Hz)					
Lower Limit	Upper Limit	Center Frequency (Hz)	Octave Band Pressure Levels (dB)	*A*-Weighting Correction Factors[a] (dB)	*A*-Weighted Octave Band Pressure Levels (dBA)
44	88	63	79	−28	51
88	177	125	86	−18	68
177	355	250	89	− 9	80
Overall sound pressure level			91		80.3

[a]Correction factors are from Rau and Wooten (1980, p. 4–28).

[6]The literature on noise distinguishes community noise from noise in work environments (industrial noise) and from particular sources (e.g., airport noise).

[7]The center frequency of an octave band is computed as $\sqrt{2}$ multiplied by the lower frequency limit of the band. As shown by Cunniff (1977, p. 39), the center frequency corresponds to the average of the logarithms of the upper and lower band limits.

Figure 14.3 plots the center frequencies and octave band pressure levels in Table 14.1 and approximates the spectrum of the incoming sound. An overall sound pressure level is obtained by substituting the octave band pressure levels (L_1 = 79 dB, L_2 = 86 dB, and L_3 = 89 dB) in equation 14-3, which gives a combined value of 91 dB. Table 14.1 also shows the results from adding *A*-weighting correction factors to the recorded pressure levels. When the *A*-weighted octave band pressure levels in the final column are entered into equation 14-3, the result is an overall level of 80.3 dB*A*. The *A*-weighted sound pressure level is much lower than the unweighted value (91 dB), because the human ear is not sensitive to the low-frequency sound in this example.

Community noise is often reported in *A*-weighted decibels. Table 14.2 indicates sound pressure levels (in dB*A*) for some familiar noise sources.

STATISTICS USED TO CHARACTERIZE COMMUNITY NOISE

Noise levels vary continually with time. Indeed, 10 min spent at a busy intersection in a residential suburb can be long enough to hear a background noise level of 50 dB*A*, an aircraft overhead at 65 dB*A*, and a sports car at 70 dB*A*.

Histograms of Sound Pressure Levels

Variations in noise levels over time are often represented using a graph of the percentage of time the sound pressure level falls into different intervals (in dB*A*). The resulting "histogram" is used to compute L_x, the sound pressure

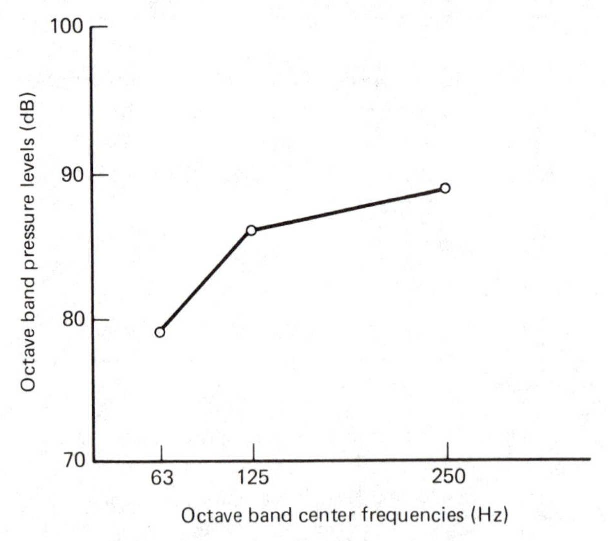

FIGURE 14.3 Octave band sound spectrum.

TABLE 14.2 Sound Pressure Levels of Common Noise Sources[a]

Sound Press Level (dBA)	Typical Source
120	Jet takeoff at 200 ft
110	Riveting machine
100	Pneumatic peen hammer
90	Subway train at 20 ft
80	Pneumatic drill at 50 ft
70	Vacuum cleaner at 10 ft
60	Large store
50	Light traffic at 100 ft
40	Residential area at night
30	Soft whisper at 5 ft
20	Studio for sound pictures

[a]Based on information in Pallett et al. (1978, p. 15).

level that is equaled or exceeded $x\%$ of the time.[8] Values of L_{10}, L_{50}, and L_{90} are commonly interpreted as peak, median, and background noise levels, respectively.

Consider an example: Suppose a sound level meter is taken into a noisy classroom and continuous sound pressure level readings (in dB*A*) are obtained for 10 min. Assume this continuous record is converted into 600 dB*A* readings corresponding to each second in the time period. (The use of 1 sec is arbitrary; any unit of time could be employed.) A histogram for sound pressure levels during this 10 min period is constructed by counting the number of seconds that the recorded dB*A* values fall in different intervals. Intervals of 5 dB*A* are used in this example, but they could have been longer or shorter. Table 14.3 has the results from the counting exercise.

Figure 14.4 is a histogram based on the data in Table 14.3. By counting the percentages associated with each interval in the figure (starting from the *right*), L_{10} and L_{90} can be determined by inspection. For the two intervals on the extreme right, the percentages add up to 10. Therefore, the sound pressure level is at or above 70 dB*A* during 10% of the time, which means $L_{10} = 70$ dB*A*. Using the same reasoning, the sound pressure level is at or above 50 dB*A* during 90% of the time and $L_{90} = 50$ dB*A*.

Because L_{50} does not fall at an endpoint of an interval, it must be calculated by interpolation. By inspection, the area under the histogram to the right of 60 dB*A* accounts for 35% of the total time. Similarly, the area to the right of 55 dB*A* accounts for 60% of the time. Therefore L_{50} lies between 55 and 60 dB*A*.

[8]More generally, a histogram for measurements of a random variable is a bar graph having a horizontal axis showing the intervals in which the variable can fall. The bar heights represent the relative frequencies with which the variable's measurements fall in the intervals. Figure 14.4 is an example in which the variable is sound pressure level.

TABLE 14.3 Sound Pressure Level Data Used for Histogram Construction

Intervals of A-Weighted Sound Pressure Levels (dBA)	Center of Interval (dBA)	Number of Seconds in Which dBA is within Interval	Percentage of Time dBA Is within Interval
$45 \leq L < 50$	47.5	60	10
$50 \leq L < 55$	52.5	180	30
$55 \leq L < 60$	57.5	150	25
$60 \leq L < 65$	62.5	90	15
$65 \leq L < 70$	67.5	60	10
$70 \leq L < 75$	72.5	30	5
$75 \leq L < 80$	77.5	30	5
Total		600	100

Starting from the right side of the figure, the area under the histogram representing 50% of the time is

$$25y + (15 + 10 + 5 + 5)(5)$$

To determine L_{50}, this expression is set equal to one half of the total area under the histogram, that is, $0.5 \times 5 \times 100$. Solving the resulting equation yields

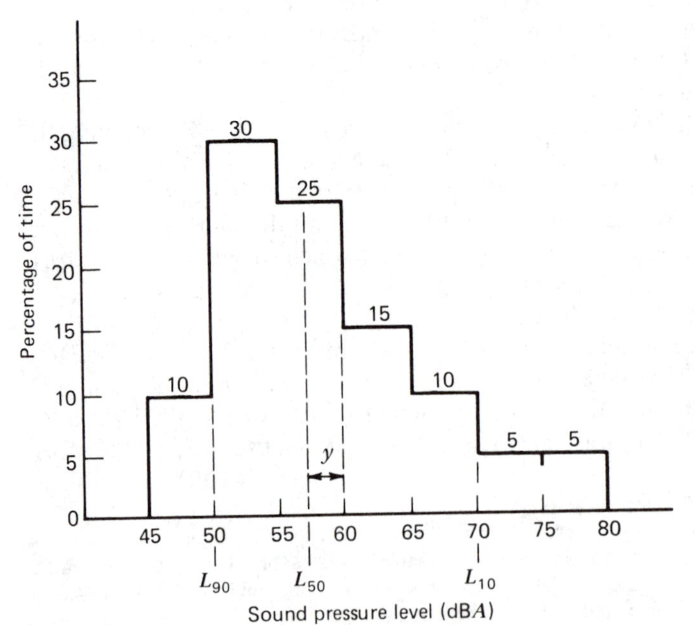

FIGURE 14.4 Determination of L_x from a histogram.

y = 3 dBA, and thus

$$L_{50} = 60 - 3 = 57 \text{ dB}A$$

Energy Equivalent Sound and the Day–Night Sound Level

Another statistic characterizing the noise level during a time interval is the *energy equivalent sound level,* L_{eq}. For any fluctuating sound occurring during a particular time period, L_{eq} is the constant sound pressure level that has the identical acoustic energy over the period.

The basis for computing L_{eq} is that a sound's acoustic energy is directly proportional to the square of its rms pressure. Consider, for example, a 1-hr interval during which the sound has an rms pressure of p_1 for the first 30 min and an rms pressure of p_2 during the second 30 min. If the equivalent sound, which is constant for 1 hr, is to have the same acoustic energy, it follows that

$$p_{eq}^2 = \tfrac{1}{2}p_1^2 + \tfrac{1}{2}p_2^2 \qquad \textbf{(14-4)}$$

where p_{eq} is the rms pressure of the energy equivalent sound. The right-hand side of the equation represents the average squared rms pressure of the two actual sounds. Using equation 14-4, together with the definitions of logarithms and sound pressure levels, it can be shown that

$$L_{eq} = 10 \log (\tfrac{1}{2}\, 10^{L_1/10} + \tfrac{1}{2}\, 10^{L_2/10}) \qquad \textbf{(14-5)}$$

Cunniff (1977) derives the following generalization of this relationship:

$$L_{eq} = 10 \log (f_1 10^{L_1/10} + f_2 10^{L_2/10} + \cdots + f_n 10^{L_n/10}) \qquad \textbf{(14-6)}$$

where f_i is the fraction of time during which the sound pressure level is constant at L_i, and $i = 1, \ldots, n$.

Equation 14-6 may be used to determine L_{eq} from noise histogram data. Consider, for example, Table 14.3. The table entries may be treated as if they resulted from seven sounds with pressure levels given by the "center intervals."[9] The appropriate time fractions are in the Percentage of Time column. Substituting these values into equation 14-6 yields L_{eq} = 66.7 dBA.

A variation of L_{eq} known as the *day–night sound level* (L_{dn}) is sometimes used to characterize community noise. It is computed in the same way as L_{eq} over a 24-hr period, except all sound pressure levels occurring between 10:00 P.M. and 7:00 A.M. are adjusted by adding 10 dBA. The adjustment reflects the increased annoyance, particularly interference with sleep, caused by night-time noise.

[9]The noise in an interval is assumed to be adequately represented by the sound pressure level at the midpoint of that interval.

A typical computation of L_{dn} first determines the L_{eq} for each hour in a 24-hr period. Then 10 dB*A* is added to each hourly value between 10:00 P.M. and 7:00 A.M. Finally, the equivalent sound pressure level for this adjusted set of 24 hourly values is calculated using equation 14-6.

The indicators L_{dn}, L_{eq}, and L_x (for x = 10, 50, and 90) are widely used both to describe and to regulate community noise. Projected values of these statistics are employed to assess the impacts of physical development projects.

EFFECTS OF NOISE ON PEOPLE

Noise influences people adversely in several ways.[10] One negative effect is impaired hearing, an important concern for some workers such as construction equipment operators and rock musicians. Hearing loss is generally reported as shifts in the "threshold of hearing," the sound pressure level at which a tone of a given frequency is barely audible. The threshold is different at different frequencies. Aspects of hearing loss are illustrated by an experiment involving one person who was exposed to a "broad-band" sound of 103 dB*A* for 2 hrs. Just prior to the experiment, the individual's hearing threshold was measured at several frequencies. The person's hearing thresholds were measured again at different times after the noise source was removed. For the individual being tested, there was a temporary threshold shift of 50 dB*A* (at 4000 Hz) soon after the noise was eliminated. In other words, compared to the preexposure condition, it took an additional 50 dB*A* before a tone at 4000 Hz was barely audible. The threshold shift was about 20 dB*A* after 5 hrs. One day later the person's hearing ability was almost back to normal.[11] Exposure to very high noise levels over long periods can lead to permanent threshold shifts.

In contrast to places like construction sites and rock concert halls, noise levels within the community at large do not generally pose a threat to hearing. To estimate the sound pressure levels at which community noise could impair hearing, the Environmental Protection Agency (1974) examined data on hearing loss in work environments. It concluded that a daily average L_{eq} of less than or equal to 70 dB*A* would protect the public from hearing damage.

Interference with speech communication is a common effect of community noise. Research has clarified the specific noise conditions under which speech communication can be carried out. Illustrative results are in Table 14.4. It indicates maximum sound pressure levels at which speech is intelligible in different situations.

Interference with sleep is another noise impact that has received attention by researchers. The extent to which different sound pressure levels awaken people

[10]Much information is available concerning the effects of noise on people, and Bugliarello et al. (1976) provide a detailed treatment of the subject.

[11]Experiments of this type are common. The figures reported here are from a graph of experimental results in Purdom and Anderson (1980, p. 384).

TABLE 14.4 Conversation Levels Required to Transmit Speech Outdoors under Various Conditions[a]

Distance (m)	Conversation Level		
	Relaxed[b]	Normal Voice	Raised Voice
1	45	65	72
2	40	60	65
3	36	56	62

Note. Table entries are the maximum steady sound pressure levels (in dB*A*) at which speech is intelligible under the conditions indicated.

[a]Based on information in Environmental Protection Agency (1974, p. D-5).

[b]Relaxed conversation dB*A* values are for 100% "sentence intelligibility"; this represents an ideal environment for speech communication.

[c]Normal and raised voice conversation figures are for 95% sentence intelligibility, which is often considered adequate for communication outdoors.

has been studied in a variety of settings. It has been found, for example, that the probability of a person being awakened by a peak sound pressure level of 40 dB*A* is only 5%, but it increases to 30% if the level is 70 dB*A* and to about 90% for an 80-dB*A* noise.[12] It is, of course, common knowledge that noise can awaken people or interfere with their ability to fall asleep. What is less well known is that noise can cause a shift in the stage of sleep (for example, from deep sleep to light sleep) without causing a person to wake up. The implications of these more subtle effects of noise are being investigated by sleep research specialists.

In addition to hearing loss and interference with speech and sleep, other impacts have been linked to community noise. One effect concerns the ability to perform complex tasks. It has been found that noise is more likely to reduce the accuracy of work than to reduce the quantity of work performed.[13] Other areas of investigation center on hypothesized linkages between noise and psychological stress and physiological disorders.[14]

Instead of investigating a particular noise impact, some researchers have surveyed communities to determine the extent to which people become annoyed at different sound pressure levels. It is difficult to compare results from different surveys because they often use dissimilar measures of noise (for example, L_{50} versus L_{dn}). Moreover, rating scales defining degrees of annoyance differ from one study to the next. Schultz (1979) attempted to put the data from 11 community surveys of various forms of transportation noise on a common footing, and his results are summarized in Figure 14.5. For each survey, annoyance levels

[12]These probability estimates are from Cunniff (1977, p. 111) and Rau and Wooten (1980, p. 4–31).

[13]Other effects of noise on task performance are reported by Cunniff (1977, p. 113).

[14]Cohen et al. (1981) review the literature on how noise affects various aspects of human physiology.

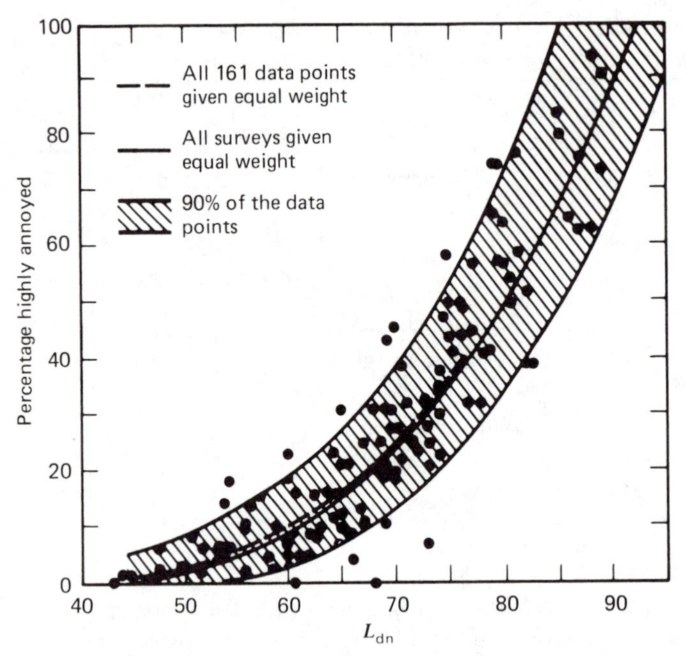

FIGURE 14.5 Data from 11 community surveys of annoyance with transportation noise. From Schultz (1979). Copyright ©, American Society for Testing and Materials, Philadelphia, Pa. Reprinted with permission.

were described in terms of *percentage highly annoyed* and sound pressure levels were put in the form of L_{dn}. The results show a clustering around the curves in Figure 14.5. Although Schultz's analysis is controversial, it represents an interesting attempt to combine disparate research results and make them more generally useful.[15]

NOISE IMPACT ASSESSMENT PROCESS

A typical assessment of noise impacts from a proposed project involves six steps:

1. Establish background noise levels.
2. Determine which land uses are potentially affected by project noise.
3. Identify applicable noise criteria.
4. Forecast future noise "with and without" the proposed project.
5. Compare predicted noise with applicable criteria.
6. Modify plans, if necessary, to deal with potential noise problems.

[15]The extent of controversy over this analysis is reflected in the discussion published with Schultz's (1979) paper.

The first step, establishing background noise levels in the vicinity of the proposed project, can sometimes be carried out by consulting governmental units such as city and county planning departments and transportation agencies. They may have results from recent noise surveys. Existing conditions can also be estimated by metering community noise directly. If the proposed project is near roadways or airports, background noise can be estimated using mathematical noise forecasting models. Much information about the sound pressure levels adjacent to transportation facilities is available, and it can be used with forecasting models to calculate sound pressure levels under existing conditions. Illustrative forecasting models are presented later in the chapter.

The second step in the noise assessment process is to determine land uses potentially influenced by noise from a proposed project. Existing (and expected) land uses in the vicinity of the project can usually be ascertained from local planning agencies. Special efforts are generally made to identify noise-sensitive land uses, such as schools, hospitals, and convalescent homes.

Several governmental sources can be consulted in carrying out the third step, determining noise criteria applicable to the potentially affected land uses. In the United States, limits on the maximum sound pressure level, or a statistical measure like L_{10}, are frequently specified in local noise ordinances. Sometimes statewide regulations of this type exist. In addition, a few federal agencies have issued recommendations that are used in noise assessments. A widely cited guide is the Environmental Protection Agency's (1974) report on "Information on Levels of Environmental Noise Requisite to Protect Public Health and Welfare with an Adequate Margin of Safety." The numerical limits in this report are only advisory, since EPA is not empowered to impose standards governing community noise.

Forecasting noise impacts, the fourth step in the assessment process, involves delineating scenarios of future conditions with and without the project. For example, if the proposal is to widen an existing highway, the scenarios would include information on average daily traffic, mix of vehicles (cars, diesel trucks, and so forth), and speed of vehicles. Conditions for both the existing highway and the proposed widened highway would be projected. For each scenario, mathematical models could be used to forecast future sound pressure levels at various nearby locations.

Once noise impacts are predicted, they can be compared to applicable noise criteria. If the criteria are satisfied, further analysis may be unnecessary. However, if they are not met, then steps to meet the criteria are investigated. Actions to lessen noise problems can be grouped in three categories: source, path, and receptor. For example, suppose a proposed highway would violate acceptable noise levels for a nearby elementary school. The noise *source* could be reduced by decreasing the capacity of the highway. The noise *path* could be modified, for instance, by putting a 10-ft.-high solid wall between the proposed highway and the school. In addition, the *receptor* could be changed: for example, the school building could be purchased and converted to a commercial facility and the school could be relocated.

Although modifications to both source and receptor are sometimes made, a typical solution to noise problems caused by physical development projects is to modify the sound's path. This is often done by increasing the distance between the source and receptor and by using sound barriers. Rau and Wooten (1980) summarize information for predicting noise reductions obtainable from barriers, including vegetation and solid walls of different heights and materials. They observe, for example, that 24 dB is a practical upper limit for reducing sound pressure levels with a straight, solid, thin wall placed between a source and receptor. If the barrier barely obstructs the line of sight between the source and receptor, the reduction might be as low as 5 dB.

Since construction activities and transportation facilities are among the principal sources of community noise, they deserve special attention in a discussion of noise impact assessment. The following sections on construction and transportation facilities illustrate aspects of the six-part assessment procedure introduced above.

NOISE FROM CONSTRUCTION ACTIVITIES

Virtually all physical development projects, from the extension of a driveway to the implementation of public works, involve a construction phase. One need only contemplate what it is like to converse in the vicinity of a jackhammer to imagine the possible social disruption.

Forecasting Construction Noise

Noise from construction equipment is often characterized by a sound pressure level (L_1) at a particular distance (r_1) from the source.[16] Forecasting construction noise typically involves using known values of L_1 and r_1 to find the sound pressure level (L_2) at a specified distance (r_2) from the source. In the simplest case, the equipment is on a flat site free of obstructions that influence sound transmission. Under these conditions, sound pressure decreases with distance primarily because of "wave divergence." As the distance from the noise source increases, the emitted sound energy is distributed over larger and larger surface areas. The sound diminution is described using

$$L_2 = L_1 - 10 \log \left(\frac{r_2}{r_1}\right)^2 \qquad \textbf{(14-7)}$$

This relationship holds for a "point source" like a jackhammer. A different equation is used for a "line source" such as a highway containing a steady stream

[16]Listings of sound pressure levels for typical pieces of construction equipment are given by Magrab (1975, p. 155).

of traffic. Equation 14-7, sometimes referred to as the "inverse square law," indicates that the sound pressure level for a point source in an obstructed, "loss-free" environment decreases by 6 dB each time the distance from the source is doubled.[17] If the noise source is obstructed by nearby walls or buildings, equation 14-7 must be modified to account for reflections of sound waves off the obstructing surfaces.

The inverse square law is demonstrated by considering an unobstructed portable air compressor on a flat construction site. (Air compressors are used to operate construction equipment such as pavement breakers and rock drills.) The compressor in this example causes a sound pressure level of 75 dB*A* at a distance of 50 ft. Suppose it is necessary to predict the sound pressure level at a property line of the construction site, a point 200 ft from the compressor. Using equation 14-7 with $r_1 = 50$ ft, $r_2 = 200$ ft, and $L_1 = 75$ dB*A* yields the required estimate:

$$L_2 = 75 - 10 \log \left(\frac{200}{50} \right)^2 = 63.0 \text{ dB}A$$

Wave divergence is not the only factor causing sound pressure levels to decrease with distance. Physical barriers, atmospheric absorption effects, and climatic influences combine to further diminish the sound. As shown by Rau and Wooten (1980), the effects of this "excess attenuation" can be estimated empirically and represented as factors (in units of dB) to be subtracted from the sound pressure levels given by equation 14-7. Of the causes of excess attenuation, physical barriers receive the greatest attention in impact assessments because they are often proposed as ways of reducing noise problems.

Regulations Controlling Construction Noise

Construction-related noise has been regulated at both federal and local levels in the United States. EPA restricts noise from newly manufactured construction equipment under the Noise Control Act of 1972. For example, consider the regulation for portable air compressors promulgated by EPA in 1978: compressors with rated capacities at or above 75 ft^3/min must have an average sound pressure level of 76 dB*A* at a point 7 m (approximately 23 ft) from the compressor.[18] Another federal effort to control construction noise is the General Service Administration's limits on noise from equipment at federal construction projects.

[17]To see this, use equation 14-7 with $r_2 = 2r_1$. The inverse square law does not apply when the locations involved are very close to the noise source. Magrab (1975, p. 4) gives a precise definition of "very close." Hothersall and Salter (1977, p. 106) indicate that for motor vehicles, the 6-dB reduction applies only for distances greater than about 7 m from a vehicle.

[18]This compressor regulation and the excerpt below from the Chicago noise ordinance are quoted by Cunniff (1977, p. 197).

At the local level, the principal mechanism for restricting noise from construction and other sources is the community noise ordinance. As an example, consider a portion of a 1971 Chicago ordinance:

> *It shall be unlawful for any person to use any pile driver, shovel, hammer, derrick, hoist tractor, roller or other mechanical apparatus operated by fuel or electric power in building or construction operations between the hours of 9:30 p.m. and 8:00 a.m., except for work on public improvements and work of public service utilities, within 600 feet of any building used for residential or hospital purposes.*

In addition to prohibiting selected construction activities during various periods, community noise ordinances can also set a maximum value of L_{10} or L_{eq} at particular locations such as property lines.

FORECASTING NOISE FROM HIGHWAYS

Methods exist to forecast how highway traffic affects sound pressure levels at different locations. The A-weighted sound pressure level is often employed in these procedures, since it is considered a reasonable predictor of human response to noise from motor vehicles.

Some highway noise forecasting models are based on physical principles such as the law of conservation of energy. The simplest of these models represents a highway as an infinitely long source emitting a constant sound intensity. It assumes the road is on level terrain with no obstructions providing excess attenuation of the sound. As shown by Lyons (1973), the effects of wave divergence will cause the sound pressure level to decrease by 3 dBA each time the distance from the source is doubled. For example, a highway noise of 70 dBA at 50 ft would diminish to 67 dBA at 100 ft, assuming wave divergence is the cause of sound diminution.

Although models based on physical laws provide important insights, they are not used extensively in practice. Mathematical models derived from both empirical studies and theoretical principles are more commonly employed. A widely used forecasting approach developed for the National Cooperative Highway Research Board (NCHRP) is introduced below.[19] The NCHRP procedure has distinct models for passenger cars and trucks. Separate models are appropriate because the key source of automobile noise is typically tire–roadway interaction, whereas the engine is the dominant noise source for most trucks moving at moderate speeds.

The NCHRP procedure finds reference values of L_{50} for a given vehicle type

[19]This discussion is based on Hothersall and Salter's (1977) summary of the NCHRP procedure. A full description of the NCHRP approach is given by Gordon et al. (1971).

(auto or truck). The *reference conditions* consider the roadway as an infinitely long line source "at grade" on flat, unobstructed terrain; vehicles are assumed to be distributed uniformly along the lane. Empirical analyses indicate that under these circumstances L_{50} can be reasonably well predicted using the following variables:

V = volume of traffic in vehicles per hour (vph)
S = average vehicle speed in miles per hour (mph)
D = distance from centerline of roadway to sound receptor in feet

The NCHRP studies show that, for both autos and trucks, L_{50} increases as V increases, and decreases as D increases. However, the relationship between L_{50} and S depends on vehicle type. For passenger cars, L_{50} increases approximately in proportion to the third power of speed, whereas for trucks L_{50} typically decreases in direct proportion to speed. Analyses of field measurements and theoretical models of traffic noise yield equations relating L_{50} to V, S, and D. An example is the following formula for autos:[20]

$$L_{50} = 10 \log V - 15 \log D + 30 \log S + 10 \log\left[\tanh\left(1.19 \times 10^{-3}\frac{VD}{S}\right)\right] + 29 \quad \textbf{(14-8)}$$

This relationship, and an analogous one for trucks, was used to prepare sets of easy-to-use curves. Figure 14.6 contains a plot of equation 14-8 with D = 100 ft. It applies only to autos on roadways satisfying the above reference conditions and gives L_{50} for different combinations of V and S. For example, if there were 1000 autos per hour passing continually on a long, straight, flat stretch of highway at an average speed of 50 mph, the forecast of L_{50} at D = 100 ft would be approximately 63 dB*A* based on the curves in Figure 14.6.

A noise forecasting procedure that could only be used for highways meeting the reference conditions would be very limited. Most highways have complex vertical and horizontal configurations, and they are typically surrounded by barriers in the form of walls, buildings, and vegetation. In addition, traffic is not distributed evenly. The NCHRP method treats these departures from the reference conditions by adjusting the reference values of L_{50}.

The overall NCHRP procedure is as follows. The roadway under consideration is broken up into elements having similar slope, road surface type, and other features. It is assumed that V and S are constant on any one segment. Figure 14.7 shows a road that has been divided into elements. For each road segment, estimates are made of V and S for both autos and trucks. Reference values of

[20]The term *tanh* stands for hyperbolic tangent, a trigonometric function given in handbooks of mathematical tables and functions.

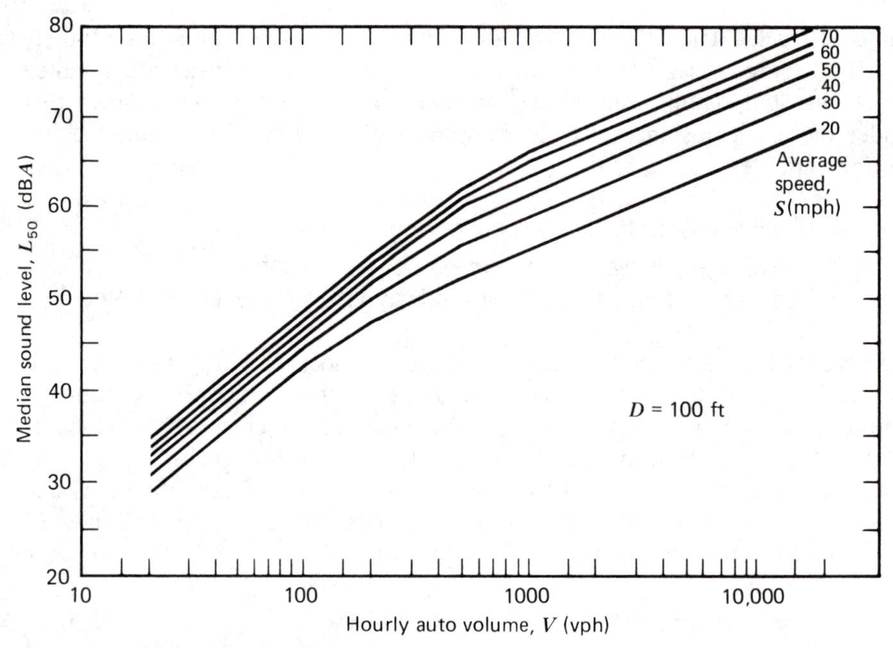

FIGURE 14.6 Relationship between L_{50} for automobiles and vehicle speed and volume of traffic. From Gordon et al. (1971).

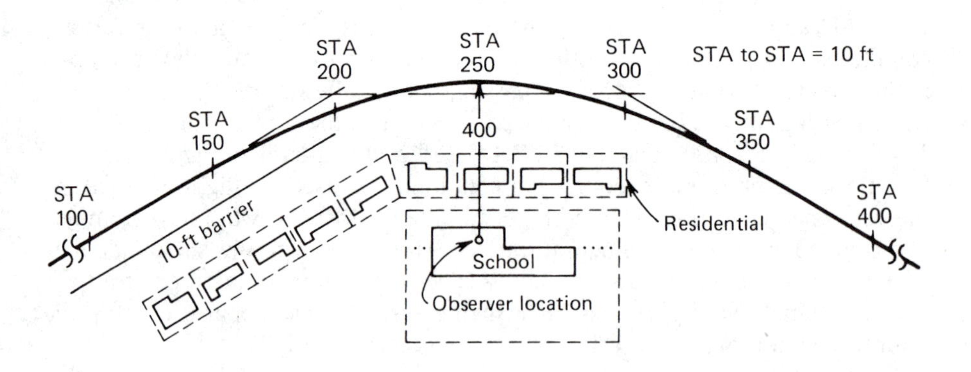

Road element No.	Description
1	Stations 100–200 on level roadway
2	Stations 200–300 on depressed roadway
3	Stations 300–400 on depressed roadway

FIGURE 14.7 Plan view of proposed highway and surroundings. Adapted from Gordon et al. (1971).

L_{50} are determined using the curves in Figure 14.6 for autos and analogous curves for trucks. Empirical and theoretical relationships are used to adjust the reference L_{50}'s based on

- Roadway width and number of lanes.
- Extent of vertical elevation or depression.
- Existence of "noninfinite" roadway lengths.
- Shielding by roadside barriers, structures, and landscaping.
- Roadway slope and road surface materials.
- Interruptions in traffic flow due to stop signs and signals.

To demonstrate how adjustments are made, suppose the road element in question had an average slope of 5%. In this case, the NCHRP procedure adjusts the reference L_{50} for trucks by adding 3 dB*A* to it. No adjustment is made for the reference L_{50} for autos, since noise does not increase significantly when cars travel uphill.

Once the adjusted L_{50}'s are computed for both autos and trucks, the NCHRP procedure determines L_{10}'s. Again, empirical and theoretical relationships are employed for this purpose. The last step of the NCHRP approach combines the sound pressure levels obtained separately for autos and trucks. This is done using logarithmic addition as in equation 14-3.

The NCHRP procedure is just one of many highway noise prediction methods in use. It forecasts L_{50} and L_{10}. Other procedures, such as the model developed by Kugler et al. (1977), rely on L_{eq} and L_{dn} as descriptors of highway noise. Goldstein (1979) provides equations for converting from one set of indicators to another.

For several highway noise forecasting models, comparisons have been made between predicted noise levels and corresponding observed noise levels. An example is given by Cohn and McColl (1979). They determined values of L_{10} using noise measurements adjacent to highways in the Albany, New York, metropolitan area. Fifty-two sets of short-term (15 to 20 min each) noise data were employed in the analysis. Cohn and McColl used the NCHRP procedure to predict L_{10}'s for each of the 52 cases. The average difference between predicted and measured values of L_{10} was 2.44 dB*A*.[21] The analysis also indicated traffic flow conditions under which other forecasting procedures gave better predictions than the NCHRP method.

FORECASTING NOISE FROM AIRPORTS

The unique characteristics of aircraft noise have led to the development of special noise indicators. They fall into two related categories: (1) measures of

[21]The standard deviation of the differences between the observed and predicted values of L_{10} was 2.92 dB*A*. Observed values of L_{10} ranged from 52 to 80 dB*A*. Comparative assessments were also conducted for L_{50} and L_{eq} as predicted using the method in Kugler et al. (1977).

the "perceived noisiness" of the takeoff or landing of a particular aircraft type and (2) measures of "community annoyance" resulting from the combined effects of all aircraft using an airport in a typical time period.

Noise from a single aircraft flyover is often described by *perceived noise level* (PNL). The design of this indicator rests on experiments in which individuals indicated their reactions to the "noisiness" or annoyance of different sounds. The experiments indicated that at higher frequencies lower sound energies are required to produce a given level of annoyance or noisiness.[22] In other words, high-frequency noise is judged to be noisier than equally loud noise of low frequency. Observe that noisiness is different from loudness.[23]

Perceived noise level, which is based on experimental tests measuring noisiness, is recorded in "perceived noise decibels" (PNdB). A doubling of "perceived noisiness" is equivalent to an increase of 10 PNdB.[24] Because the procedure for determining PNL is complex, it is common to approximate PNL using easy to measure parameters such as the *D*-weighted decibel. Sound level meters can automatically apply "*D*-weighting correction factors." Sound pressure levels in different frequency bands are corrected to approximate results from experiments testing the human perception of noisiness. The "*D*-weighting network" involves the same kind of electronic addition of adjusted sound pressure levels used in computing the *A*-weighted sound pressure level in Table 14.1. An approximate relationship between *D*-weighted sound pressure levels (L_d) and the perceived noise level is

$$\text{PNL} \approx L_d + 7 \tag{14-9}$$

where L_d is the maximum *D*-weighted sound pressure level recorded during an aircraft overflight. A rough estimate of PNL can also be obtained by adding 13 to the maximum *A*-weighted sound pressure level recorded during an aircraft's takeoff or landing. The accuracy of these approximations to PNL depends on the spectrum of the measured sound.

Although PNL is considered a useful measure of noisiness from aircraft flyovers, it does not fully account for the following: (1) as the *duration* of a single-event sound increases in time, ratings of perceived noisiness increase in magnitude; and (2) if a complex sound has an identifiable *pure tone,* such as the whining of a jet aircraft, it is judged to be noisier than a similar sound without such a tone. The *effective perceived noise level* (EPNL) accounts for these two factors by "correcting" the PNL readings

$$\text{EPNL} = \text{PNL} + C_1 + C_2$$

[22]The basis for the PNL is described more fully by Goldstein (1979).

[23]Loudness is more closely related to experiments in which listeners indicate their ability to hear sounds at different frequencies. These experiments are the basis for the *A*-weighted decibel.

[24]This interpretation of a doubling of perceived noisiness is explained by Magrab (1975, p. 63).

with EPNL measured in EPNL decibels, and

C_1 = a factor reflecting the existence of identifiable pure tones in the sound spectrum ($0 \leq C_1 \leq 3$)

C_2 = a factor accounting for the duration of a flyover[25]

Manufacturers estimate contours of EPNL while testing new aircraft under various government certification programs. These contours, referred to as *noise footprints,* represent the loci of points on the ground where EPNL is constant for a given aircraft type performing under a particular set of conditions. Figure 14.8 shows noise footprints for a takeoff and landing. Contours of EPNL depend on operating procedures followed in takeoff and landing and on conditions of temperature and relative humidity.

Indicators of the noisiness of a particular takeoff or landing are used in constructing measures of community annoyance from an airport's operations on a typical day. The *noise exposure forecast* (NEF), a widely used measure of community annoyance from airports, provides an illustration.[26] The NEF uses contours of EPNL together with "corrections" to account for the increased community annoyance caused by night-time aircraft operations. Procedures for computing NEF generally require computers and the following information:[27]

- Airport runway configurations.
- Number of aircraft operations by aircraft type.
- Aircraft flight plans and arrival and departure data.
- Percentage use of each flight path (by aircraft type) for takeoffs and landings.

Results from an assessment of airport noise can be presented as contours of NEF (see Figure 14.9). An analysis of NEF contours indicates particular land areas likely to be affected by a proposed airport or by changes in operations at an existing airport. Sometimes L_{dn} contours are used. The two measures are related approximately by

$$L_{dn} \approx \text{NEF} + 35$$

The EPA (1974) has issued guidelines indicating the fraction of people likely to be annoyed when L_{dn} or NEF values exceed various levels.

In general, theoretically derived noise forecasting models, such as the inverse square law, are not used alone in forecasting airport noise. Predictions are more commonly based on both empirical and theoretical findings. Although computers are often needed, there are methods for making simple, approximate predictions

[25]The factor C_2 is computed as $C_2 = 10 \log (\Delta t/20)$, where Δt is the time interval (in seconds) during which the noise level is within 10 PNdB of the maximum instantaneous value of the perceived noise level (Cunniff, 1977, p. 157).

[26]Goldstein (1979) discusses several other indicators of community annoyance due to airport noise.

[27]For a summary of a well-documented Federal Aviation Administration computer program for predicting NEF and L_{dn} contours, see Cohn and McVoy (1982, pp. 178–185).

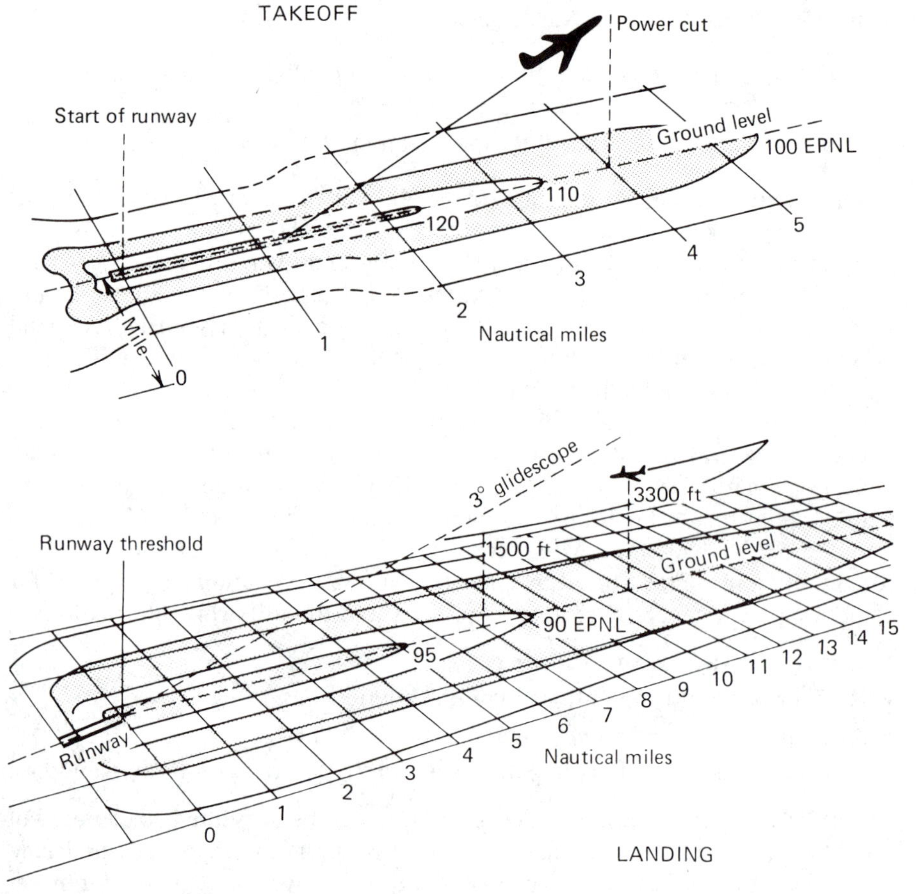

FIGURE 14.8 Contours of EPNL for aircraft takeoff and landing. From U.S. Department of Transportation (1972).

without them.[28] The development of noise forecasting procedures is a complex and active field of research, and acoustical specialists must be consulted for the most up-to-date approaches.

KEY CONCEPTS AND TERMS

THE NATURE OF SOUND AND ITS MEASUREMENT
- Air pressure fluctuations
- Pure tones
- Frequency of oscillation
- Cycles per second, Hz
- Amplitude of oscillation
- Root-mean-square pressure, Pa
- Sound pressure level, dB
- Sound spectrum
- Octave bands
- *A*-Weighted sound pressure level, dB*A*

[28]An example is the report by Pallet et al. (1978) containing worksheets and background information for making rough estimates of airport noise; the report also contains worksheets for predicting noise from highways.

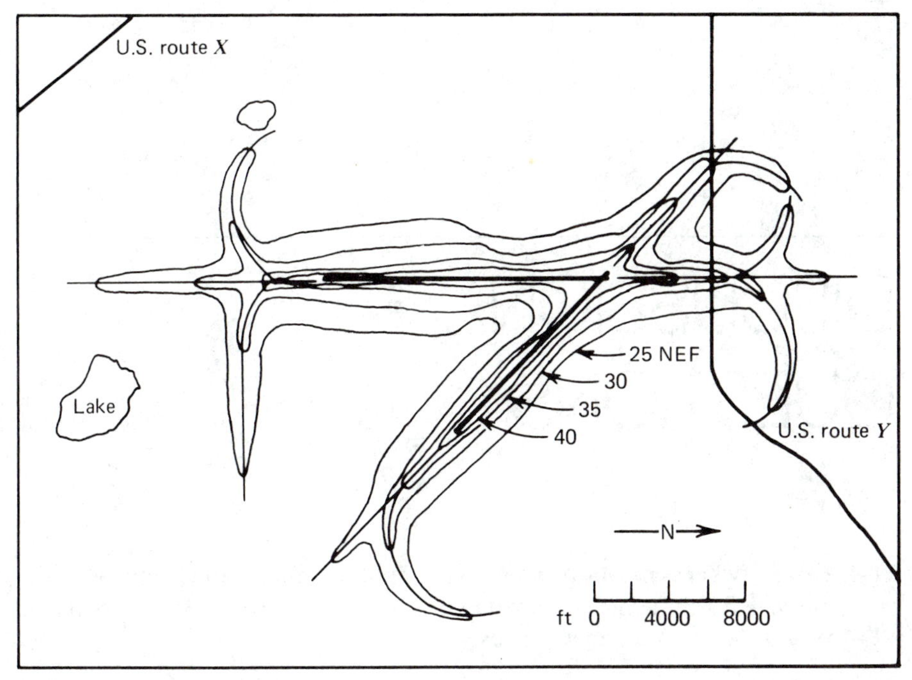

FIGURE 14.9 Typical NEF contours around an airport. From U.S. Department of Transportation (1972).

STATISTICS USED TO CHARACTERIZE COMMUNITY NOISE
- Histogram of sound pressure levels
- Sound pressure level exceeded *x*% of time
- Energy equivalent sound level
- Day–night sound level

EFFECTS OF NOISE ON PEOPLE
- Shift in threshold of hearing
- Interference with speech communication
- Shifts in stages of sleep
- Community annoyance

NOISE IMPACT ASSESSMENT PROCESS
- Community noise survey
- Noise-sensitive land use
- Noise criteria
- Local noise ordinances
- Modifications in source, path, and receptor

NOISE FROM CONSTRUCTION ACTIVITIES
- Point source versus line source
- Inverse square law
- Wave divergence
- Excess attenuation
- Physical barriers

FORECASTING NOISE FROM HIGHWAYS
- Empirical models
- Tire–roadway interaction
- Engine noise
- NCHRP procedure

FORECASTING NOISE FROM AIRPORTS
- Noisiness versus loudness
- Perceived noise level
- *D*-weighted sound pressure level
- Effective perceived noise level
- Noise footprints
- Noise exposure forecast

DISCUSSION QUESTIONS

14–1 Explain what is meant by loudness, amplitude, pitch, and frequency using musical instruments you are familiar with. Use two instruments having very different frequency ranges.

14–2 Suppose a sound with the following characteristics was received by a sound level meter that recorded A-weighted decibels:

Frequency (Hz)	*Sound Pressure Level (dB)*
125	50
250	55

Estimate the meter reading in dBA.

14–3 Consider the following data for a particular street corner from 4:00 P.M. to 6:00 P.M. on a weekday.

Sound Pressure Level Intervals	*Number of Minutes in Which dBA Is within Interval*
$50 \leq L < 55$	20
$55 \leq L < 60$	40
$60 \leq L < 65$	35
$65 \leq L < 70$	15
$70 \leq L < 75$	10

Compute L_{10}, L_{50}, and L_{90}.

14-4 The following data characterize the distribution of sound pressure levels at a particular point on a typical day:

Time Period	*dBA*
Midnight–6 A.M.	40
6 A.M.–noon	50
noon–6 P.M.	55
6 P.M.–midnight	45

Compute L_{eq} and L_{dn}.

14-5 Use equation 14-6 to verify that $L_{eq} = 66.7$ dB*A* for the data in Table 14.3.

14-6 In the United States, community noise ordinances control maximum sound pressure levels at property lines. Cunniff (1975) reports that a typical daytime limiting value for commercial areas is 60 to 65 dB*A*. Explain why this regulatory approach would not be effective for controlling construction noise. Why would a limit on L_{10} or L_{eq} be more useful?

14-7 Characterize the relative difficulty in forecasting noise from construction activities, highways, and airports.

REFERENCES

Bugliarello, G., A. Alexandre, J. Barnes, and C. Wakstein, 1976, *The Impact of Noise Pollution, A Socio-Technological Introduction.* Pergamon, New York.

Cohen, S., D. S. Krantz, G. W. Evans, and D. Stokols, 1981, Cardiovascular and Behavioral Effects of Community Noise. *American Scientist* **69 (5),** 528–535.

Cohn, L. F., and W. McColl, 1979, L_{10} versus L_{eq}—A User's Perspective, *in* R. J. Pepper, and C. W. Rodman (eds.), *Community Noise,* pp. 237–246. ASTM Special Tech. Pub. 692. American Society of Testing Materials, Philadelphia, Pa.

Cohn, L. F., and M. R. McVoy, 1982, *Environmental Analysis of Transportation Systems.* Wiley, New York.

Council on Environmental Quality, 1979, *Environmental Quality, the Tenth Annual Report of the Council on Environmental Quality.* U.S. Government Printing Office, Washington, D.C.

Cunniff, P. F., 1977, *Environmental Noise Pollution.* Wiley, New York.

Environmental Protection Agency, 1974, *Information on Levels of Environmental Noise Requisite to Protect the Public Health and Welfare with an Adequate Margin of Safety.* Technical Document 550/9-74-004, EPA, Washington, D.C.

Goldstein, J., 1979, Description of Auditory Magnitude and Methods of Rating Community Noise, *in* R. J. Pepper and C. W. Rodman (eds.), *Community Noise,* pp. 38–72. ASTM Special Tech. Pub. 692, American Society of Testing Materials, Philadelphia, Pa.

Gordon, C. G., W. J. Galloway, B. A. Kugler, and D. L. Nelson, 1971, *Highway Noise: A Design Guide for Highway Engineers.* Report No. 117, National Cooperative Highway Research Program, Washington, D.C.

Hothersall, D.C., and R. J. Salter, 1977, *Transportation and the Environment.* Granada, Hertfordshire, Great Britain.

Kugler, B. A., et al., 1977, *Highway Noise, A Design Guide for Prediction and Control.* Report No. 174, National Cooperative Highway Research Program, Washington, D.C.

Lyons, R. H., 1973, *Lectures in Transportation Noise.* Grozier, Cambridge, Mass.

Magrab, E. B., 1975, *Environmental Noise Control.* Wiley, New York.

Pallett, D. S., R. Wehrli, R. D. Kilmer, and T. L. Quindry, 1978, *Design Guide for Reducing Transportation Noise in and Around Buildings.* Building Science Series 84, U.S. Department of Commerce, National Bureau of Standards, Washington, D.C. (April).

Purdom, P. W., and S. H. Anderson, 1980, *Environmental Science, Managing the Environment.* Charles G. Merrill, Columbus, Ohio.

Rau, J. G., and D. C. Wooten (eds.), 1980, *Environmental Impact Analysis Handbook.* McGraw–Hill, New York.

Schultz, T. J., 1979, Community Annoyance with Transportation Noise, *in* R. J. Pepper and C. W. Rodman (eds.), *Community Noise,* pp. 87–103. ASTM Special Tech Pub. 692, American Society of Testing Materials, Philadelphia, Pa.

U.S. Department of Transportation, 1972, *Transportation Noise and Its Control.* Publication DOT P 5630.1, Washington, D.C. (June).

CHAPTER 15

ESTIMATING AIR QUALITY IMPACTS

Since the 1940s, much attention has been devoted to analyzing the causes and consequences of air-borne residuals. The literature on the subject is voluminous and it crosses several academic disciplines. This chapter draws from that literature and introduces techniques for conducting an air quality impact assessment.

THE NATURE OF AIR POLLUTION

Air pollution is the presence of air-borne residuals such as dust, fumes, and smoke at levels causing injury to life and property. To protect public health and welfare, U.S. air quality laws mandate controls on emissions of carbon monoxide (CO), hydrocarbons (HC), oxides of nitrogen (NO_x), oxides of sulfur (SO_x), total suspended particulates (TSP), and lead (Pb). Although many other pollutants can be singled out, these have received the most attention in the United States.

Carbon monoxide, a colorless, odorless, and tasteless gas, is a product of incomplete combustion. The majority of the CO emissions in the United States during the 1970s were from motor vehicles. Carbon monoxide combines with hemoglobin in the human blood stream and reduces hemoglobin's ability to carry oxygen to body tissues. Depending on the degree of exposure, effects of inhaling CO range from headaches, dizziness, and nausea to unconsciousness and death. Coffin and Stokinger (1977, p. 305) report that these effects are not "to be expected at even the highest concentrations found in the ambient air . . . . However, great concern has been expressed regarding the possibility that covert effects may decrease physical or mental acuity or interfere with the function of organs already suffering oxygen deficiency, such as the heart in coronary disease."

Hydrocarbons, chemical compounds of hydrogen and carbon, are present in an enormous number of forms. A major source of air-borne hydrocarbons is the combustion exhaust from motor vehicles. In addition, industrial operations such as dry cleaning contribute large quantities of hydrocarbons to the ambient air through evaporation. Although hydrocarbons produce some adverse effects directly, they are of special concern because of their role in chemical reactions yielding photochemical smog. Reaction products, referred to collectively as *photochemical oxidants,* cause a variety of ailments (such as difficulty in breathing) and damage to vegetation and property. Ozone is the principal reaction product and is often used as an overall indicator of photochemical oxidants.

Oxides of nitrogen, especially nitrogen dioxide and nitric oxide, are characteristic products of high-temperature combustion processes. The most significant sources of NO_x emissions in the United States are motor vehicles and coal- and gas-fired power plants. Nitrogen oxides can cause a variety of negative effects including nose and eye irritation and atmospheric discoloration. In addition, they play an important role in the production of photochemical oxidants.

Emissions of sulfur oxides, mainly sulfur dioxide, also result from combustion processes, in this case the incomplete combustion of fuels containing sulfur impurities. Although much of the concern over SO_x emissions is directed toward coal-fired power plants, industrial facilities such as copper smelters are also an important source. Sulfur dioxide in the ambient air has been linked to a variety of human health problems, especially respiratory disorders, and to the formation of acid precipitation.

The term *particulates* refers to the multitude of solid and liquid particles present in the atmosphere. Somewhat more than half of the air-borne particulates are estimated to originate from natural sources such as dust storms and volcanic eruptions.[1] The main sources of man-made particulates are coal combustion and various industrial processes. Forest fires, both natural and man-made, are also a major contributor. Particulates exist in a range of sizes and can adsorb many chemicals present in the ambient air. High concentrations of particulates have been linked to increases in respiratory diseases and gastric cancers, poor atmospheric visibility, and the soiling of buildings and other materials.

Lead in the atmosphere results primarily from motor vehicles fueled with leaded gasoline. A variety of human health problems, including liver and kidney damage and nervous system disorders, have been linked to atmospheric lead. In the United States, concern over these problems inspired a nationwide effort to eliminate the use of lead additives in gasoline. By the late 1980s, atmospheric lead from motor vehicles is not expected to be a significant problem in the United States.

Combustion processes, which occur in motor vehicle engines, steam electric power plants, home heating units, and various industrial operations, constitute

[1]For a breakdown of man-made and natural sources of air-borne particulates, see Perkins (1974, pp. 30–32).

the major sources of modern air pollution problems. The materials used as fuels and the way fuels are burned are key elements in determining air quality. Some experts feel that air pollution problems can be significantly reduced only if there is a major change in sources used for energy production.[2]

AIR QUALITY IMPACT ASSESSMENT PROCESS

A comprehensive assessment of a proposed project's air quality impacts involves the following steps:

1 Establish background air quality levels.
2 Identify applicable air quality criteria and standards.
3 Forecast future air pollutant *emissions* with and without the project.
4 Forecast future ambient air pollutant *concentrations* with and without the project.
5 Compare predicted air quality with applicable standards.
6 Modify plans, if necessary, to deal with potential air quality problems.

The first step, establishing background levels of air quality, is carried out only for air quality indicators likely to be influenced by the proposal. For example, if the proposed project were a highway, the main indicators of interest would be hydrocarbons, nitrogen oxides, carbon monoxide, and photochemical oxidants. In the United States, data on background air quality levels are available from local, regional, and state air quality management agencies, various state implementation plans proposed in response to federal air quality laws, and EPA's computerized data retrieval systems. The EPA data include measurements from the federal Continuous Air Monitoring Program (CAMP) stations established in the early 1960s.

Applicable air quality criteria and standards are determined in the second step of the assessment process. The EPA has issued criteria relating levels of air pollutants to human health and welfare.[3] In the United States, standards used in impact assessments include the national ambient air quality standards and pertinent state or local standards.

The NAAQS, as illustrated by carbon monoxide standards in 1980, stipulate both concentration and time of exposure. The CO standards require that the average concentration for any 1-hr period be less than 9 parts of CO per million parts of ambient air. These CO measurements are said to have an "averaging time" of 1 hr. The NAAQS also require average CO concentrations during any 8-hr period to be less than 35 parts per million (ppm). Limits on CO are not

[2]Chambers (1976) provides further discussion of the linkage between energy sources and air quality.

[3]"Criteria documents" have been issued by EPA for all pollutants included in the NAAQS. See, for example, Environmental Protection Agency (1971).

to be exceeded more than once per year. As of 1980, the NAAQS applied to particulates, SO_x, CO, NO_2, hydrocarbons, ozone, and lead.[4] The national standards are revised periodically and up-to-date versions are obtainable from EPA.

In the third step of the assessment process, the proposed project's emissions are estimated in units of weight (or mass) per time period. Procedures for estimating emissions are reviewed in the next section.

The fourth step, predicting changes in the ambient *concentrations* of air quality indicators due to a new discharge, is often complex. In fact, sometimes it is not carried out and the assessment considers only the increased *emissions* from the proposed project. Reasons for not estimating concentrations include (1) inadequate understanding of the underlying physical and chemical processes and (2) unwillingness to commit the time and money needed to utilize existing forecasting procedures.

The final steps in a comprehensive impact assessment are to compare forecasted concentrations with applicable standards and to modify the proposed project if expected air quality degradation is unacceptable. Air-borne residuals are commonly reduced by changing combustion processes and using emission control devices such as scrubbers and filters.[5] However, there are many options for mitigating adverse air quality effects that do not involve control devices. Examples include reducing the scale of a facility or changing the locations of discharges (see Table 3.1).

Forecasts of air pollutant emissions and concentrations are carried out at various levels of sophistication. Some impact assessments are limited to quick and simple estimates of increases in emissions. Others use elaborate computer-based mathematical models to translate increases in emissions into changes in concentrations at various times and places.

ESTIMATING EMISSIONS OF RESIDUALS

Emission factors are often used to project amounts of air-borne residuals resulting from a proposed action. An emission factor is the estimated quantity of residuals discharged per unit of activity. For instance, the Environmental Protection Agency (1973) indicated that when coffee beans are processed in a direct-fired roaster, 7.6 lb of particulates is released per ton of coffee processed. EPA periodically publishes updated discharge data in its "Compilation of Emission Factors."

The use of factors to project emissions is illustrated by EPA data for power plants fired with bituminous coal. The main emissions from such plants are particulates and sulfur oxides. Other substances discharged include carbon monoxide, hydrocarbons, nitrogen oxides, and aldehydes. The EPA compilations

[4]A listing of the NAAQS in 1980 is given in Council on Environmental Quality (1980, p. 172).

[5]Traditional technology for controlling emissions is described in detail by Stern (1977).

include emission factors for each of these materials and instructions for adjusting factors to reflect variations in the sulfur and ash content of coal. For example, EPA indicates that $16 \cdot A$ lb of particulates are emitted per ton of coal burned, where A is the percentage of ash in the coal.[6] If a proposed power plant burns bituminous coal with a 10% ash content at a rate of 200 tons/hr, the estimated emission of particulates is

$$(16 \times 10) \frac{\text{lb}}{\text{ton of coal}} \times 200 \frac{\text{tons of coal}}{\text{hr}} = 32{,}000 \text{ lb/hr}$$

This estimate assumes no emission controls are employed. Using devices known as electrostatic precipitators, as much as 99.5% of the particulates can be removed.

Emission factors for mobil sources must account for variations in operating conditions. For example, carbon monoxide releases from an automobile depend significantly on the vehicle's speed and whether it has been warmed-up. A car operating at 20 mph might emit twice as much CO per mile as it would at 45 mph.[7] Also, emissions of CO (per mile) during the first 3–5 min after starting a cold vehicle can be four times higher than emissions after the vehicle is warmed up.[8] Variables such as speed and whether a vehicle starts cold or hot are accounted for in EPA's computer-based procedures for estimating motor vehicle emission factors. According to Cohn and McVoy (1982, p. 201), average "fleet" emission factors given by these procedures are especially sensitive to year of analysis, average vehicle speed, "start mode" (hot versus cold), ambient temperature, and vehicle mix (percentage of autos, motorcycles, diesel trucks, and so forth).

Once average emission factors are estimated for a fleet of vehicles, they are combined with traffic projections to yield forecasts of total emissions. For instance, suppose all vehicles on a proposed highway segment consist of passenger cars. For operating conditions of the average car in 1990, assume EPA estimating procedures indicate a CO emission of 7 g/mile. Total daily CO emissions expected in 1990 equal 7 g/vehicle-mile multiplied by the number of vehicle-miles-traveled (VMT) per day by automobiles using the highway in 1990. Projections of VMT are generally available from applicable highway planning studies.[9]

[6]This factor applies to pulverized bituminous coal in a "general-type furnace" of size greater than 10^8 British thermal units/hr. (Environmental Protection Agency, 1973, p. 1.1–3).

[7]A graph showing variations in average CO emissions with vehicle speed is given by Environmental Protection Agency (1973, p. 3.1.1–7).

[8]Horowitz (1982, pp. 167–171) discusses how CO emissions vary depending on whether the car is started in a cold or warmed-up state.

[9]Realistic examples demonstrating computer-based procedures for estimating mobile source emission factors are given by Horowitz (1982). His applications concern air quality impacts of mass transit improvements, car pool incentives, and other transportation management schemes.

SIMPLE MODELS RELATING EMISSIONS TO CONCENTRATIONS

Most methods for predicting how emissions influence ambient air quality rely on mathematical models. Rudimentary models providing rough estimates of how emissions affect concentrations are introduced below. Before proceeding, however, a warning is in order. Forecasting air pollutant concentrations is a specialized endeavor that can easily be carried out incorrectly. Atmospheric pollutant transport processes are complex, especially when emissions are chemically reactive. Except for very simple cases, forecasting concentrations is best left to air quality modeling specialists.

Box Models

A *box model* assumes that a constant emission rate, P (units of mass/time), enters a volume of ambient air moving in one direction at a constant speed, U. The air in motion is confined from above by a layer of stable air at an elevation, h. The moving air is also confined in the direction perpendicular to the wind speed (see Figure 15.1). This model represents a valley, in which air passes through a zone of width, w, formed by two rows of hills. Air-borne residuals are diluted and carried from the area by lateral motion in the direction of the wind.

The simplest box model assumes steady-state conditions: the emissions, wind speed, and characteristics of air available for dilution do not vary over time. It

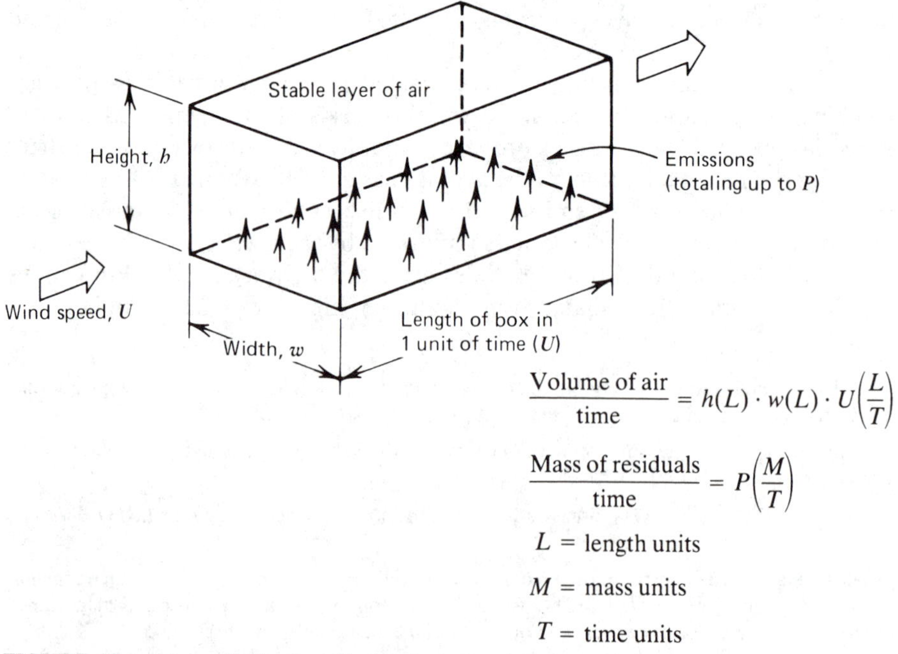

FIGURE 15.1 Air available for dilution in a simple box model.

also assumes discharges mix completely and instantaneously with the air available for dilution and the released material is chemically stable and remains in the air. Under steady-state conditions, the air volume passing over the area in a unit of time equals the product of wind speed, U, and cross-sectional area, hw (see Figure 15.1). This relationship for volume per time is based on the law of conservation of mass applied to steady fluid motion. The concentration of the discharged material in the ambient air equals the mass of substance emitted per time divided by the corresponding air volume available for dilution. In mathematical terms,

$$c = \frac{P}{Uhw} \tag{15-1}$$

where c is the average concentration under steady-state conditions, and other terms are as previously defined. The numerator has units of mass per time and the denominator is in volume per time. Therefore, c has units of mass of substance emitted per volume of air. Commonly employed units yield concentrations in micrograms (10^{-6} g) per cubic meter, abbreviated as $\mu g/m^3$.

Box models are used to estimate average air pollutant concentrations in regions with many small sources that are distributed uniformly. Kohn's (1978) application of equation 15-1 for sulfur dioxide in the St. Louis airshed during 1970 is an example. Average parameter values for the box model as applied to St. Louis in 1970 were estimated as

$$\begin{aligned} U &= 15{,}400 \text{ m/hr, average annual wind speed} \\ h &= 1210 \text{ m, ``mixing height''} \\ w &= 10^5 \text{ m, ``width of the airshed''} \end{aligned}$$

Approximately 1375 million lb of sulfur dioxide was discharged in the airshed during 1970. This corresponds to an emission rate of

$$P = \frac{1375 \times 10^6 \text{ lb/year} \times 454 \text{ g/lb} \times 10^6 \ \mu\text{g/g}}{8760 \text{ hr/year}} = 7.126 \times 10^{13} \ \mu\text{g/hr}$$

Direct substitution into equation 15-1 yields

$$c = \frac{P}{Uhw} = \frac{7.126 \times 10^{13}}{15{,}400 \times 1210 \times 10^5} = 38 \ \mu\text{g/m}^3$$

which is equivalent to a volumetric concentration of 0.014 part of SO_2 per million parts of ambient air.[10] During 1970, the average SO_2 concentration observed at

[10]The factor used by Kohn to convert from units of $\mu g/m^3$ to ppm is 0.000369. Bibbero and Young (1974, p. 20) explain how this conversion constant can be derived using the ideal gas law. The constant depends on the molecular weight of SO_2 and on the temperature and pressure of the ambient air.

the federal Continuous Air Monitoring Program station in St. Louis was 0.03 ppm. The box model result is thus only about half the observed concentration. One explanation for the discrepancy is that the CAMP station was located in a zone where SO_2 concentrations were much higher than average; the box model gives only a regional average concentration.

Box models have several limitations. As illustrated by the SO_2 figures above, the concentrations at some times and places within a region can be very different from the average yielded by a box model. In addition, box models do not account for atmospheric dispersion of materials in vertical and lateral directions. They also assume that emissions are chemically nonreactive during the time periods used in the analysis. While it is often reasonable to consider CO, SO_x, and particulates as chemically stable, this assumption is inappropriate for emissions such as hydrocarbons and nitrogen oxides that contribute to photochemical smog.

Rollback Models

Rollback (or "proportional scaling") models provide another simple approach for estimating how emissions influence ambient air quality. These models *assume* a linear relationship between P, the total quantity of a substance discharged in a region during a particular period, and c, the concentration of the substance at a *specified point.* In mathematical terms, the assumption is

$$c = kP + b \tag{15-2}$$

where b ("background concentration") represents the concentration when emissions are zero, and k is an empirically determined constant (see Figure 15.2). The parameters k and b are for a particular pollutant and location.

Kohn's (1978) application of a rollback model to calculate concentrations at the CAMP station in St. Louis is illustrative. Emission and concentration data for 1963 were used to determine k for various pollutants. For example, 300×10^6 lb of particulates was discharged in the St. Louis region in 1963, and the concentration of particulates near the CAMP station was 128 $\mu g/m^3$. Kohn estimated the background level of particulates as 31 $\mu g/m^3$. Based on this information, k was computed as

$$k = \frac{c - b}{P} = \frac{128 - 31}{300 \times 10^6} = 0.32 \times 10^{-6}$$

This value of k was employed in calculating the particulates concentration near the CAMP station in St. Louis for 1970, a year when regional particulates emissions were 233×10^6 lb. Assuming b remained constant at 31 $\mu g/m^3$, Kohn computed the 1970 concentration using equation 15-2:

$$c = kP + b = (0.32 \times 10^{-6})(233 \times 10^6) + 31 = 106\ \mu g/m^3$$

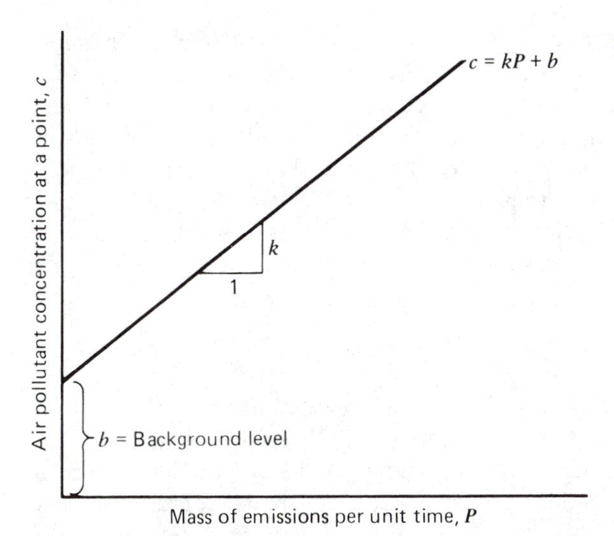

FIGURE 15.2 Linear relationship between emissions and concentrations in rollback models.

The measured concentration of particulates at the CAMP station in 1970 averaged 130 μg/m^3. Between 1963 and 1970, discharges of particulates decreased from 300×10^6 to 233×10^6 lb, but the average concentration at the CAMP station actually *increased* from 128 to 130 μg/m^3. Since it assumes a linear relationship between concentration and emissions, the rollback model predicted a *decrease* in concentration.

The model's failure to predict the increase in particulates concentration results partly because variations in meteorological conditions can cause k to change over time. For example, if average wind speeds were different in 1963 and 1970, that difference itself would cause an erroneous estimate of the 1970 particulates concentration. In addition, k can change if the locations of sources shift over time. Suppose that compared to 1963, a greater portion of the regional discharge was close to the CAMP station in 1970. The CAMP station concentration might then show an increase even though total regional emissions decreased.

Rollback models are also employed to determine discharge reductions needed to meet an ambient air quality standard, c_s. Suppose equation 15-2 is written for two time periods, a recent year in which the standard was violated, and a future year when the standard is met. Combining the two equations algebraically and then simplifying yields

$$R = 100\left(\frac{P_0 - P_s}{P_0}\right) = 100\left(\frac{c_0 - c_s}{c_0 - b}\right) \qquad \textbf{(15-3)}$$

where R is the requisite percentage reduction in emissions. The subscript 0 refers to when the ambient standard was violated, and s refers to the period in which the standard is met.

Equation 15-3 has been used to compute required reductions in emissions from motor vehicles. This application has been disputed because the linear form of rollback models cannot account for photochemical reactions that occur among vehicular emissions of HC and NO_x. Another objection is that rollback models have not been widely tested with observed data, and thus they can yield incorrect, misleading results. Despite these criticisms, rollback models are often used because of their simplicity. De Nevers and Morris (1975) have worked to develop "modified rollback models" that eliminate some of the shortcomings of the models above.

Statistical Models

Rollback models treat the concentration of an air pollutant at a point in terms of an *average* value. Using results from Larsen's (1971) statistical analysis of air quality records, rollback models can be extended to consider fluctuations in concentration at a given location. Larsen's statistical model represents short-term variations in air pollutant concentrations at a point as a "log-normal distribution."

Log-Normal Distributions To understand Larsen's use of log-normal distributions, consider a more common statistical form, the "normal distribution." Figure 15.3 represents the variability of hourly temperature readings at the CAMP station in Washington, D.C. for a 7-year period. It is interpreted in a similar way as the noise histogram in Figure 14.4. The horizontal axis indicates hourly temperatures and the vertical axis gives the frequency of occurrence of those temperatures. In this case, the temperature intervals used for purposes of preparing the graph are so small that the results are approximated as a smooth curve instead of the collection of bars in the noise histogram.

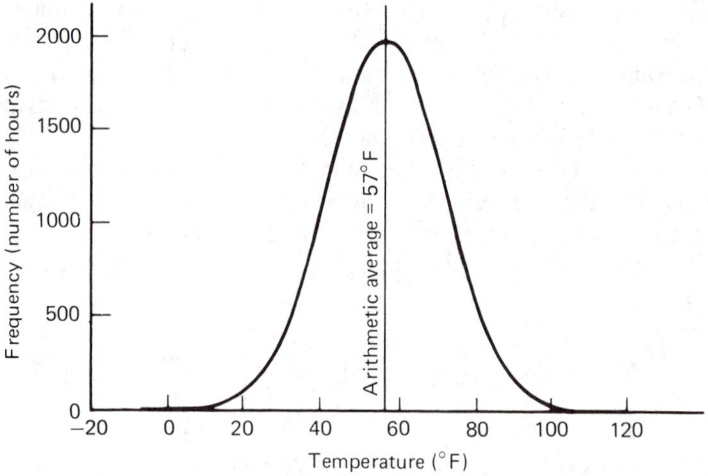

FIGURE 15.3 Estimated frequency of hourly temperatures at CAMP site, Washington, D.C., December 1, 1961, to December 1, 1968. Adapted from Larsen (1971).

The curve in Figure 15.3 is bell shaped and symmetric about 57°F, the average hourly temperature observed during the 7-year period. Values above 57°F occurred with the same frequency as values below it. Temperature is a "continuous variable." Its values are not limited to integer values like 60°F or 61°F; an enormous number of intermediate temperatures can be observed. When a frequency analysis of a continuous variable yields the bell-shaped pattern in Figure 15.3, it is common to assert that the variable, in this case hourly temperature, follows a normal distribution.[11] Statisticians are fond of normal distributions because they can be manipulated easily to yield interesting information.

Larsen's analysis of air quality data for eight cities established that the frequency of occurrence of a pollutant's concentration at a particular point did *not* have a normal distribution. Consider, for example, his 7 years of data for the 1-hr average sulfur dioxide concentrations at the CAMP station in Washington, D.C. The frequency analysis for SO_2 did not yield a bell-shaped curve like the one in Figure 15.3. Instead, SO_2 concentrations lower than the arithmetic average occurred much more frequently than values above the average. Based on preliminary findings, Larsen reexamined the data by taking the logarithm of each 1-hr SO_2 measurement. His frequency analysis of the transformed data indicated that variations in the logarithms of the SO_2 concentrations could be represented using a normal (bell-shaped) curve. In the language of statisticians, the original SO_2 concentrations follow a log-normal distribution. Larsen's data for CO, NO_x, HC, and photochemical oxidants also followed a log-normal distribution.

Use of log-normal probability paper simplifies the data analysis for an air quality indicator presumed to follow a log-normal distribution. The vertical axis on such paper represents the logarithm of concentration; the horizontal axis is the percentage of time a particular concentration is equaled or exceeded. Figure

[11]As explained by Larsen (1971, p. 9), the normal (or "Gaussian") distribution curve in Figure 15.3 is described using

$$Y = \frac{n}{s\sqrt{2\pi}} \exp\left[-\frac{1}{2}\left(\frac{t-m}{s}\right)^2\right]$$

where

Y = frequency of occurrence (hr)
t = hourly temperature (°F)
n = total number of hourly temperature readings (61,320)
m = arithmetic average of readings (57°F)
s = "standard deviation" of readings (15° F)

The curve has two parameters, m and s, and they are computed using formulas available in introductory statistics textbooks. The normal equation indicates, for example, that the number of samples with a t value of 40°F (more precisely, $39.5 < t < 40.5$), is

$$Y = \frac{61{,}320}{15\sqrt{2\pi}} \exp\left[-\frac{1}{2}\left(\frac{40-57}{15}\right)^2\right] = 858 \text{ hr}$$

15.4 shows results from using log-normal probability paper to perform a data analysis exercise suggested by Larsen. The measurements consist of average sulfur dioxide concentrations assumed to be taken in Washington, D.C. for each month of 1 year. Monthly average values for January through December, in micrograms per cubic meter, are 300, 250, 180, missing, 150, 60, 120, 100, 140, 160, 190, and 220, respectively.

Figure 15.4 is constructed as follows: First the 11 monthly average readings are ranked from highest to lowest as in Table 15.1. The symbol r represents the rank order of a measurement, where $r = 1$ corresponds to the maximum observed monthly SO_2 concentration. Next, f, the percentage of time a particular concentration is equaled or exceeded, is determined. Larsen calculated f using a formula based on the theory of log-normal distributions:

$$f = 100\% \left(\frac{r - 0.4}{n} \right) \qquad \textbf{(15-4)}$$

where n is the number of observations ($n = 11$ in this example). Equation 15-4 is for data points higher than the median (or middle value) of the observations. The median (which corresponds to $r = 6$ in this case) is plotted at the 50% frequency. Values of f for points below the median are found using equation 15-4, but with 0.4 replaced by 0.6. Figure 15.4 shows f values plotted against corresponding SO_2 concentrations.

Combining Rollback and Statistical Models Rollback models are sometimes combined with a log-normal representation of the distribution of air pollutant concentrations. This is demonstrated in EPA's (1975) impact assessment for a proposed wastewater treatment plant expansion in the Livermore-Amador Val-

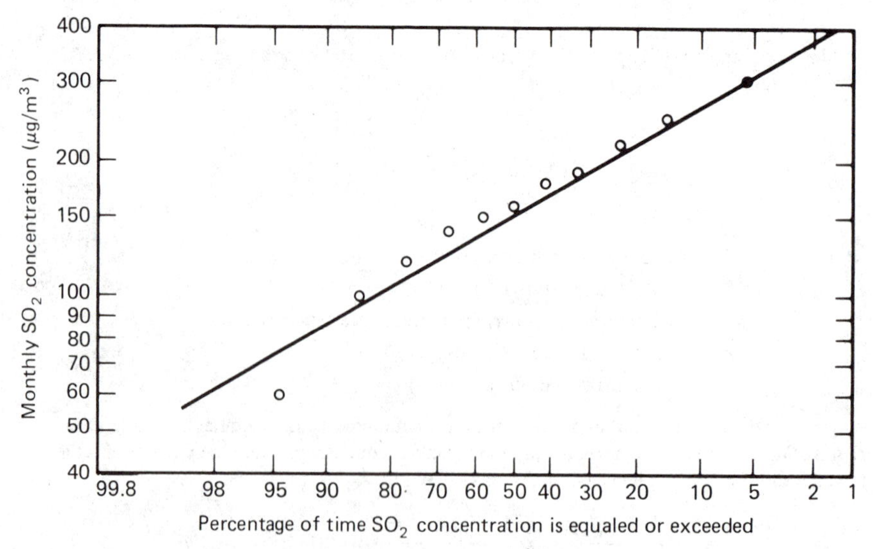

FIGURE 15.4 Frequency analysis of monthly average SO_2 concentrations.

TABLE 15.1 Frequency Analysis of Monthly Average SO_2 Concentrations

Month	Monthly SO_2 ($\mu g/m^3$)	Rank Order r	Percentage Equaled or Exceeded f (%)
January	300	1	5.5
February	250	2	14.5
March	180	5	41.8
April	Missing	Missing	Missing
May	150	7	58.2
June	60	11	94.5
July	120	9	76.4
August	100	10	85.5
September	140	8	67.3
October	160	6	50.0
November	190	4	32.7
December	220	3	23.6

ley in northern California. EPA concluded that the proposed facility would allow more people to reside in the Valley, and new residents would substantially increase the vehicle miles traveled during commuting hours. Since the NAAQS for photochemical oxidants were violated in the valley, EPA was very concerned about air quality impacts of the proposed expansion.

The impact assessment assumed that daily maximum concentration of photochemical oxidants at a station in Livermore-Amador Valley follows a log-normal distribution. Figure 15.5 is from EPA's frequency analysis of daily high oxidant concentrations occurring in 1973, a year in which meteorological conditions were average. Point A in the figure is the *average* daily high concentration of oxidants during 1973 plotted at $f = 30\%$. For a variable that follows a log-normal distribution, the average will be equaled or exceeded only 30% of the time.

Rollback models assume that ambient air pollutant concentrations at a point are proportional to total emissions of the pollutant. Since photochemical oxidants are not discharged directly, additional assumptions were needed before the rollback concept could be applied. EPA assumed a one-to-one correspondence between average daily high concentrations of oxidants at the monitoring station and average emissions of reactive hydrocarbons (RHC) in the valley. The term *reactive* means these are hydrocarbons that take part in photochemical reactions. The complexities and nonlinearities involved in these reactions make this a tenuous assumption.

EPA predicted the frequency distribution of daily high concentrations of oxidants in Livermore-Amador Valley for 1990. Their forecasting approach required three items of information. One was the distribution of daily high concentrations of oxidants for a base year; this is the line labeled "1973" in

Figure 15.5. Another was the total 1973 RHC emission in the valley, which was estimated to be 9.3 tons. The final required item was the RHC emissions expected in 1990. Based on an analysis of projected commuter travel in 1990, a total RHC discharge of 4.8 tons was predicted. The 1990 RHC figure is substantially lower than the 1973 value because of expected improvements in motor vehicle emission controls. EPA's application of the rollback model took the form

$$C_{f=30\%}\,(1990) = C_{f=30\%}(1973) \times \frac{\text{RHC emissions (1990)}}{\text{RHC emissions (1973)}} \qquad \textbf{(15-5)}$$

where C represents the concentration of photochemical oxidants. The value of $f = 30\%$ is used because the rollback model is applied for the *average* concentration of oxidants. Using the above-mentioned RHC emissions together with the 1973 average concentration (point A in Figure 15.5), equation 15-5 yields a 1990 average daily high concentration of 0.038 ppm. This is point B in the figure. To obtain the distribution of concentrations for 1990, a line parallel to the 1973 line is drawn through point B. Fluctuations in concentration, relative to the average, are assumed to be the same in 1990 as they were in 1973.

Figure 15.5 was employed to estimate the frequency with which the NAAQS for photochemical oxidants would be violated in 1990 in the Livermore-Amador

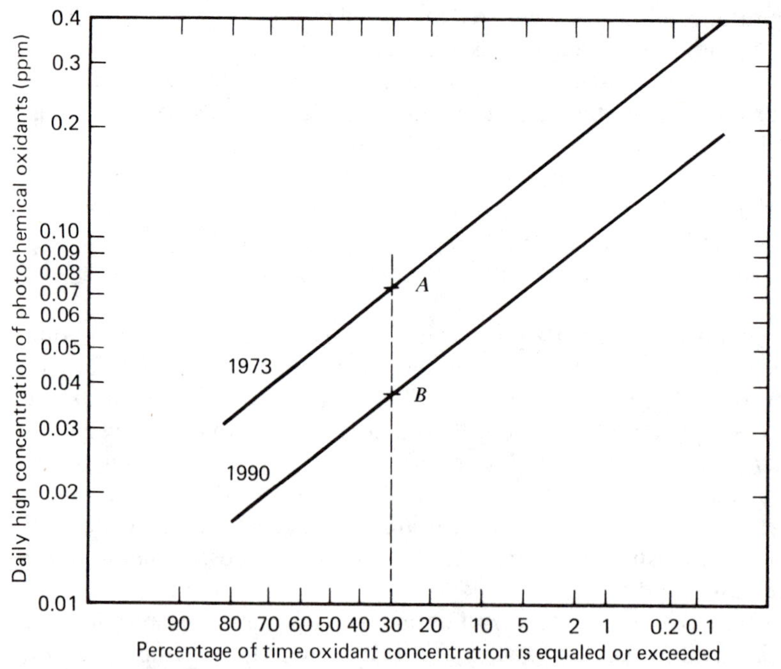

FIGURE 15.5 Analysis of oxidant concentration distributions in the Livermore-Amador Valley. Adapted from Environmental Protection Agency (1975).

Valley. This information was used to analyze alternative ways of meeting the national ambient standards for oxidants. Possible actions included reducing the proposed treatment plant expansion and adopting transportation management strategies to decrease the number of vehicle miles traveled.

ATMOSPHERIC DISPERSION MODELS

Much effort has been spent analyzing how an air pollutant disperses downwind from its point of discharge. During the 1920s, the dispersion process began to be examined systematically on both theoretical and experimental levels. Mathematical models of atmospheric turbulence were used to predict the motion of particles in a puff of smoke released into the ambient air. Experimental work indicated that, under some conditions, materials downwind from an emission were distributed according to a bell-shaped (or "normal") curve. Extensions of this early research yielded "atmospheric dispersion models" for point, line, and area sources. A smokestack is a typical point source, and a highway is a common line source. An area source is illustrated by the collection of space heating units in a 1000-unit suburban housing development.[12]

Point Source Gaussian Models

The most widely discussed atmospheric dispersion model is the point source Gaussian model. In contrast to box and rollback models which use average concentrations, the Gaussian model describes concentrations at numerous points in space. It typically incorporates the following assumptions:

- Rate of pollutant emission is constant.
- Average windspeed and direction are constant.
- Emitted substance is chemically stable and does not "fall out" from ambient air.
- Area surrounding source is flat, open country.

Movement of the substance in the direction of the wind is due to the wind's average motion, a form of transport termed *advection*. Random fluctuations within the wind cause material to be distributed in the plane perpendicular to the wind's direction; this is referred to as *diffusion*.

The point source Gaussian model is derived by applying the law of conservation of mass to the emitted substance under the conditions outlined above. The law is written as a differential equation that includes terms representing both advection and diffusion. A solution to the equation (under steady-state

[12]Distinctions between point, line, and area sources are made by examining a source's length and width relative to the size of the area of analysis. A point source is small in both length and width, whereas a line source has one dimension (for example, a highway's length) that is large compared to the dimensions of the study area. An area source is relatively large in both length and width.

conditions) is the normal distribution equation used in statistics. The normal equation is also known as the Gaussian equation, and that is the basis for naming the model. The Gaussian model describes the concentration of the substance downwind from the source by a series of bell-shaped curves like those in Figure 15.6. The figure's coordinate system has the x axis extending horizontally in the direction of the wind. The y axis is in the horizontal plane and perpendicular to the x axis, and the z axis extends vertically. The overall height of the plume in Figure 15.6 is somewhat higher than the source height. This occurs because the emission of a gaseous material from a point source such as a smokestack generally experiences an extra rise (due to buoyancy and momentum effects) when it hits the atmosphere. The height, h, in the figure, termed the *effective emission height,* is determined using standard equations.[13]

In its general form, the point source Gaussian model predicts concentration at a downwind location (x, y, z). A commonly used simplification of the model forecasts pollutant concentration at downwind points along the ground $(z = 0)$:

$$c(x, y) = \frac{Q}{\pi \sigma_y \sigma_z U} \exp\left(-\frac{y^2}{2\sigma_y^2} - \frac{h^2}{2\sigma_z^2} \right) \quad \textbf{(15-6)}$$

where

Q = constant emission rate (μg/sec)
U = constant wind speed (m/sec)
h = effective emission height of the stack (m)
σ_y = horizontal dispersion coefficient (m)
σ_z = vertical dispersion coefficient (m)

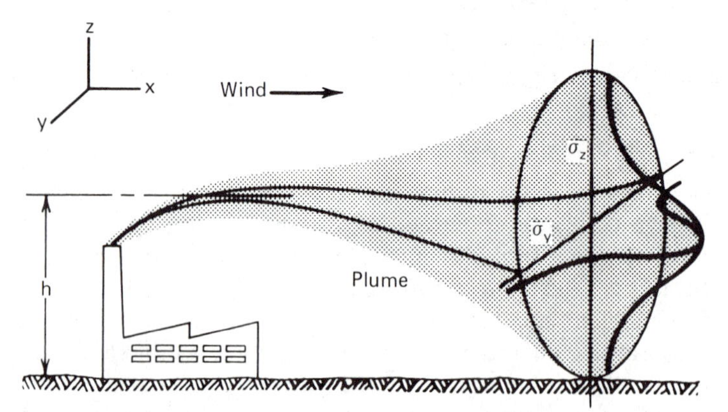

FIGURE 15.6 Gaussian (bell-shaped) distribution of a substance from a point source. From R. J. Bibbero, "Systems Approach Toward Nationwide Air Pollution, III, Mathematical Models," in *IEEE Spectrum,* Vol. 8, No. 12, pp. 47–58. Copyright © 1971 IEEE.

[13]Methods for computing effective emission height are reviewed by Perkins (1974).

Equation 15-6 is the two-dimensional form of the Gaussian distribution equation used in statistics, and the terms σ_y and σ_z are analogous to "standard deviations" in the y and z directions. A standard deviation is a statistic related to the character of the peak in the bell-shaped curve. If the standard deviation is low, the curve has a relatively sharp peak. Values of σ_y and σ_z generally increase with x, the distance downwind from the source. This indicates that the bell-shaped curves of concentration get flatter as the distance from the source increases.

To predict the concentration at any ground location (x, y) it is first necessary to determine the dispersion coefficients, σ_y and σ_z, based on meteorological conditions to be used in making the forecast. One of several available procedures for estimating σ_y and σ_z uses empirical observations in conjunction with an appropriate "stability category." The category is determined from surface wind speed, daytime incoming solar radiation, and night-time cloud cover. Table 15.2 shows six stability categories (labeled A through F) defined by these factors. Once the category is established, the empirically derived graphs in Figures 15.7 and 15.8 are employed to estimate σ_y and σ_z, respectively. Using these curves, only the stability category and distance downwind from the source are needed to obtain dispersion coefficients.[14]

The computation of $c(x, y)$ is demonstrated with Turner's (1969) example concerning a petroleum refinery to be located in an open, unobstructed area. A smokestack at the proposed refinery would have an effective height $h = 60$

TABLE 15.2 Key to Stability Categories[a]

	Day			Night	
	Incoming Solar Radiation			Cloud Cover	
Surface Wind Speed at 10 m (m/sec)	Strong	Moderate	Slight	Thinly Overcast or ≤4/8 Cloudiness[b]	≤3/8 Cloudiness
<2	A	A–B	B		
2–3	A–B	B	C	E	F
3–5	B	B–C	C	D	E
5–6	C	C–D	D	D	D
>6	C	D[c]	D	D	D

[a]Adapted from Turner (1969). This source should be consulted for definitions of strong, moderate, and slight incoming solar radiation.

[b]The degree of cloudiness is defined as the fraction of sky above the local apparent horizon that is covered by clouds (Bibbero and Young, 1974, p. 319).

[c]The neutral class, D, should be assumed for overcast conditions during day or night. Class A is the most unstable and class F is the most stable.

[14]This discussion of procedures for estimating σ_y and σ_z is from Turner (1969).

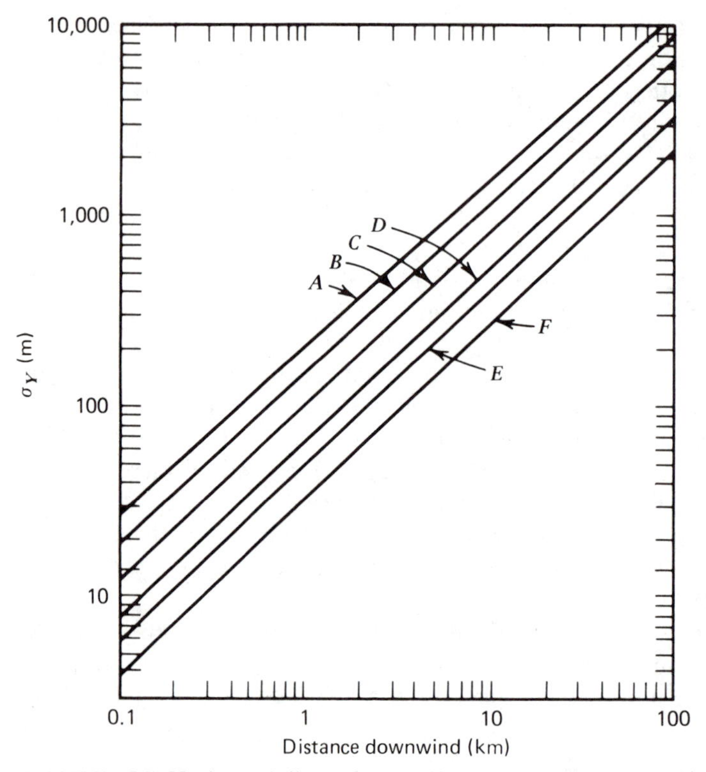

FIGURE 15.7 Horizontal dispersion coefficient as a function of downwind distance from the source. From Turner (1969).

m and emit sulfur dioxide at a rate $Q = 80 \times 10^6$ μg/sec.[15] Suppose it were necessary to forecast SO_2 concentrations at a distance 500 m downwind from the refinery stack. The conditions used in forecasting are for an overcast winter morning with a ground level wind speed $U = 6$ m/sec.

Dispersion coefficients are estimated as follows: For overcast days with wind speeds of 6 m/sec, Table 15.2 indicates "D" as the applicable stability category. Figures 15.7 and 15.8 show that for category D and a downwind distance of 500 m, the dispersion coefficients are $\sigma_y = 36$ m and $\sigma_z = 18.5$ m, respectively.

Equation 15-6 is used to construct a curve showing the concentration distribution at 500 m downwind. This is done by setting $x = 500$ and letting y take on different values in equation 15-6. For example, at $x = 500$ and $y = 0$ (the center line of the plume), substitution of the above values for x, y, Q, U, h, σ_y, and σ_z into equation 15-6 yields

$$c(x, y) = \frac{80 \times 10^6}{(3.14)(36)(18.5)(6)} \exp\left[-½\left(\frac{0}{36}\right)^2 - ½\left(\frac{60}{18.5}\right)^2\right] = 33\ \mu\text{g/m}^3$$

[15]The example assumes SO_2 is chemically nonreactive and any SO_2 fallout is reflected back into the ambient air when it reaches the ground.

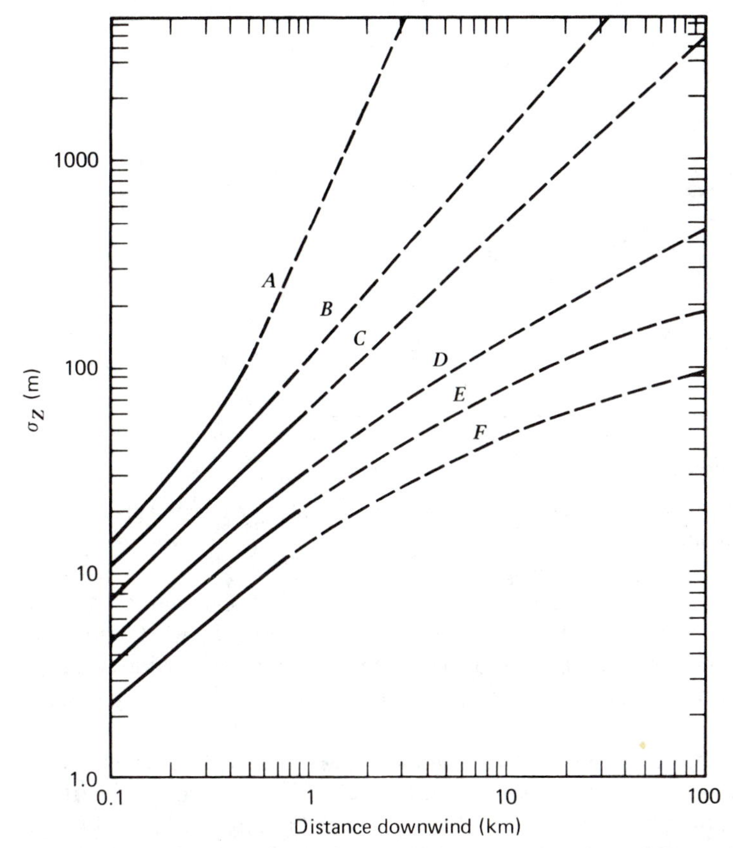

FIGURE 15.8 Vertical dispersion coefficient as a function of downwind distance from the source. From Turner (1969).

Performing this computation with $y = 10$ instead of 0 and with all other values constant, the SO_2 concentration is 32 $\mu g/m^3$. The concentration distribution in Figure 15.9 is found by repeating the procedure using $y = 20$, $y = 30$, and so forth. The figure is one half of the bell-shaped distribution of SO_2 concentration at $x = 500$; the mirror image of the curve applies for negative values of y.

Many empirical investigations have been undertaken to test or "validate" the point source Gaussian model. Typically, such studies begin by estimating the model parameters, σ_y and σ_z, and appropriate values of Q, U, and h. Forecasts are then made with equation 15-6 and the results are compared with field measurements. Numerous validation studies indicate that when the point source Gaussian model is applied for CO, SO_2, and particulates, forecasts of concentration are reasonably close to observed values.[16] In this context, *reasonably close* means that a substantial majority of predicted values are within ±50% of observed values. Accuracy of forecasts may decrease significantly if there are

[16]Bibbero and Young (1974, pp. 331–345) summarize results from several studies undertaken to validate the point source Gaussian model.

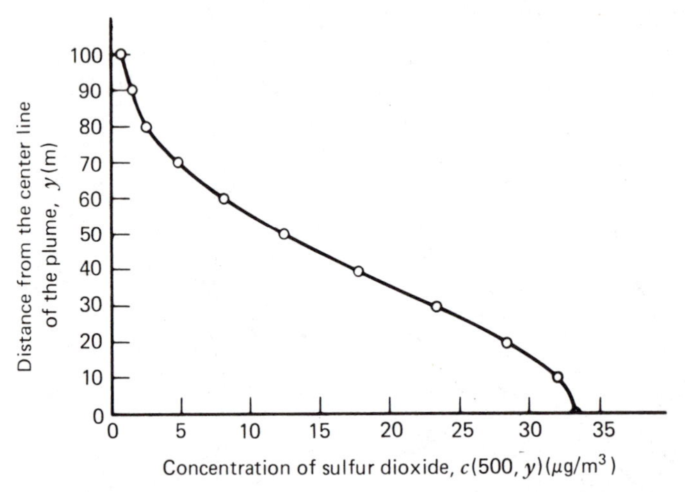

FIGURE 15.9 Estimated distribution of SO_2 concentration along a line 500 m downwind from a refinery smokestack.

departures from conditions used in formulating the Gaussian model: for example, the assumption that wind speed is constant. Because the model does not account for the chemical reactions that occur between nitrogen oxides and hydrocarbons, it is not useful in predicting concentrations of photochemical oxidants.

An extension of the Gaussian point source model is used to predict effects of chemically stable discharges from line sources. Consider an example involving emissions of carbon monoxide from motor vehicles on a long, straight highway. The road is treated as an infinite number of very small point sources, the individual motor vehicle exhausts. For the case in which a steady wind blows perpendicular to the highway, the CO concentration resulting from the numerous point sources is

$$c(x) = \frac{2q}{\sqrt{2\pi}\,\sigma_z U} \exp - \left(\frac{h^2}{2\sigma_z^2}\right) \qquad \textbf{(15-7)}$$

where $c(x)$ is the concentration x m downwind, q equals the "line source strength" (mg/m-sec) and U, s_z, and h are as previously defined.

Multiple-Source Models for Chemically Stable Substances

Complex models have been created to forecast the concentration of chemically stable substances released from a large number of point, line, and area sources. Some of these are combinations of Gaussian models. Others rest on differential equations based on the law of conservation of mass. These equations are solved using numerical analysis methods to yield ambient concentrations. Digital computers are frequently used in operating multiple-source forecasting models.

The need for models to predict ambient concentrations increased dramatically in the United States during the 1970s. The stringent federal air quality regulations in this period led to an upsurge in modeling activity. Models were developed to treat time scales ranging from 1 hr to 1 year. In addition, numerous physical circumstances were analyzed including hilly terrain and urban areas containing complex building configurations. Efforts were also made to model non-steady-state conditions, such as those in early morning when auto emissions gradually build to their rush hour peak.

During the mid-1970s, EPA set up a system that made multiple-source models easy to obtain. A collection of computer programs referred to as UNAMAP, the User's Network for Applied Modeling of Air Pollution, was made available for a nominal charge.[17]

The Climatological Dispersion Model (CDM) is one of several models included in UNAMAP. It relies on extensions of the Gaussian models above and treats both point and area sources in flat terrain. CDM forecasts concentrations of chemically stable substances under steady-state conditions with averaging times ranging from 1 month to 1 year.

A test of CDM by Prahm and Christensen (1977) indicates the model's ability to provide reasonable predictions. Their study involved a 500-km^2 flat, urban area that included the city of Copenhagen, Denmark. The period of analysis was the first 3 months of 1971. Information gathered for testing CDM included SO_2 discharge rates from 252 point sources and 454 area sources during the 3-month period. Measurements of average wind speed and direction were collected to estimate dispersion coefficients. These emission and metereological data provided a basis for using CDM to calculate SO_2 concentrations at 22 monitoring stations in the Copenhagen area. Calculated values were compared to average SO_2 concentrations at the 22 stations during the 3-month analysis period.

Figure 15.10, which summarizes some of the test results, indicates that calculated SO_2 concentrations were close to measured values at many of the 22 stations.[18] Although the model's performance was good in many respects, it systematically underestimated observed concentrations. The line in Figure 15.10 intercepts the vertical axis at 39; the intercept would be at zero if there were no systematic underestimation. Prahm and Christensen explain the underestimation in terms of rural background concentrations and emission sources not accounted for in the CDM calculations. Based on their overall analysis, Prahm and Christensen felt that CDM provides forecasts adequate for planning purposes when steady-state conditions and long-term averaging periods are used.

[17]For additional information on UNAMAP, see Turner (1979).

[18]A statistical measure of the closeness of observed and computed concentrations is the coefficient of linear correlation, R^2. For the data in Figure 15.10, $R^2 = 0.82$. This is interpreted to mean that CDM accounts for more than 80% of the variation in the observed SO_2 concentrations during the period of analysis.

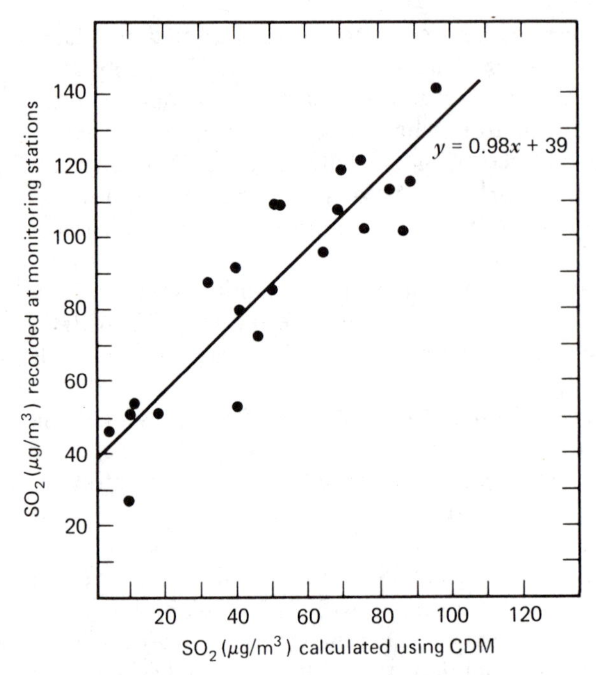

FIGURE 15.10 Measured and computed 3-month average SO_2 concentrations at 22 receptor points (test No. 3c). Modified from L. P. Prahm, and M. Christensen (1977, p. 793), copyright © Pergamon Press, New York.

Models for Chemically Reactive Substances

The dispersion models above forecast how releases of a chemically stable pollutant influence ambient concentrations of the pollutant. It is much more difficult to relate discharges to concentrations when the emitted material undergoes chemical transformations in the atmosphere. Sometimes the substance of interest is a reaction product and is not released directly. For example, when forecasts of ambient concentrations of ozone are made, the important emissions are of hydrocarbons and nitrogen oxides. Ozone itself is not discharged directly. It is a product of photochemical reactions involving HC and NO_x discharges.

Difficulties in modeling atmospheric dispersion for chemically reactive substances are demonstrated using photochemical smog. In this case, sunlight triggers a complex set of reactions involving numerous chemical compounds. The reaction products, referred to collectively as photochemical oxidants, include ozone, peroxyacetyl nitrate, and peroxybenzoyl nitrate. Collectively, photochemical oxidants have been linked to harmful effects on humans including eye irritation, coughing, and difficulty in breathing. Ozone has produced significant damage to plants. Photochemical smog has a characteristic odor and produces a haze that decreases visibility.

Figure 15.11 shows the time pattern of pollutant formation on a typical smoggy day in Los Angeles. For the first hour or so after sunrise there is a buildup of

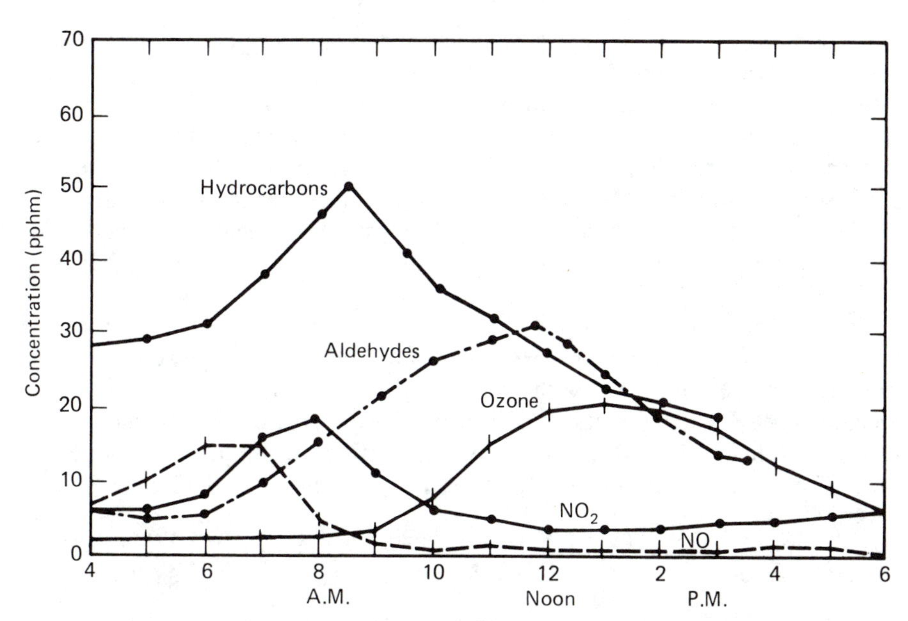

FIGURE 15.11 Average concentration during days of eye irritation in downtown Los Angeles, California. Hydrocarbons, aldehydes, and ozone for 1953–1954. Nitric oxide and nitrogen dioxide for 1958. From Haagen-Smit and Wayne (1976, p. 254), copyright © Academic Press, New York.

direct discharges of hydrocarbons and nitric oxide (NO). During this period, the concentrations of reaction products, represented by ozone, aldehydes, and nitrogen dioxide (NO_2), remain low. Formation of nitrogen dioxide from nitric oxide is well underway shortly after sunrise and by about 8 A.M. the NO_2 levels peak out. By late morning, ozone concentrations begin to increase and there is a further reduction in NO. The ozone level decreases in late afternoon and reaches zero late in the night.

Relationships between photochemical oxidants and NO_x and HC emissions are difficult to understand because of the enormous number of chemicals involved. For example, the simple process of gasoline evaporation yields hundreds of different organic compounds. After the reaction products are accounted for, the number of organic compounds in photochemical smog formation is in the thousands.[19] Moreover, the particular hydrocarbon molecules present influence the extent to which oxidants are formed. Some hydrocarbons (e.g., tri- and tetraalkyl ethylenes) are much more chemically reactive than others.

The formation of smog involves thousands of chemical species in thousands of different photochemical reactions. To cope with the multitude of chemical species and reactions, air quality modeling specialists group large numbers of different hydrocarbons into classes and then examine representative reactions for each class. There are many alternative ways of categorizing individual hy-

[19]For information on the types of compounds involved, see Haagen-Smit and Wayne (1976).

drocarbons and describing average reactions. A typical photochemical smog model involves between 10 and 30 reactions.

Because of the complexities in photochemical smog formation, there is no direct relation between emissions of HC and NO_x and concentrations of photochemical oxidants. For example, a 10% decrease in NO_x discharge does not imply a corresponding 10% reduction in ozone concentration. In fact, there are circumstances in which a reduction in NO_x emissions can lead to an *increase* in the ozone concentration. This counterintuitive result can be explained using mathematical models of smog formation.[20]

Models of photochemical smog are emphasized here since they have received much attention since the 1950s. However, air quality modeling specialists have recently become concerned with the formation and transport of aerosols, especially those involving sulfates and nitrates. Aerosols consist of microscopic solid or liquid particles dispersed in a gas; common examples are smoke and haze. There are two reasons for the increased interest in modeling aerosols:

1 Visibility reduction caused by the light scattering properties of aerosols.
2 Acidification of watercourses due to deposition of sulfate and nitrate aerosols.

In many cases, aerosols are transported from sources several hundred miles upwind of regions experiencing visibility or acidification problems.

The role of aerosols in decreasing visibility is well known. Light scattered into an observer's line of sight by aerosols in the ambient air is perceived as a haze that obscures visible objects. Efforts are being made to model visibility changes associated with emissions of SO_x and NO_x. For example, an analysis of the Los Angeles air basin linked visibility reductions to nitrate aerosols caused by motor vehicle emissions and sulfate aerosols caused by point sources of SO_x.[21]

Models are also being developed to predict effects of SO_x and NO_x emissions in one region on acidification of watercourses in distant downwind regions. Although this is commonly referred to as the *acid rain problem,* acidification is also associated with snow and with *dry deposition.* The latter occurs when gaseous materials are adsorbed by particles which eventually settle out of the atmosphere. In addition to aerosol formation and transport, models of acidification processes include chemical reactions involved when SO_x or NO_x are transformed to acidic forms and "washed out" of the atmosphere by precipitation.[22] A complication occurs when acid precipitation is caused by both NO_x *and* SO_x emissions. In this case, chemical transformations involving SO_x are

[20]Chock et al. (1981) demonstrate how this outcome can be explained using mathematical models.

[21]The Los Angeles analysis is presented by White and Roberts (1977). Numerous other visibility modeling studies are reported by White, Moore, and Lodge (1981).

[22]An approach to incorporating the washout process into atmospheric dispersion models for SO_2 is given by Barrie (1981).

combined with complex models of photochemical smog caused by NO_x and HC emissions.[23]

Despite the attention devoted to modeling chemically reactive substances, models involving reactions are not often used in routine impact assessments because they are expensive to apply in real settings. A substantial amount of field data are needed to determine a model's parameters and test the accuracy of its predictions. Because many chemical compounds in these models are present in very low concentrations, they are often difficult and expensive to measure. Data-related problems are such that few dispersion models involving complex reactions have been subject to tests of the type noted above in connection with the Climatological Dispersion Model. Until such tests are made, chemical reaction models cannot be considered fully reliable and model forecasts need to be used with caution. The potential value of these models in planning and decision making is very great.

KEY CONCEPTS AND TERMS

THE NATURE OF AIR POLLUTION
- Carbon monoxide
- Hydrocarbons
- Nitrogen oxides
- Sulfur oxides
- Particulates
- Atmospheric lead
- Photochemical smog
- Photochemical oxidants
- Air quality and energy sources
- Air pollution and human health

AIR QUALITY IMPACT ASSESSMENT PROCESS
- Data sources for background air quality
- National ambient air quality standards
- Concentration units, ppm and $\mu g/m^3$
- Averaging time
- Emissions versus ambient concentrations

ESTIMATING EMISSIONS OF RESIDUALS
- Emission factors
- Stationary versus mobile sources
- Vehicle miles traveled

SIMPLE MODELS RELATING EMISSIONS TO CONCENTRATIONS
- Chemically stable pollutants
- Steady-state conditions
- Box models
- Rollback models
- Frequency analysis
- Log-normal distributions
- Reactive hydrocarbons

ATMOSPHERIC DISPERSION MODELS
- Point, line, and area sources
- Gaussian models
- Advection and diffusion
- Effective emission height
- Dispersion coefficients
- Stability categories
- Line source strength
- Multiple-source models
- Photochemical smog models
- Aerosol formation
- Visibility reduction
- Acid precipitation and dry deposition

[23]An example of an aerosol model involving SO_x and NO_x is given by Kasahara, Takahashi, and Tohno (1980).

DISCUSSION QUESTIONS

15–1 What considerations are important in determining emissions from automobiles? How would you expect emission factors for the *average* auto to vary over the next decade?

15–2 Indicate the strengths and weaknesses of rollback and box models. Why are they widely used? What variables would these models have to include to reduce their shortcomings?

15–3 Suppose a metropolitan area has a mass of stable air 500 ft above the ground and surface wind speed is steady at 5 mph. The width of the air basin in the direction perpendicular to the wind is 10 miles. Use a box model to estimate the maximum regional carbon monoxide emission (in tons per day) consistent with attaining a CO standard of 10,000 $\mu g/m^3$. (This was the 1980 NAAQS for carbon monoxide using an 8-hr averaging time.)

15–4 Consider a metropolitan area in which the average background concentration of particulates is 20 $\mu g/m^3$. Suppose the annual geometric average concentration of particulates in 1980 was 200 $\mu g/m^3$, and the total regional emission of particulates for that year was 400×10^6 lb. The 1980 NAAQS for particulates was set at an annual geometric average concentration of 75 $\mu g/m^3$. What is the percentage by which the 1980 emissions of particulates would have to be "rolled back" to meet the NAAQS? How might your rollback estimate be criticized if it were used to establish an air quality management plan?

15–5 An open, burning dump emits nitrogen oxides at the rate of 2 g/sec on an overcast night when the wind speed averages 5 m/sec. Assume the dump can be represented as a point source with an effective stack height of zero. Find the ground level NO_x concentrations at the centerline of the plume for points located 1, 2, and 5 km downwind of the source. For purposes of this exercise, assume NO_x is chemically stable.[24]

15–6 Suppose a proposed highway is expected to accommodate a maximum of 8000 vehicles per hour during late afternoon rush hours. Overcast conditions are expected on a typical afternoon. The average emission factor for CO is estimated at 10 g/vehicle mile and it is assumed that a motor vehicle exhaust has an effective height of zero. Calculate the "line source strength" (in μg/m – sec) as the product of the emission factor, the number of vehicles per hour, and appropriate units conversion factors. Use the line source Gaussian model, equation 15-7, to predict the CO concentration 300 m downwind if the wind blows perpendicular to the highway at a constant speed of 4 m/sec.

15–7 Equation 15-7 applies when wind blows perpendicular to a line source. Suppose the angle between the wind direction and the line source is ϕ, where ϕ is greater than or equal to 45°. For this case, a line source model to forecast the effects of chemically nonreactive emissions on concentrations has the form

$$c(x) = \frac{2q}{\sin\phi\sqrt{2\pi}\,\sigma_z U}\exp\left(-\frac{h^2}{2\sigma_z^2}\right)$$

where all terms except ϕ are as defined in the discussion of equation 15-7.[25] Using the conditions associated with Question 15-6, construct a graph of CO concentration 500 m downwind of the highway versus the angle ϕ (where $45° \leq \phi \leq 90°$).

[24]This problem (and others below) is adapted from Turner (1969, Chapter 7).

[25]This modeling result is presented by Turner (1969, p. 46).

REFERENCES

Barrie, L. A., 1981, The Prediction of Rain Acidity and SO_2 Scavenging in Eastern North America. *Atmospheric Environment* **15,** 31–41.

Bibbero, R. J., and I. G. Young, 1974, *Systems Approach to Air Pollution Control.* New York.

Chambers, L. A., 1976, Classification and Extent of Air Pollution Problems, *in* A. C. Stern (ed.), *Air Pollution,* 3rd ed., Vol. 1, pp. 3–22. Academic Press, New York.

Chock, D. P., A. M. Dunker, S. Kumar, and C. S. Sloane, 1981, Effect of NO_x Emission Rates on Smog Formation in the California South Coast Air Basin. *Environmental Science and Technology* **15** (8), 933–939.

Coffin, D. L., and H. E. Stokinger, 1977, Biological Effects of Air Pollutants, *in* A. C. Stern (ed.), *Air Pollution,* 3rd ed., Vol. II, pp. 231–360. Academic Press, New York.

Cohn, L. F., and G. R. McVoy, 1982, *Environmental Analysis of Transportation Systems.* Wiley, New York.

Council on Environmental Quality, 1980, *The Eleventh Annual Report of the Council on Environmental Quality.* Council on Environmental Quality, Washington, D.C.

de Nevers, N., and J. R. Morris, 1975, Rollback Modeling: Basic and Modified. *Journal of the Air Pollution Control Association* **25** (9), 943–947.

Environmental Protection Agency, 1971, *Air Quality Criteria for Nitrogen Oxides,* Pub. No. AP-84. EPA, Washington. D.C.

Environmental Protection Agency, 1973, *Compilation of Air Pollutant Emission Factors,* Pub. No. AP-42, 2nd ed. EPA, Research Triangle Park, N.C.

Environmental Protection Agency, 1975, "Draft Environmental Impact Statement, Proposed Wastewater Management Program, Livermore-Amador Valley, Alameda County, CA," EPA Region IX Office, San Francisco, Calif.

Haagen-Smit, A. J., and L. G. Wayne, 1976, Atmospheric Reactions and Scavenging Processes, *in* A. C. Stern (ed.), *Air Pollution,* 3rd ed., Vol. I, pp. 235–288. Academic Press, New York.

Horowitz, J. L., 1982, *Air Quality Analysis for Urban Transportation Planning.* MIT Press, Cambridge, Mass.

Kasahara, M., Takahashi, K., and S. Tohno, 1980, Photochemical Aerosol Formation in Multi-Component Systems Containing Preexisting Particles, *in* M. M. Benarie (ed.), *Atmospheric Pollution, 1980, Proceedings of the 14th International Colloquium, Paris,* pp. 221–226. Elsevier, Amsterdam.

Kohn, R. E., 1978, *A Linear Programming Model for Air Pollution Control.* MIT Press, Cambridge, Mass.

Larsen, R. I., 1971, *A Mathematical Model for Relating Air Quality Measurements to Air Quality Standards,* Pub. No. AP-89, EPA, Research Triangle Park, N.C.

Perkins, H. C., 1974, *Air Pollution.* McGraw–Hill, New York.

Prahm, L. P., and M. Christensen, 1977, Validation of a Multiple Source Gaussian Air Quality Model. *Atmospheric Environment* **11,** 791–795.

Stern, A. C. (ed.), 1977, *Air Pollution,* 3rd ed., Vol. IV. Academic Press, New York.

Turner, D. B., 1969, *Workbook of Atmospheric Dispersion Estimates,* revised edition, Pub. No. 999-AP-26, Public Health Service, National Center for Air Pollution Control, Cincinnati, Ohio.

Turner, D. B., 1979, Atmospheric Dispersion Modeling, a Critical Review. *Journal of the Air Pollution Control Association* **29** (5), 502–519.

White, W. H., and P. T. Roberts, 1977, On the Nature and Origin of Visibility Reducing Aerosols in the Los Angeles Air Basin. *Atmospheric Environment* **11,** 803–812.

White, W. H., Moore, D. J., and J. P. Lodge, Jr. (eds.), 1981, Plumes and Visibility: Measurements and Model Components. *Atmospheric Environment* **15** (10/11), 1785–2406.

CHAPTER 16

ASSESSING IMPACTS ON WATER RESOURCES

Water resources have been degraded unintentionally for centuries. During the past several decades, many techniques have been developed to forecast how human actions affect natural bodies of water. Some of these methods predict impacts on the *quantity* of water present in both surface water and groundwater. Others focus on changes in water *quality*. The hydrologic cycle provides a basis for understanding how water quantity and quality are inextricably related.

REPRESENTATIONS OF THE HYDROLOGIC CYCLE

Elements of the Cycle

The hydrologic cycleis the circulation of water from the atmosphere to the earth and back (see Figure 16.1). Water's movement from the atmosphere to earth, referred to as *precipitation,* generally consists of either rain or snow. Water molecules follow many paths after reaching the earth's surface. For example, if precipitation takes the form of snow and the weather is sufficiently cold, the snow may simply accumulate. As the weather warms, the snowmelt either moves along the land as surface runoff or infiltrates into the ground and percolates down to the groundwater zone. In this zone, the spaces between soil particles or rocks are completely filled with water. Groundwater itself moves very slowly in lateral directions, eventually discharging into streams, lakes, or coastal waters.

If precipitation takes the form of rain, the path it follows after reaching the earth depends on the surface it hits. If rain falls on an impermeable area such as a parking lot, it will flow across the surface until it reaches a natural or man-made channel or a depression where it is stored temporarily. If rain falls on a permeable area such as an open field, its path will depend on surface wetness. The rain will either infiltrate into the ground, become temporarily stored in

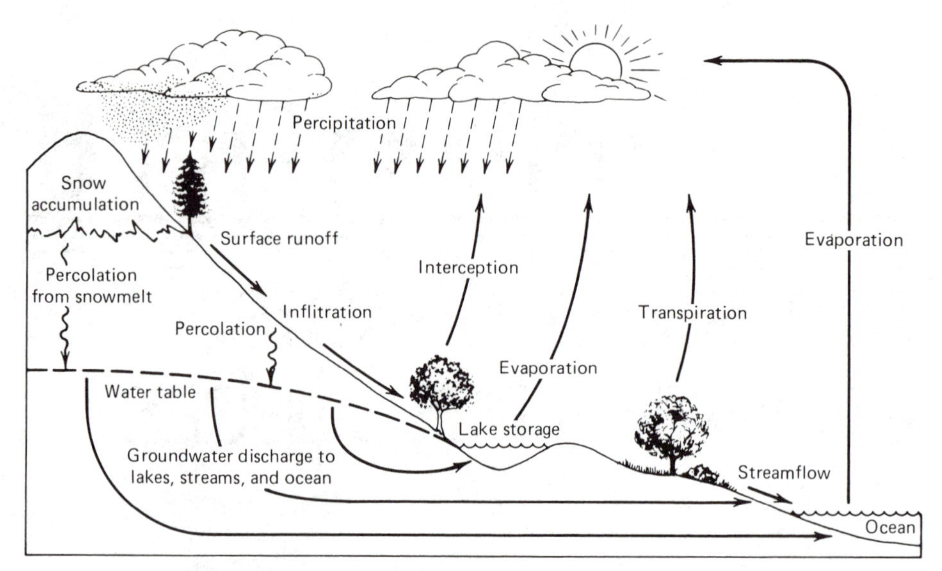

FIGURE 16.1 Schematic diagram of the hydrologic cycle. Adapted from *Water in Environmental Planning* by Thomas Dunne and Luna B. Leopold. Copyright © 1978, Freeman, San Francisco.

surface depressions, or become surface runoff. If the rain falls on vegetation, it will either drip off or evaporate from the leaves to the atmosphere. The latter process is termed *interception.*

As indicated in Figure 16.1, the return of water from the earth to the atmosphere occurs by either evaporation or transpiration. In the latter process, water diffuses out from plant cells and vaporizes into the atmosphere. Because it is hard to differentiate between evaporation and transpiration from croplands and vegetated areas, the processes are sometimes viewed collectively as *evapotranspiration.*

Systematic analysis of parts of the hydrologic cycle has been going on for centuries, and an historic account of these analytic efforts is given by Biswas (1970). Since the seventeenth century, components of the cycle have been described using equations. These relationships provide the bases for "hydrologic simulation models" that are applied in forecasting the effects of human actions on elements of the hydrologic cycle.

Hydrologic Simulation Models

Computer-based models representing the land phase of the hydrologic cycle have been widely used since 1960. A typical model requires information characterizing (1) the geometric configuration of the river basin, including locations of reservoirs and stream channels; (2) the land surfaces, including their potential for infiltration and evapotranspiration; and (3) the soil system and physical attributes of the groundwater zone. Once this information is specified, a simulation model can produce a forecast of streamflows and reservoir levels based on anticipated rainfall patterns. The model itself consists of equations representing processes in Figure 16.1. Some equations are based on physical principles such as the laws of conservation of mass and energy, and others rest on empirical studies. The primary model "outputs" are hydrographs of flow at selected stream locations in the basin. A *hydrograph* is a plot of how streamflow at a particular point (measured in volume of water per unit time) varies over time.

There are dozens of hydrologic simulation models.[1] One of the earliest, the Stanford Watershed Model (SWM) developed by Crawford and Linsley (1966), has been adapted for use in a variety of circumstances. Figure 16.2 delineates SWM's structure. Equations in the model provide a running account of how water entering a river basin undergoes the various processes (such as interception, infiltration, and runoff) shown in Figure 16.1. Formulas based on mass conservation keep track of all water entering and leaving the basin. Other theoretical and empirical equations route surface runoff and groundwater flows into stream channels. Information on streamflow is used to develop hydrographs at selected points.

The SWM requires much data for *calibration,* the process of estimating model parameters. Between 20 and 40 parameters characterizing the river basin under study must be specified. Most of these can be determined from maps and hydrologic records. A small number are estimated using a trial-and-error procedure. Trial values of parameters are based on the analyst's judgment and past experience in using SWM. Once all parameter values are set, the model uses historical precipitation data to produce hydrographs at various locations. These "simulated hydrographs" are then compared with corresponding hydrographs constructed from field measurements ("observed hydrographs"). If there are significant differences between simulated and observed hydrographs, trial parameters are reestimated in an effort to reduce discrepancies. The model is then run again to produce another set of hydrographs and the comparison with measured hydrographs is repeated. This iterative process continues until the model is considered calibrated.[2] Figure 16.3 demonstrates simulated and observed hydrographs obtained in calibrating SWM in a study of Morrison Creek in California. The close agreement between simulated and measured hydrographs in the figure can only be expected when much data are available for calibration.

Hydrologic simulation models are used to predict how surface runoff and groundwater flows respond to water resources development projects and weather modification (for example, using cloud seeding). These models can also describe hydrologic impacts that accompany land surface modifications affecting infiltration and evapotranspiration. Examples of such changes include suburban housing tracts that lower infiltration rates by covering permeable land, and forest harvesting practices that decrease rates of evapotranspiration.[3]

[1]For a review of numerous hydrological simulation models, see Overton and Meadows (1976).

[2]James (1965) presents criteria for judging whether simulated and observed hydrographs are sufficiently close to stop the trial-and-error calibration process.

[3]These comments on how mathematical models are used in forecasting hydrologic impacts are based on Linsley (1976).

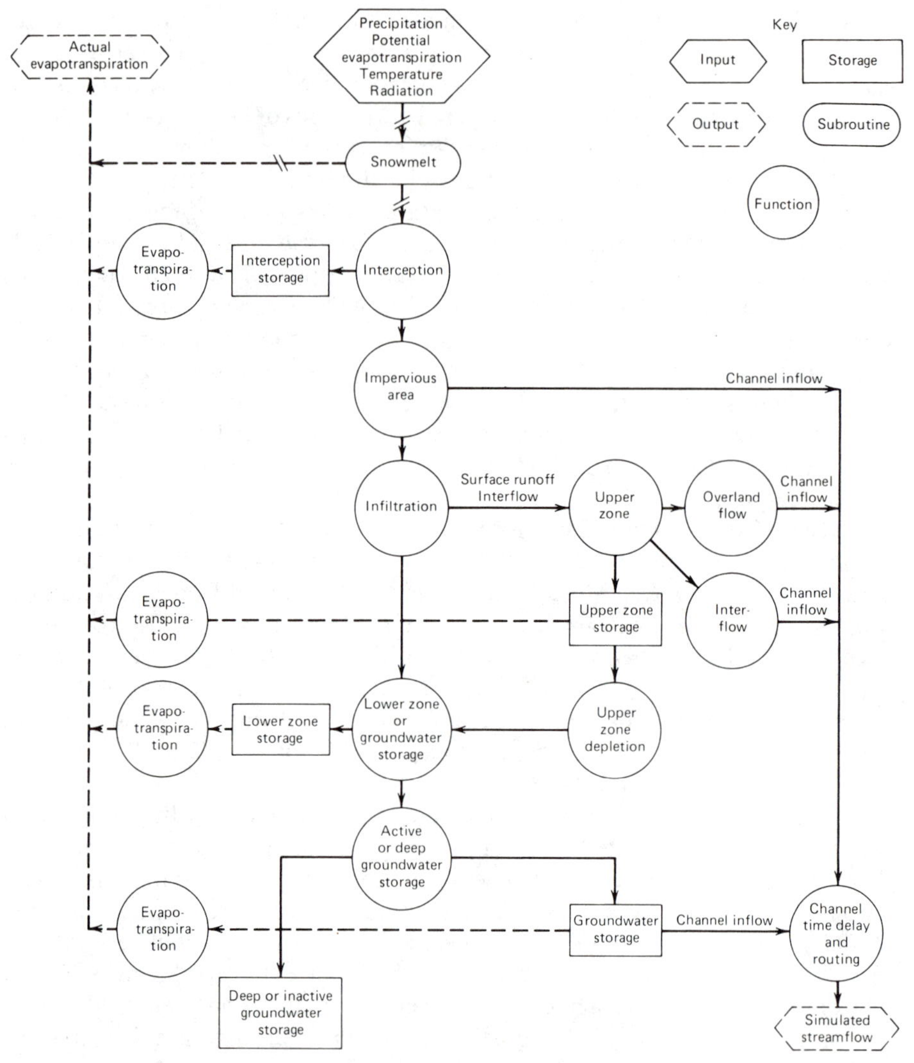

FIGURE 16.2 Flowchart for the Stanford Watershed Model IV. From N. H. Crawford and R. K. Linsley (1966).

INFLUENCE OF LAND USE CHANGES ON FLOOD FLOWS

Land use changes, especially those associated with residential, industrial, and commercial developments, can greatly influence the timing and magnitude of flood flows. This can occur if the impermeable land area is increased or the water conveyance network is modified to remove storm water more rapidly. The latter occurs when gutters, drains, and storm sewers are installed or when natural stream channels are widened and lined with impervious materials.

Figure 16.4 depicts the impacts of urban development on a flood hydrograph.

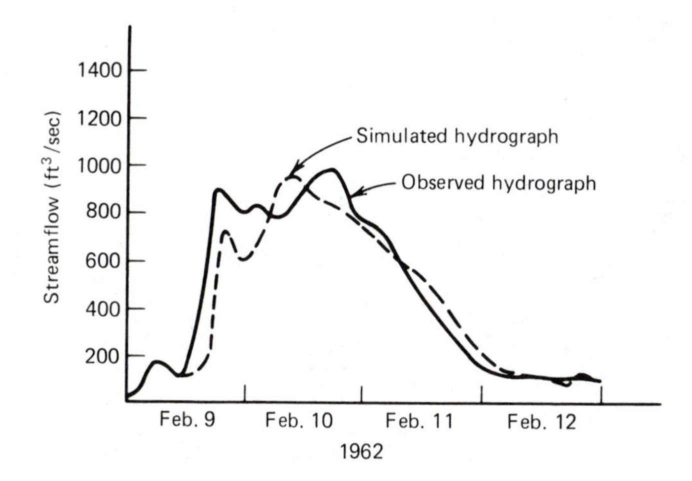

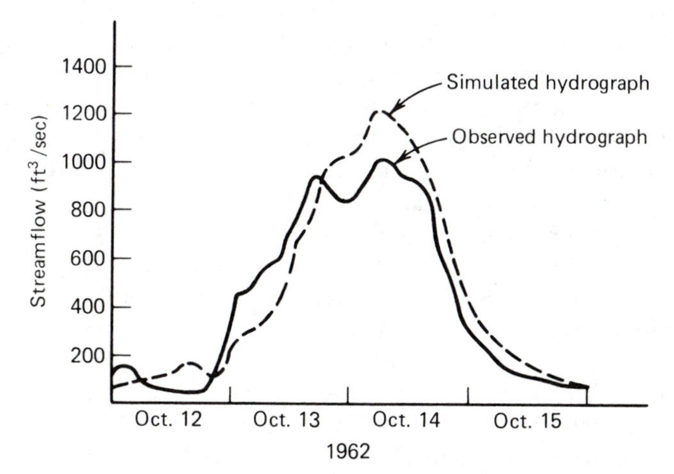

FIGURE 16.3 Comparisons between simulated and observed hydrographs using the Stanford Watershed Model on Morrison Creek, California. Adapted from L. D. James, *Water Resources Research,* Vol. 1, No. 2, pp. 223–234, American Geophysical Union, copyright © 1965.

The bar graph on the left side of the figure shows the time distribution of a rainstorm over a particular river basin. The two curves are hydrographs at a stream location that receives surface runoff produced by the storm. The dashed curve represents the basin in its natural state. The solid curve is for conditions following urban development in the basin. Because of development, the peak rate of streamflow is higher and occurs sooner. These effects are due to increased impermeability of the land surface and conveyance system modifications using gutters, storm sewers, and so forth.

To be useful in analyzing how shifts in land use influence flood flows, a hydrologic simulation model must include parameters representing river basin

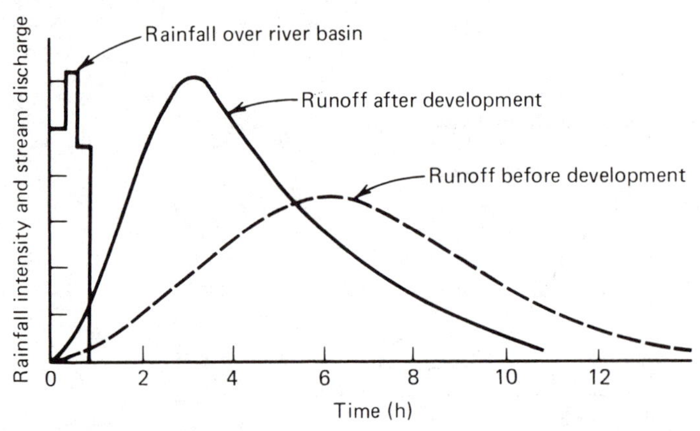

FIGURE 16.4 Effects of urban development on surface runoff at a particular stream location.

features affected by land development: rates of infiltration and evapotranspiration, and hydraulic characteristics of a basin's water transport network. A typical analysis determines how particular rainfall events are converted to surface runoff under different land development scenarios.

James' (1965) investigation of how development affects flood flows in Morrison Creek in California is illustrative. He examined two variables related to land use change: channelization and urbanization. *Channelization* was defined as the fraction of the natural stream channel system modified by either installing storm sewers or straightening, widening, or lining channels. The fraction of the study area devoted to residential, commercial, and industrial development, and to associated public services was used to measure *urbanization.*

James employed the Stanford Watershed Model to simulate streamflows that would have occurred had the Morrison Creek basin been at each of 13 different levels of urbanization and channelization. Each level was represented by suitable values of the models parameters. For any particular combination of urbanization and channelization, hydrographs associated with different rainfall events were computed using SWM. Since James was concerned with flood flows, he used rainfall events that occur infrequently and produce flood conditions.

Figure 16.5 is one of several graphs that summarizes the Morrison Creek study results. It indicates how the flood peak expected once in 10 years increases with rising levels of urbanization and channelization in the basin. Flood peaks are given as cubic feet per second of streamflow at a particular location on the creek. With intense urbanization and channelization, the flood peak could be three times as high as the corresponding peak under completely rural conditions. Similar outcomes were obtained for other frequencies of flood occurrence.

James' results include quantitative representations of the hydrograph changes sketched in Figure 16.4. The hydrograph shifts are major for the Morrison Creek basin, but this is not always the case. Studies by Beard and Chang (1979) indicate that when a basin's natural soils have low permeability, the increase in flood

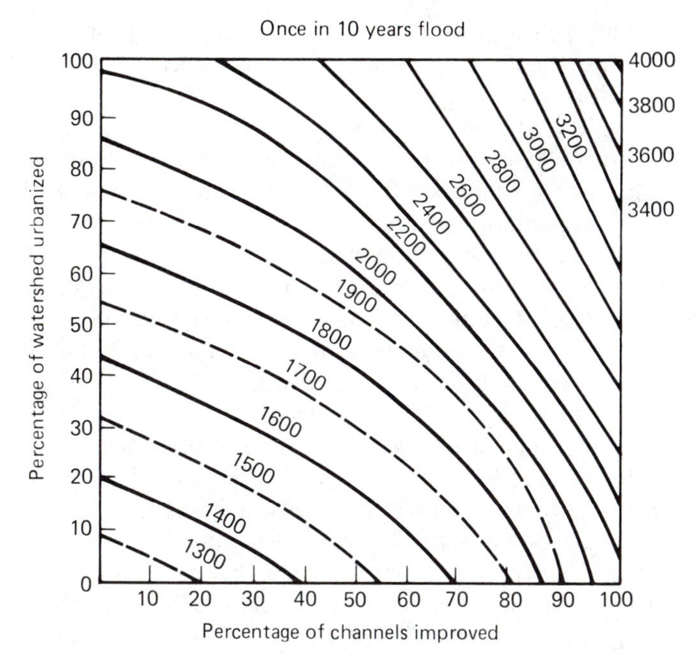

FIGURE 16.5 Flood peak versus urbanization and channelization (flood flows are in cubic feet per second). From L. D. James, *Water Resources Research,* Vol. 1, No. 2, pp. 223–234, American Geophysical Union, copyright © 1965.

peaks caused by urban development may not be large. The effect of land use changes on flood peaks depends on particular conditions in the watershed being analyzed.[4]

PARAMETERS USED TO MEASURE IMPACTS ON WATER QUALITY

There are an enormous number of constituents in water and a correspondingly large number of water quality measures. This discussion introduces indicators commonly employed in considering how human actions affect water quality.

An especially significant aspect of water is the extent to which it contains disease-causing (*pathogenic*) organisms. Examples are bacteria causing typhoid fever and cholera, and viruses responsible for hepatitis A and poliomyelitis. Also included are protozoa such as *Giardia lamblia,* which can cause gastrointestinal disorders. Although modern methods of water disinfection have greatly reduced the threat of water-borne disease epidemics, such threats are a continuing concern. The possible contamination of water by pathogenic microbes is signaled by the presence of coliform bacteria. These bacteria, although usually not pathogenic themselves, are contained in the feces of humans and other

[4]Other studies of the effects of urbanization and channelization on flood flows are summarized by Dunne and Leopold (1978, pp. 324–330).

warm-blooded animals. Although coliform bacteria are imperfect indicators of fecal contamination, they are widely used because it is impractical to test for the presence of individual disease-causing organisms on a routine basis.

Water also contains a variety of dissolved substances (*solutes*), classified as inorganic or organic. A distinguishing feature is that organic solutes contain one or more atoms of organic carbon. Many solutes appear in water as a result of natural processes such as the dissolution of rocks by surface runoff.

Inorganic solutes in water exist as chemically charged substances known as *ions.* Common negatively charged ions include carbonate, bicarbonate, sulfate, nitrate, phosphate, and chloride. Among the usual positively charged ions are calcium, magnesium, sodium, and potassium. *Total dissolved solids* (TDS) is an aggregate measure of these various ions. In recent years much attention has been given to a group of positively charged ions in water known, collectively, as *heavy metals.* These metals, which are toxic to humans and other organisms at very low concentrations, include lead, mercury, and cadmium. The significance of trace quantities of heavy metals in water bodies is demonstrated by the mercury pollution in Minamata Bay in Japan between 1953 and 1961. Mercury had been discharged to the bay in wastewaters from industrial facilities. It gradually appeared in the form of methylmercury in fish and shellfish caught for human consumption from Minamata Bay. There were 121 reported cases of methylmercury poisoning caused by eating fish and shellfish from the bay. Of these, 46 cases resulted in death.[5]

It is difficult to fully characterize the *organic solutes* in water because thousands of different compounds are involved, the majority of which have not been identified. Recent developments in gas chromatography and mass spectrometry have led to remarkable advances in measuring organic solutes. They allow volatile organic compounds to be identified at concentrations as low as nanograms per liter (10^{-6} mg/1). However, the majority of organic compounds in water are nonvolatile and not yet subject to identification.

A committee of the National Academy of Sciences (1977) reviewed the health effects of 309 volatile organic solutes that had been found in drinking water supplies. A few of the substances (for example, vinyl chloride and benzene) are known or suspected causes of cancer in humans. Many others have caused cancer in laboratory animals. It was difficult for the committee to make definitive statements on cancer risks associated with human consumption of the 309 solutes. In many cases the laboratory research needed to characterize risks of long-term exposure to trace quantities of these materials had not yet been done.

Although most nonvolatile organic solutes in water are not identified individually, they are measured collectively. A widely used aggregate measure is biochemical oxygen demand (BOD), the quantity of dissolved oxygen in a given volume of water used by bacteria to oxidize "decomposable" organic matter to

[5]This discussion of methylmercury poisoning in Minamata Bay is from the National Academy of Sciences (1977). Förstner and Wittman (1981, pp. 18–26) summarize numerous incidents of poisoning from water polluted by arsenic, cadmium, chromium, lead, and mercury.

carbon dioxide, water, nitrate–nitrogen, and other stable end products.[6] These bacteria are *aerobic;* they must have dissolved oxygen to live. BOD is widely used as a measure of potential depletions in a stream's dissolved oxygen (DO). High concentrations of BOD may cause decreases in dissolved oxygen and thereby disrupt fish and other aquatic organisms. When there is no oxygen dissolved in a stream, decomposition of organic matter often yields gases with offensive odors. In this case, decomposition is carried out by bacteria that use oxygen bound up in the organic matter itself.

Some nitrogen compounds associated with biochemical oxidation processes in natural waters are plant nutrients that can cause excessive growth of algae and other aquatic plants. Compounds of phosphorus, such as those originating in some household soaps and detergents, are also aquatic plant nutrients.

In addition to dissolved substances, water also contains particles that remain in suspension. Land erosion and wastewater discharges are common sources of suspended matter. In flowing streams, material transported can include everything from fine clay to rocks. When a stream's velocity is reduced to zero, heavy material settles out, and the remaining particles are termed *suspended solids.* The light-scattering properties of the suspended material provide a basis for measuring *turbidity,* another common indicator of water quality.

Suspended particles can pick up (or, more precisely, "adsorb") organic substances, heavy metals, and other chemicals dissolved in water. When the particles settle out, the resulting "bottom deposits" become a storehouse for future releases of adsorbed chemicals. This phenomenon occurred in Lake Michigan. Prior to 1970, there were substantial discharges of polychlorinated biphenyls (PCBs) to the lake, and some of this material accumulated in sediments on the lake's bottom. Polychlorinated biphenyls are organic solutes known to cause cancer in laboratory animals and suspected of causing cancer in humans. Direct discharges of PCBs to Lake Michigan were curtailed in 1970. For several years thereafter, the concentrations of PCBs in the lake remained nearly constant at the pre-1970 level. As explained by Neely (1980), this concentration remained high because of PCBs released from the sediments to the lake.

Of the many other descriptors of water quality, temperature, pH, and radioactivity are notable because they are frequently influenced by human actions. The release of heated water from the cooling systems of electric power plants and various industrial facilities can increase stream temperatures and harm aquatic life. The term *pH* refers to a logarithmically scaled measure of the hydrogen ion concentration of a solution. For most natural surface waters, pH varies from slightly acidic (pH 6.5 on a scale from 0 to 14) to slightly "basic" (pH 8.5). The pH of natural waters is influenced by discharges from industrial facilities such as metal plating firms, by waters draining from coal mines, and by acid precipitation and dry deposition. Acidification of surface waters is a

[6]The laboratory technique to determine BOD does not measure all nonvolatile organic solutes, since some are resistant to bacterial oxidation during the short time periods used in the procedure. Although other aggregate measures of organic solutes have been developed to overcome its shortcomings, BOD continues to be widely used.

serious concern, since potential consequences include massive fish kills and other damage to aquatic ecosystems. Radioactivity is another important measure of water quality. Principal man-made sources include fallout from nuclear weapons testing, and residuals associated with nuclear energy production and other non-military uses of radioactive materials.[7]

POINT SOURCES: MUNICIPAL AND INDUSTRIAL EFFLUENTS

Water-borne residuals originate from both point and nonpoint sources. Examples of point sources are the direct discharges to water bodies from municipal and industrial wastewater treatment plants. Nonpoint sources include residuals carried to streams, lakes, and estuaries by surface runoff and to groundwater zones by infiltration and percolation. A parking lot is a common nonpoint source. When it rains, the resulting runoff picks up residuals deposited on the lot's surface and carries them to a natural drainage system.

Residuals in municipal effluents have been studied scientifically for over a century. The character of municipal wastewater is affected by industrial wastes that enter a municipal sewer system. Thus, individual municipal discharges can be quite different from each other. A substantial portion of a typical municipal effluent consists of domestic wastewaters containing many organic compounds, most of which are biodegradable. The effect of municipal wastewater on water quality depends on the treatment provided prior to discharge and on the volume of "receiving water" available to dilute the effluent.

Industrial discharges vary greatly in their physical and chemical characteristics and are generally discussed in terms of four classes of industrial water use: sanitary, cooling, cleaning, and process.[8] The first of these, *sanitary,* consists of water used by employees for personal sanitation. The residuals generated are similar to those in domestic wastewater.

The largest quantity of water used by industry is for *cooling,* as illustrated by the production of electricity. In a typical steam–electric power plant, large quantities of cooling water are used for condensing steam. Water leaving a plant's condenser may be 10 or 15°F higher in temperature than the water entering. The heated waters may be either discharged to a natural water body or recirculated in a loop containing a device such as a cooling tower to reduce water temperatures.[9]

[7]The measurement and distribution of radioactive materials in water bodies is a complex topic; Velz (1970, pp. 254–269) provides an introductory discussion. For a review of radioactive waste treatment procedures, see Nemerow (1978, pp. 662–721).

[8]This discussion of industrial wastewater is based on information from Nemerow (1978); he provides detailed characteristics of wastewaters from a variety of industrial sources.

[9]Small quantities of water high in total dissolved solids are frequently discharged from power plants with recirculating cooling water systems. This "blowdown water" is released from the recirculatory cooling system to control TDS. Fresh water (low in TDS) is added to make up for the water released. Frequently, blowdown also contains chemicals used to control aquatic organisms in the cooling water system.

An example of *cleaning* water use is the washing of "soft-drink" bottles that have been returned for reuse. Wastewaters from bottle washing contain high concentrations of BOD and suspended solids due to the left-over soft drinks and debris contained in returned bottles. Because alkaline detergent baths are used in bottle cleaning, the wastewaters are highly alkaline.

Process water includes water used for product formation, waste and by-product removal, and product transport. Four steps in the production of printed cotton fabric demonstrate how residuals enter process water: desizing, scouring, dyeing, and printing. The desizing operation removes starch from cotton cloth soon after it has been woven. (Starch is added early in the production process to give new cotton thread the tensile strength it needs to be woven into cloth.) The starchy wastewater from desizing has high concentrations of BOD and suspended solids. In the scouring operation, desized woven cotton fabric is cooked in a hot, caustic solution to remove impurities. The resulting wastewater is high in BOD and suspended solids and it is very alkaline. Fabric dyeing and printing that occur near the final stages of cloth production yield colored wastewaters that can also be high in BOD.

NONPOINT SOURCES OF WATER-BORNE RESIDUALS

Nonpoint sources of water-borne residuals do not involve continual or direct releases from a pipe or channel and they are often difficult to describe and control.[10] Discharges from nonpoint sources are sometimes similar in magnitude and quality to that of nearby point sources. As municipal and industrial effluents are controlled by treatment and other means, the relative importance of nonpoint sources in causing water pollution is expected to increase. This discussion emphasizes the principal nonpoint sources in the United States: surface runoff from urban, agricultural, and commercial forest areas, discharges from solid waste disposal sites, and wastewaters from septic tanks, cesspools, and industrial lagoons.[11]

Urbanization and Road Construction

Substances deposited on roads and other impervious surfaces created during urbanization are carried away in surface runoff. Residuals accumulating on urban land include street litter, pesticides used in lawn care, fecal deposits of animals, and substances settled out from the atmosphere. In addition, many residuals are deposited from motor vehicles. Examples are fibers from brake linings and hydrocarbons and heavy metals from engine exhausts. Residuals also include sodium chloride and other chemicals used to keep roads free of ice and

[10]The discussion of nonpoint sources relies on Wanielista (1978) and Krenkel and Novotny (1980); they provide detailed information on specific sources and methods of control.

[11]Evidence of the significance of these nonpoint sources in the United States is given by the Council on Environmental Quality (1980, pp. 87–97).

snow. Many drinking water supplies have had to be abandoned because of contamination by runoff containing these "deicing" chemicals.[12]

Soil erosion is a major source of residuals in areas where roads and other physical facilities are being constucted. Wolman and Schick (1967) cite cases where sediment yields from erosion during construction (measured in tons per square mile per year) were 100 times higher than yields in comparable undisturbed areas. Rising sediment loads are reflected in turbidity and suspended solids in surface runoff.

Suspended solids in runoff from urban areas ("urban runoff") are important because various organic compounds and heavy metals are frequently adsorbed on their surfaces. Loadings of suspended solids in urban runoff may be much greater than those of nearby point sources. This is illustrated by Pirner and Harms' (1978) studies of runoff from Rapid City, South Dakota. They found the amount of suspended solids in the runoff from one rainstorm in 1975 to be equal to that from the Rapid City wastewater treatment plant during 7 months. Moreover, the suspended solids yielded by this one storm were delivered to the local stream in only 3 hr.

For small areas, the time distribution of residuals in urban runoff is often similar to that of the runoff itself. This is shown in Figure 16.6 for a 2.2- mile^2

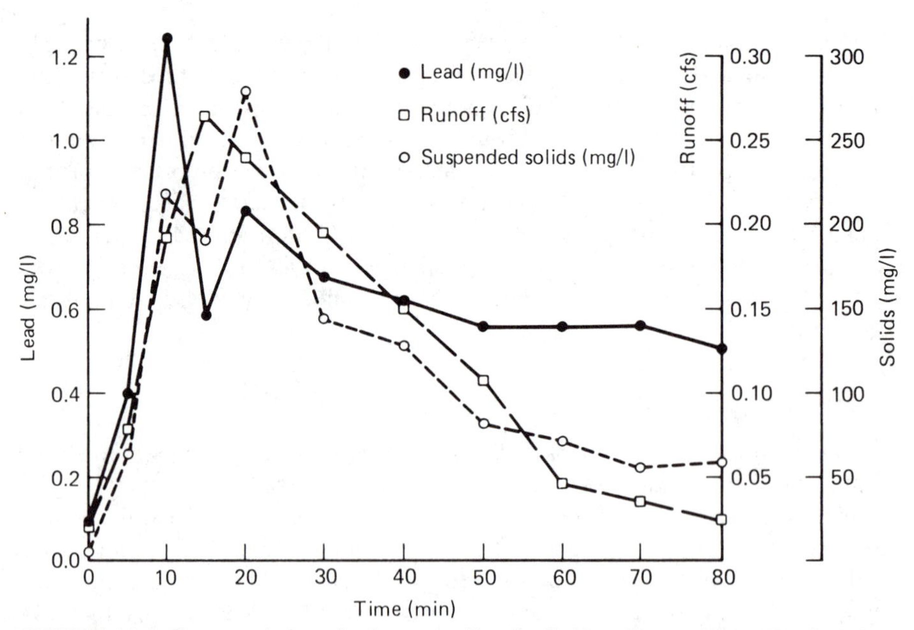

FIGURE 16.6 Characteristics of urban runoff at Lodi, New Jersey. From W. G. Wilber and J. V. Hunter in W. Whipple, Jr. (ed.), *Urbanization and Water Quality Control,* American Water Resources Association, pp. 45–54. Copyright © 1975.

[12]Field et al. (1975) elaborate on the contamination of drinking water supplies by deicers.

urbanized watershed at Lodi, New Jersey. The figure demonstrates a "first flush effect," in which concentrations of residuals (lead and suspended solids) reach their maxima shortly after the runoff begins. The peak concentrations result early because the initial volume of runoff carries away the high loadings of residuals that accumulated since the last rainstorm or street sweeping.

Several measures can be taken to reduce water pollution caused by urban runoff. Some involve structural works such as detention basins to store surface runoff and allow some solids to settle out. Others involve changes in operations such as the use of more efficient devices to sweep litter from streets.

Agriculture, Forestry, and Mining

Table 16.1 indicates residuals commonly associated with runoff from agricultural areas ("farm runoff"). Sediments from eroding farmlands frequently carry pesticides, fertilizers, and other chemicals adsorbed onto soil surfaces. Fertilizers often contain compounds of nitrogen and phosphorus that can stimulate the growth of algae and other aquatic plants. Nitrates in farm runoff are of special concern since they can percolate into groundwater zones. High concentrations of nitrates in groundwater used for drinking have been linked to methemoglobinemia, a serious disease in infants. Irrigated farms can degrade water quality because flows remaining after irrigation often have high concentrations of total dissolved solids. High TDS results because much water enters the atmosphere during the evapotranspiration that accompanies irrigation. Since smaller quan-

TABLE 16.1 Residuals in Surface Runoff from Agriculture, Forest, and Mining Areas

Activity	Water Quality Parameters Affected by Surface Runoff
Agriculture	
Croplands	Sediments Pesticides Compounds of phosphorus and nitrogen Total dissolved solids
Animal feedlots	Biodegradable organic matter Pathogenic organisms Compounds of phosphorous and nitrogen
Commercial forests	Sediments Pesticides Water temperature
Mining	Sediments Heavy metals Acidity

tities of water remain to dilute solids dissolved in the water prior to irrigation, the TDS concentration increases. This solids concentration also rises because water dissolves mineral salts in the soil it irrigates. Concentrations of TDS are sometimes quintupled due to combined effects of salt dissolution and evapotranspiration.

Other nonpoint sources associated with agriculture are animal feedlots, areas containing high densities of animals raised for slaughter. The buildup of animal manure on feedlot surfaces is often massive. When it rains, the resulting runoff is high in BOD and contains various nitrogen and phosphorous compounds. Pathogenic organisms, as indicated by high concentrations of coliform bacteria, may also exist in this runoff.

Table 16.1 also lists residuals in runoff from commercially harvested forests. As in the case of farmlands, runoff from commercial forests often contains high concentrations of sediments and pesticides. Sediment yields from eroding logging roads are especially high. In addition, notable increases in water temperature may result when trees shading a stream are harvested.

Surface runoff from mining areas is also rich in sediments. Problems can be caused by sediments from either strip mines or open pit mines because the exposed bare land is easily eroded by rainfall. The eroded soils often contain adsorbed metals and other substances associated with mining. When water comes in contact with mined materials, chemical reactions leading to acidification may take place. This occurs frequently in water draining from coal mines.

Leachates from Landfills and Wastewater Disposal Sites

When water infiltrates soil it dissolves some substances and transports them. This process is called *leaching,* and the water carrying the material is termed a *leachate.* Significant quantities of residuals enter leachates passing through "landfills" and dumps containing solid wastes or containerized liquid wastes, and pits and lagoons receiving industrial wastewaters. Contamination of groundwater by such leachates was observed with alarming frequency in the United States during the 1970s. This resulted, in part, because firms relied more heavily on pits and lagoons to dispose of small volumes of wastewater. Low-volume effluents were sometimes diverted to the ground to avoid stringent point source controls mandated by federal water quality laws.

A 1977 assessment of 50 industrial waste disposal sites found that groundwater below the sites was contaminated by leachates in nearly all cases.[13] Heavy metals were discovered in the groundwater at 49 sites, and various organic solutes (including PCBs and benzene) were measured at 40 sites. It is difficult to prevent groundwater contamination by leachates from dumps and landfills because these facilities are frequently unlined. An additional complication is that many sites are not equipped to monitor groundwater quality.

[13]For details on this assessment, see Council on Environmental Quality (1980, p. 89).

Another nonpoint source of groundwater pollution is the "individual" wastewater disposal system, typically a septic tank or cesspool. In 1980, nearly 20 million households in America relied on individual systems. Groundwater contamination by septic tanks and cesspools typically consists of domestic wastes. However, problems are sometimes caused by chemicals used to clean septic tanks. These cleaners, which frequently contain trichloroethylene, benzene, or methylene chloride, have contaminated numerous drinking water wells on Long Island, New York.[14]

Water Resources Projects

Because water resources projects such as dam construction often influence water quality, they are sometimes viewed as nonpoint sources of pollution. This discussion introduces some of the many water quality issues commonly associated with reservoirs and dredging projects.[15]

Consider first the creation of water storage reservoirs. When a stream is dammed, its velocity decreases and sediments formerly held in suspension settle out. The water's turbidity is reduced and a pool of sediments is created behind the dam. Heavy metals and organic substances originally adsorbed on sediments may become dissolved at the sediment–water interface. Sediment trapping also affects water quality below the dam. Because water released from the reservoir has a reduced turbidity, the downstream pattern of bank erosion and sediment deposition will be different from what it was before the dam was built.

Reservoir development also influences water quality by means of thermal stratification, a process that occurs seasonally in reservoirs located in temperate climates. In this process, layers of water are formed because of the way the temperature, and density, of the stored water varies with depth. The cold, bottom layer (the "hypolimnion") is especially important since withdrawals are often made from a reservoir's lower levels to maximize the use of stored water. Releases from the hypolimnion are often low in dissolved oxygen and high in iron and manganese concentrations.[16]

Dredging to develop navigation channels is another type of project influencing water quality. Coastal dredging in highly populated regions is especially problematic because the water bodies frequently contain bottom sediments that have

[14]The Long Island case and numerous other incidents of groundwater contamination are summarized by the Council on Environmental Quality (1981).

[15]In addition to reservoirs and dredging projects, channel modifications (e.g., stream widening and lining) also affect water quality. The literature on these subjects is voluminous and is reviewed by Ortolano (1973); for a brief overview, see Krenkel and Novotny (1980, pp. 236–239).

[16]These conditions result because the amount of dissolved oxygen in the hypolimnion is frequently less than required for the biochemical decomposition of organic matter. In the absence of DO, chemical reactions at the sediment–water interface include the release of iron and manganese from the sediments into solution. For additional information on reservior stratification, see Krenkel and Novotny (1980, pp. 462–487).

adsorbed heavy metals and synthetic organic compounds. Dredging exposes a large surface area of bottom sediments to water and this can lead to numerous chemical and biological changes. For example, the resuspension of bottom muds having a high BOD often depresses dissolved oxygen concentrations in the dredging area. In addition, complex chemical interactions at the newly created sediment–water interfaces can increase the concentrations of many solutes, some of which may be toxic to aquatic life.

FORECASTING CHANGES IN SURFACE WATER QUALITY

Mathematical models are often used to forecast how proposed projects influence water quality. The simplest cases involve substances such as chlorides, which are "conservative" in that they are not modified by chemical or biological reactions in natural waters. Materials that undergo biochemical transformations (for example, nitrogen compounds) are much more difficult to model. For many substances, the transformations that occur in natural waters are not understood well enough to be described with equations.

The degree of difficulty in modeling water quality impacts of proposed projects is different for streams, lakes, and estuaries. The simplest forecasting exercise is for a stream where the transport of residuals is in one direction and due entirely to the stream's average motion (*advection*). In this case, *dispersion* of residuals, mixing due to factors such as turbulent and molecular diffusion, wind effects, and density differences, is considered insignificant. Analyses of lakes and estuaries are often more complex because they typically include pollutant transport due to both advection and dispersion. In addition, they frequently account for flows in more than one direction.[17]

Predicting Effects of Point Source Discharges

For the residuals in Table 16.2, it is common to use mathematical models to estimate how point source discharges influence concentrations within receiving waters. Although the residuals listed have been modeled in streams, lakes, and estuaries, this introduction considers only streams. Discharges and streamflows are considered to be constant and steady-state conditions are assumed.[18]

The listing in Table 16.2 is ordered from the simple case of conservative substances to more complex cases in which biochemical transformations must be accounted for. The same ordering of topics is followed in this discussion. It is assumed throughout that only a single point source is involved.

Models for predicting how the discharge of a conservative substance affects

[17]There are exceptions: for example, lakes are sometimes modeled as uniformly mixed systems. Also, models of streams can be extended to include dispersion effects.

[18]Krenkel and Novotny (1980) introduce more sophisticated models used in analyzing non-steady-state conditions and complex water systems such as lakes and estuaries.

TABLE 16.2 Water Quality Measures for Which Forecasting Procedures Exist

Conservative Residuals
Chlorides
Total dissolved solids
Residuals that diminish according to first-order reaction kinetics
Biochemical oxygen demand
Coliform bacteria
Total phosphorus
Coupled indicators
DO and BOD
Complex systems of indicators
Compounds of nitrogen
Biomass of algae
Temperature
Suspended sediments

stream quality are invariably based on the law of conservation of mass, also referred to as the "mass balance equation." Chlorides and total dissolved solids are the indicators modeled most frequently. As shown in Figure 16.7, a mass balance analysis for conservative substances uses two equations, one for the conservation of flow and one for the conservation of residuals. Solving these

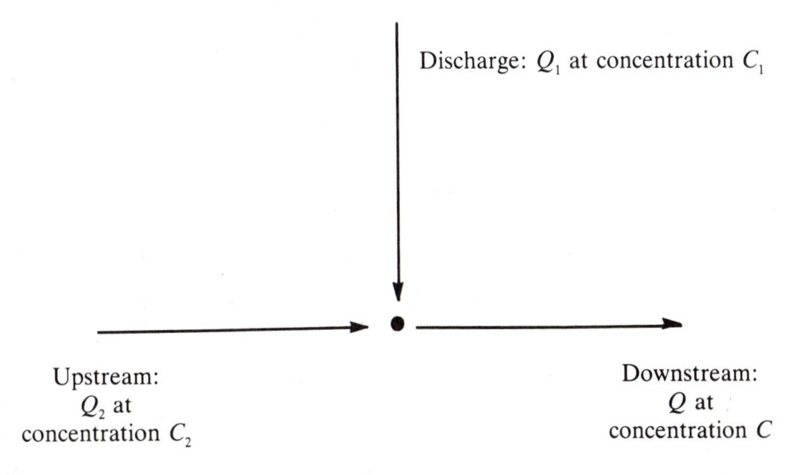

Conservation of streamflow: $Q = Q_1 + Q_2$

Conservation of mass: $QC = Q_1C_1 + Q_2C_2$

FIGURE 16.7 Steady-state mass balance analysis for a conservative substance discharged to a stream.

equations simultaneously yields C, the concentration of residuals downstream of the discharge point:

$$C = \frac{Q_1 C_1 + Q_2 C_2}{Q_1 + Q_2} \qquad \textbf{(16-1)}$$

where

Q_1 = flowrate upstream of the discharge (cfs)
Q_2 = rate of wastewater flow (cfs)
C_1 = residuals concentration upstream of the discharge (mg/l)
C_2 = residuals concentration of the wastewater (mg/l)

The units shown (cfs and mg/l) are illustrative; other flowrate and concentration units can also be used. Because the substance is conservative and streamflow is constant, the increase in concentration caused by the discharge is independent of the distance downstream. This analysis assumes the effluent is completely mixed once it enters the stream and the concentration of residuals is uniform throughout any stream cross section. Equation 16-1 is used in Chapter 3 to determine how chloride reduction by the Margarita Salt Company influences the concentration of chlorides in the Cedro River.

The next level of modeling complexity is for a substance that decreases in concentration in accordance with "first-order reaction kinetics." Underlying physical or biochemical causes of the decrease are not treated explicitly. At any instant, the rate of decay is assumed proportional to the amount of substance present; the proportionality factor is called the *rate constant.* As indicated in the discussion of equation 7-2, coliform bacteria are predicted using a first-order reaction model. More generally, a first-order model has the form

$$C(x) = C_0 e^{-k(x/u)} \qquad \textbf{(16-2)}$$

where

x = distance downstream of the discharge (miles)
$C(x)$ = residuals concentration at location x (mg/l)
C_0 = residuals concentration at $x = 0$ (mg/l)
u = velocity of streamflow (miles/day)
k = empirically determined rate constant (per day)

The concentration at $x = 0$ is calculated using the mass balance concepts yielding equation 16-1. The value of C_0 is the total mass of residuals per time just below the discharge point divided by the volume of streamflow per time at that point. Because stream velocity is assumed constant, (x/u) represents the "time of travel" below the discharge location.

Chapter 7 shows how k is estimated from observed data. It also indicates how the first-order reaction model predicts reductions in downstream coliform concentrations due to wastewater treatment. BOD and total phosphorous are also modeled by assuming they decrease according to a first-order reaction.

Sometimes the indicator of interest interacts with other water quality measures. This holds for dissolved oxygen, which depends on the concentration of biochemical oxygen demand. The simplest mass balance analysis of DO considers two effects: loss of oxygen caused by the biochemical oxidation of organic solutes and the gain due to oxygen transfers across the air–water interface ("atmospheric reaeration"). Mass balance equations for dissolved oxygen and biochemical oxygen demand are solved simultaneously to yield the "oxygen sag equation":

$$C(x) = C_s(x) - \left[\frac{k_1 L_0}{k_2 - k_1} \left(10^{-(k_1 x/u)} - 10^{-(k_2 x/u)}\right) + D_0 10^{-(k_2 x/u)} \right] \quad \textbf{(16-3)}$$

where

x = distance downstream from BOD discharge (miles)
$C(x)$ = concentration of DO at location x (mg/1)
$C_s(x)$ = "saturation concentration" of DO at location x (mg/1)
D_0 = DO "deficit" ($C_s - C$) at $x = 0$ (mg/1)
L_0 = "ultimate" BOD at $x = 0$ (mg/1)
k_1 = coefficient of deoxygenation (per day)
k_2 = coefficient of reaeration (per day)
u = velocity of streamflow (miles per day)

Equation 16-3 holds only if the oxygen consuming organic solutes are not so plentiful that they drive the DO concentration to zero.

Several terms in equation 16-3 require elaboration. The coefficient of deoxygenation, k_1, is associated with the biochemical oxidation of organic matter. Since BOD is assumed to follow a first-order reaction, k_1 is estimated using the previously discussed methods for finding decay rates for coliform (see Figure 7.5). The coefficient of reoxygenation, k_2, is computed based on stream velocity and average stream channel depth.[19] The saturation value of dissolved oxygen, $C_s(x)$, is the maximum amount of oxygen that can be dissolved per unit volume of water. Tables for determining $C_s(x)$ based on water temperature are available in water quality management textbooks.[20] The ultimate biochemical oxygen

[19]Several methods for calculating k_2 are reviewed by Krenkel and Novotny (1980, pp. 384–392).

[20]For example, saturation values of DO are tabulated by Velz (1970, p. 194). Since all streams have some background BOD, the dissolved oxygen level of an "unpolluted" stream is often lower than the saturation value given in the tables.

demand, L_0, is the amount of dissolved oxygen required to convert all the biodegradable organic matter (per unit volume) to carbon dioxide, water, and other stable end products. If the wastewater contains substantial amounts of organic nitrogen, the oxygen sag analysis can be modified to distinguish between "carbonaceous" and "nitrogenous" BOD. The former is due to organic compounds consisting only of carbon, oxygen, and hydrogen, and the latter is due to the oxidation of ammonia–nitrogen.[21] Equation 16-3 has also been extended to account for other factors affecting DO such as photosynthetic activities of aquatic plants and biochemical oxidation of organic sediments.

Some water quality experts prefer to analyze DO by maintaining separate accounts for deoxygenation and reoxygenation instead of combining them as is done in the oxygen sag equation. This approach, due to Velz (1970), is referred to as the "rational method of stream analysis." Figure 16.8 illustrates results from using Velz's approach to analyze DO in a reach of the Upper Chattahoochee River in Georgia. Several municipal wastewater discharges were accounted

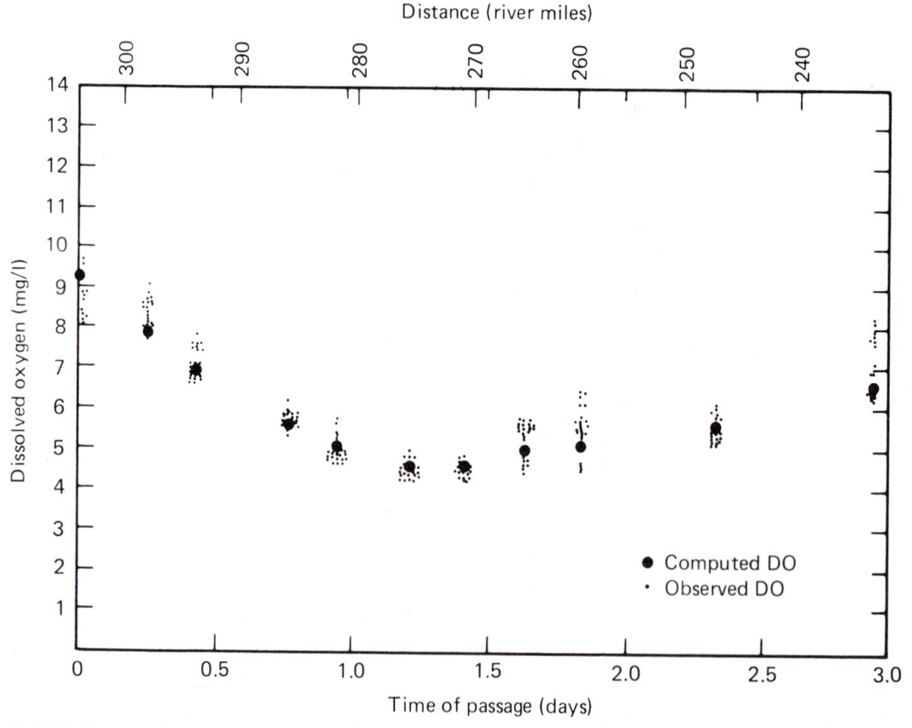

FIGURE 16.8 Computed and observed dissolved oxygen concentrations in the Atlanta to Franklin reach of the upper Chattahoochee River in Georgia. From R. N. Cherry, et al. in P. E. Greeson (ed.), *River-Quality Assessments,* American Water Resources Association, pp. 22–42. Copyright © 1977.

[21]The distinction between carbonaceous and nitrogenous BOD is discussed by O'Connor, Thomann, and Di Toro (1976).

for. The figure indicates a close correspondence between observed DO and forecasts obtained using the rational method. Dissolved oxygen decreases immediately below the principal waste sources (river miles 300 to 275) and then gradually increases as the oxygen required in stabilizing organic matter is reduced. If there were no additional effluents downstream, atmospheric reaeration would eventually raise the river's dissolved oxygen to near its saturation value.

Several mass balance equations are solved simultaneously in modeling different forms of nitrogen in surface waters: organic-nitrogen, ammonia–nitrogen, nitrite–nitrogen, and nitrate–nitrogen. Figure 16.9 depicts the nitrogen transformations that occur in the presence of dissolved oxygen. The interconnections in the figure can be described by combining a mass balance relationship for DO with mass balance equations for various nitrogen constituents. O'Connor, Thomann, and Di Toro (1976) solved simultaneously the mass balance equations for DO and various nitrogen constituents and used the results to analyze several water systems. Models treating oxygen and nitrogen have been extended to include biological parameters such as the biomass of algae. However, models including biomass require substantial amounts of data and are not used routinely.

Two other water quality indicators that have been investigated extensively are temperature and suspended sediments. Procedures used in forecasting temperature are based on the law of conservation of energy, and they have have been applied in a variety of environmental assessments.[22] In contrast, techniques used in modeling the transport of sediments are very complex.[23]

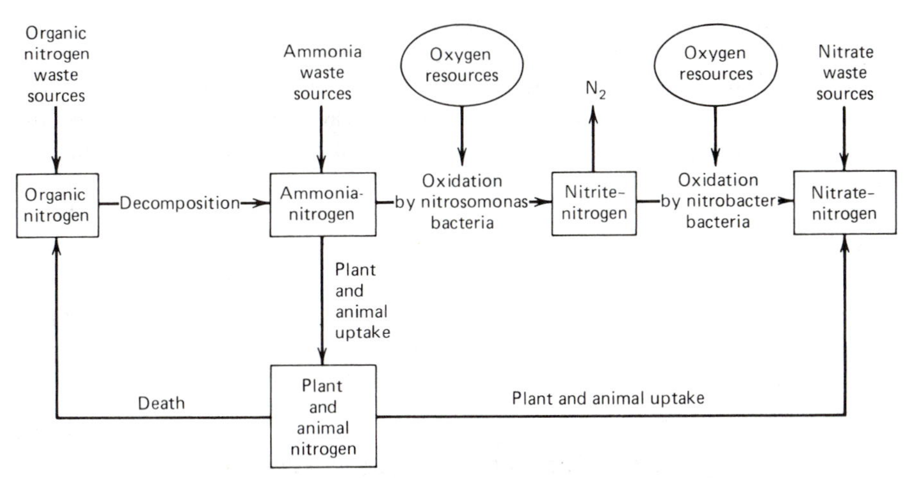

FIGURE 16.9 Major features of the nitrogen cycle in waters containing dissolved oxygen. From D. P. Loucks, J. R. Stedinger, and D. A. Haith, *Water Resource Systems Planning and Analysis,* copyright © 1981, p. 451. Reprinted by permission of Prentice–Hall, Englewood Cliffs, N.J.

[22]Velz (1970, pp. 304–337) gives several practical examples demonstrating how models are used to forecast water temperatures.

[23]An introductory discussion of sediment transport modeling is given by Dunne and Leopold (1978, pp. 672–685).

Forecasting Effects of Nonpoint Sources

To estimate water quality impacts of nonpoint sources, it is necessary to predict how residuals on land are washed away by surface runoff. This forecasting problem is discussed below. Once the water-borne residuals have been predicted, extensions of the previously described point source models are used to determine how receiving water quality is influenced. The response of natural waters to nonpoint sources cannot be fully analyzed by considering only steady-state conditions. Non-steady-state analyses are called for since large quantities of residuals in surface runoff may be delivered to receiving waters in very short time intervals.

Runoff from Urban Areas Many procedures for predicting residuals in urban runoff are extensions of the hydrologic simulation models introduced in the beginning of this chapter. In one fashion or another, the extended models treat the following questions: What quantity of solid material has accumulated on impervious surfaces? Which water quality indicators usefully characterize the solids? What portion of the accumulated residuals will be washed away in a given volume of runoff? How much of the materials adsorbed on the solids will "desorb" into solution? The quantitative responses to these questions are used in constructing *pollutographs,* plots of residuals concentration in surface runoff at a particular location versus time. The process of extending hydrologic simulation models to produce pollutographs only began in the 1970s, and thus procedures for answering the questions above are still being developed. A few illustrative approaches are introduced below.

The question of how much solid material accumulates on impermeable surfaces in urban areas is generally answered empirically. For example, Sartor and Boyd (1975) gathered data from several cities on the solids accumulating between periods of rainfall or street sweeping. Their results, summarized in Figure 16.10, have been used to derive an equation for estimating solids accumulation on streets:

$$F = \frac{G}{k}(1 - e^{-kt}) \qquad \textbf{(16-4)}$$

where

F = loading of solids on street surfaces (lb/curb-mile)
G = constant rate of solids deposition (lb/curb-mile-day)
k = empirically determined rate constant (per day)
t = time elapsed since last rainfall or street sweeping (days)

Estimates of the two parameters, k and G, have been made for many cities.

Equation 16-4 is demonstrated by a case presented by Overton and Meadows (1976) in which G = 100 lb of solids/curb-mile-day and k = 0.40/day. The

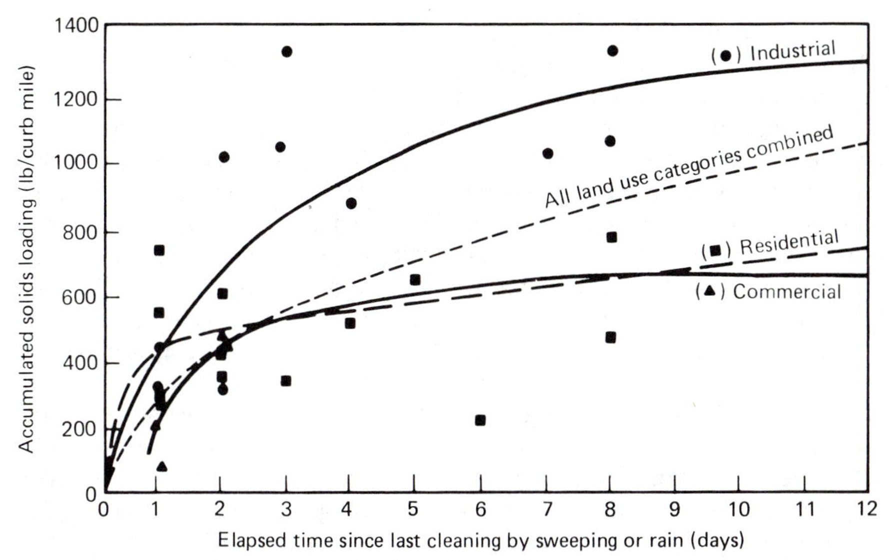

FIGURE 16.10 Solids accumulation for different land uses: Results of a study for eight U.S. cities. Reprinted from J. D. Sartor and G. B. Boyd, *in* W. I. Jewell and R. Swan (eds.), *Water Pollution Control in Low Density Areas* by permission of University Press of New England. Copyright © 1975 by Trustees of the University of Vermont.

estimated quantity of solids accumulating in 2 days after the last rainfall or street sweeping is

$$F = \frac{100}{0.4}(1 - e^{-0.4(2)}) = 138 \text{ lb/curb-mile}$$

As in Figure 16.10, this equation shows the solids buildup reaching a maximum as t, the time elapsed, becomes very large. The maximum value, G/k, is 250 lb/curb-mile in this example.

To predict the constituents in urban runoff, specific substances associated with the solids accumulating on streets must be identified. Data indicating the amount of a substance per unit of solids for different residuals and land use categories are used for this purpose. For example, the phosphate in solids accumulating on residential neighborhood streets has been estimated as 50 mg/kg of solids.[24] The mass of phosphate accumulating in such an area is calculated by multiplying 50 mg/kg by the total solids accumulation (in kilograms).

After the individual residuals in the solids accumulating on streets are determined, the question remains as to how much material will be washed away by surface runoff. Loucks, Stedinger, and Haith (1981) determine this quantity by assuming there is a runoff rate that, if equaled or exceeded, will flush away all

[24]This value, which is for neighborhoods with single-family homes, is from Loucks, Stedinger, and Haith (1981, pp. 481). Other ways of characterizing the quantities of substances in urban runoff are given by Whipple et al. (1983, pp. 84–90).

accumulated solids. If the surface runoff is less than this rate, the fraction of solids washed off is assumed equal to the actual surface runoff divided by the minimum required for a complete washout.[25]

A remaining question concerns how much adsorbed material goes into solution when the solids are flushed away. This can be examined using *partition coefficients* obtained from laboratory experiments. A partition coefficient is the concentration of a substance in solution (mg/1) divided by the concentration of the substance adsorbed on the solids (mg/kg) after the water and solids have been in contact long enough to reach a chemical equilibrium. A partition coefficient's value depends on characteristics of the solid and the adsorbed substance. Loucks, Stedinger, and Haith (1981) demonstrate how partition coefficients are used to calculate residuals concentrations in surface runoff.

Once the residuals concentrations in surface runoff at a particular point are estimated, hydrologic modeling techniques are used to construct pollutographs at downstream locations. The analysis methods frequently combine the water quality prediction techniques above with a hydrologic simulation model.[26] Indicators commonly included in models that analyze water quality in urban runoff are suspended solids, BOD, coliform bacteria, and compounds of phosphorus and nitrogen.

Runoff Outside of Urban Areas In nonurban settings, the amount of soil eroded by surface runoff is frequently the key factor in determining water quality effects of nonpoint sources. Soil erosion is often significant for activities related to construction, farming, open pit and strip mining, and forest harvesting.

Although many relationships exist for predicting soil erosion due to surface runoff, the most widely used is the "universal soil loss equation":

$$A = R\,K\,L\,S\,C\,P \qquad \textbf{(16-5)}$$

where

A = average soil loss per unit area (tons/acre/time period)
R = local rainfall factor (per time period)
K = soil erodibility factor
L = slope length factor
S = slope gradient factor
C = cropping management factor
P = erosion control factor

[25]More complicated approaches to estimating the extent of washout are discussed by Overton and Meadows (1976, pp. 318–328) and Whipple et al. (1983, pp. 104–107).

[26]Over a dozen computer simulation models have been developed to predict residuals concentrations in urban runoff, and several of them are reviewed by Overton and Meadows (1976) and Whipple et al. (1983). The models generally assume that no chemical or biological transformations of residuals take place in the time during which the runoff is routed to a stream, lake, or estuary. In addition, residuals transferred from surface runoff to the ground via infiltration are typically considered neglible.

Although equation 16-5 was originally developed for analyzing soil conservation activities on farms, it has been adapted for nonagricultural settings. Factors on the right-hand side of equation 16-5 are determined using a variety of charts and graphs, such as those summarized by Wanielista (1978).

The universal soil loss equation can be applied for each of several time periods, depending on how R is selected. Values of R have been developed to yield average annual quantities of eroded soil. They have also been developed to correspond to different frequencies of occurrence for storm events. For example, it is possible to estimate soil loss for a storm whose magnitude is normally exceeded once in 10 years, once in 20 years, and so forth.[27]

After the soil loss is determined it is necessary to analyze the distribution of various substances, such as pesticides and heavy metals, adsorbed on the eroded soil. Available analysis methods rely on empirical equations to calculate how much of the adsorbed materials will be desorbed.[28]

In addition to affecting concentrations of pesticides, heavy metals and other residuals, soil eroded from land surfaces can greatly influence the shape of streams. The increase in suspended sediments delivered to a stream alters the downstream pattern of bank erosion and sediment deposition. These effects are difficult to forecast because sediment transport processes are complex and not fully understood.

FORECASTING CHANGES IN GROUNDWATER QUALITY

Forecasting how human actions influence groundwater quality is generally more difficult than predicting how various projects affect surface water quality. One reason is that test wells and other equipment are needed to delineate the groundwater zone and determine hydraulic properties of soils and rocks ("porous media") in that zone. This type of data gathering exercise is often quite costly. The lack of uniformity in the porous media through which the water flows is another source of difficulty. Hydraulic properties of soils can vary greatly, even over small areas. If the porous media are very irregular, it may be necessary to analyze water movement in all three spatial dimensions. Further complications in predicting groundwater quality are the chemical and biological transformations at the soil–water interface.

Advection and Dispersion of Conservative Solutes

As in the case of surface waters, the simplest procedures for predicting effects of human actions on groundwater quality are those involving conservative solutes

[27]Overton and Meadows (1976, pp. 311–318) present several applications of the universal soil loss equation.

[28]Approaches for determining the disposition of substances adsorbed on eroded soil are discussed by Krenkel and Novotny (1980, p. 229) and Loucks, Stedinger, and Haith (1981, pp 484–485).

such as chlorides. By definition, conservative substances do not react with either the porous media comprising the groundwater aquifer or the flowing groundwater itself. Forecasting procedures for conservative solutes generally rest on (1) a hydraulic analysis to estimate groundwater velocity, including both speed and direction, and (2) a mass balance analysis to determine concentrations at various locations.

In developing forecasting methods, the behavior of the conservative substance is assumed to be affected by only two processes, advection and dispersion. In this context, *advection* is solute transport by the average motion of groundwater flow. (For advective transport, the average rate of solute motion equals the average velocity of the water.) *Dispersion* refers to the movement of the substance due to molecular diffusion and small scale nonuniformities in water velocity. Fluctuations in velocity occur because groundwater flows through nonuniform spaces between soil and rock particles.

Differences between advection and dispersion are clarified by a laboratory experiment simulating the flow of a conservative substance in groundwater. Figure 16.11*a* shows a cylindrical column packed with clean sand through which ordinary tap water flows at a constant average velocity. Assume a nonreactive tracer material is injected into the tap water at time, t_0, and the concentration of tracer in the tap water prior to time t_0 is zero.

After time t_0 the tracer concentration in the column inflow is C_0. A short time later, the concentration of tracer in the column *outflow* jumps from zero to C_0. If advection were the only transport process, the increase in outflow concentration from zero to C_0 would occur in an instant. The tracer injected at time t_0 would advance through the column with a uniform velocity equal to the average water velocity. Tracer concentration in the column outflow versus time is shown in Figure 16.11*b*. The figure assumes advection is the only transport mechanism. By convention, the graph's vertical axis is C/C_0, the fraction of the original solute concentration that appears in the column outflow. Use of this ratio, termed the *fractional breakthrough,* makes it possible to compare the transport of different solutes on a single graph.

When the tracer experiment is actually conducted, the graph of outflow concentration versus time does not have the form in Figure 16.11*b*. Instead, the outcome is more like the "S-shaped" curve in Figure 16.11*c*. Some of the initial tracer injection arrives at the outlet sooner than would be the case if transport were due only to advection, and some of it arrives later. The spreading of the initial tracer injection is attributed to dispersion.

Mass balance analyses have been used to forecast how the outflow concentration of a nonreactive tracer varies with time in the above experiment. The parameters in the resulting mathematical model include average path length of a unit volume of water, average velocity of water, and measures of the dispersion characteristics of the porous medium in the column.[29] For the simple experi-

[29]Mathematical models based on mass balance analyses for conservative solutes are presented by Freeze and Cherry (1979, pp. 388–397).

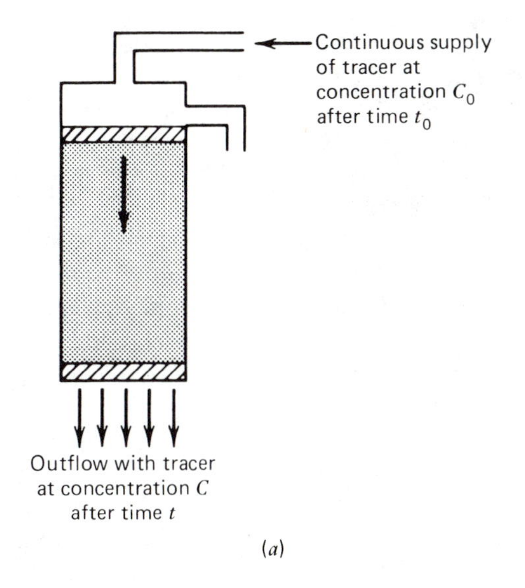

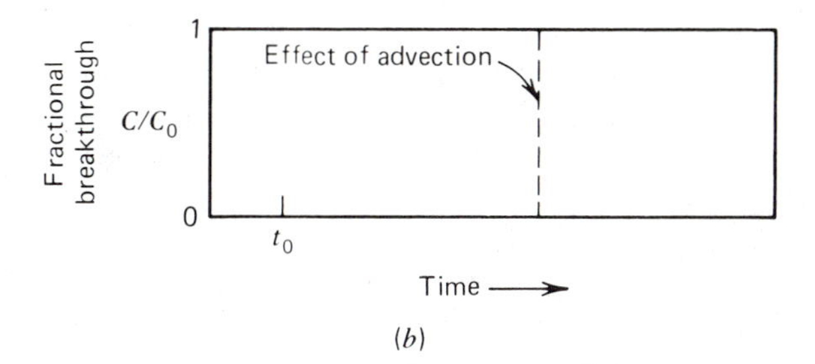

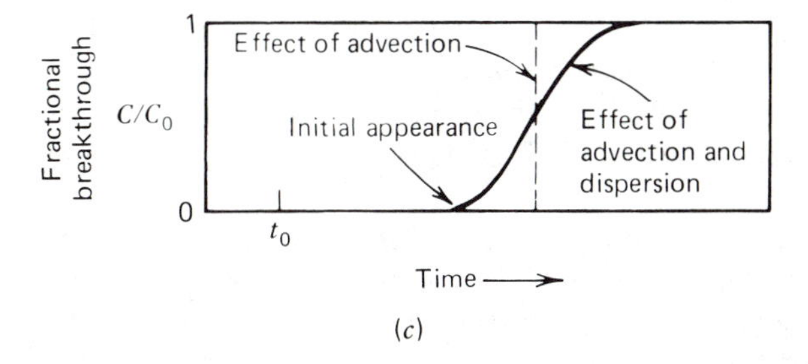

FIGURE 16.11 Dispersion of conservative tracer passing through a column of clean sand. Adapted from R. Allan Freeze and John A. Cherry, *Groundwater,* copyright, © 1979, p. 390. Adapted by permission of Prentice–Hall, Englewood Cliffs, N.J. (*a*) Column with uniform flow and continuous supply of tracer after time t_0. (*b*) Concentration of tracer in outflow if transport is due only to advection (hypothetical condition). (*c*) Concentration of tracer in outflow due to combined effects of advection and dispersion.

mental system, the concentrations predicted with mathematical models are often close to observed values. However, real groundwater aquifers are much more complex than the laboratory column, and the accuracy of forecasted solute concentrations in real aquifers is often quite low. Complications result because of nonuniformities in the porous media and the need to consider more than one dimension when analyzing paths of flowing groundwater.

Sophisticated techniques exist for solving the equations governing the movement of conservative solutes in aquifers. However, constants in the equations, especially aquifer dispersion characteristics, often cannot be estimated because there are insufficient data. Despite these difficulties, mathematical models have been used to guide data collection efforts and formulate wastewater management strategies.[30]

Transport of Reactive Solutes

Transformations influencing reactive solutes in groundwater include adsorption, ion exchange, chemical precipitation, acid–base reactions, oxidation–reduction reactions, and microbial cell synthesis.[31] For many solutes there is not sufficient information to model transformations mathematically.

Studies by Roberts et al. (1980) illustrate the complexities in predicting solute concentrations when biological and chemical reactions influence transport processes. Their investigation concerned the suitability of injecting highly treated municipal wastewaters into an aquifer as a means of replenishing the groundwater in Palo Alto, California. This discussion treats the part of their study concerning chlorides and two reactive organic solutes, chlorobenzene and naphthalene.

Roberts and his colleagues measured solute concentrations at two locations, an injection well and a test (or observation) well. The quantity of wastewater pumped into the injection well was also measured. Observations were presented on graphs of fractional breakthrough (C/C_0) versus volume of wastewater pumped into the aquifer at the injection well ("cumulative volume injected"). For any solute, the concentrations are C_0 at the injection well and C at the observation well. A comparative analysis of fractional breakthrough patterns led to hypotheses about how chemical and biological reactions influenced solute transport.

Figure 16.12*a* summarizes results for the transport of chlorobenzene and chlorides. Compared to chlorides, a much greater volume of wastewater had to be injected before any chlorobenzene was detected at the observation well. The following reasoning was used to explain the differences in transport patterns: For chlorides, transport between the injection and test wells was due only to advection and dispersion. The average time of travel for chlorides was identical

[30]This use of models is demonstrated by Konikow's (1981) efforts to predict chloride concentrations in aquifers in the Denver, Colorado, area.

[31]For a discussion of these transformations, see Freeze and Cherry (1979, pp. 402–426).

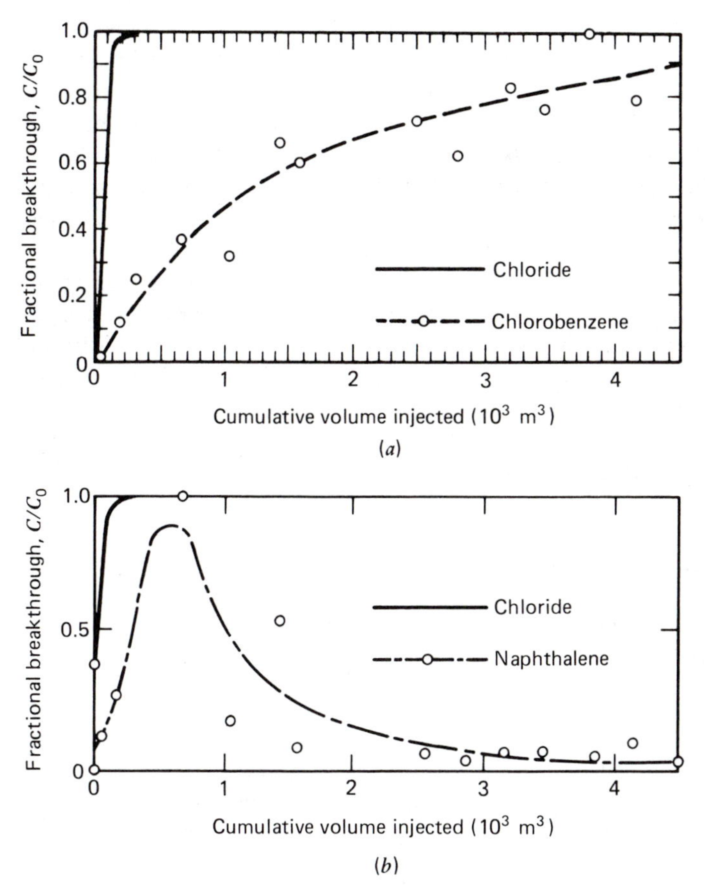

FIGURE 16.12 Concentration of organic solutes and chloride at the observation well. From P. V. Roberts, et al., *Journal Water Pollution Control Federation,* Vol. 52, No. 1, pp. 161–172. Water Pollution Control Federation. Copyright © 1980. (*a*) Concentrations of chlorobenzene and chloride. (*b*) Concentrations of naphthalene and chloride.

to that of water. For chlorobenzene, however, transport was affected by adsorption in addition to advection and dispersion. Roberts and his colleagues postulated that in the early stages of pumping, much of the chlorobenzene was adsorbed on the porous media. Eventually, the adsorption capacity of the porous media in the path between the two wells was exhausted. When this occurred, values of C/C_0 were equal to unity for chlorobenzene. Thereafter, since adsorption was no longer significant, only advection and dispersion played a role in the transport process and the chlorobenzene concentration at the test well remained at C_0.

The results for the movement of naphthalene, shown in Figure 16.12*b,* are quite different from either chlorides or chlorobenzene. In the early stages of pumping, the naphthalene concentration at the test well increased, but then it decreased without reaching the concentration at the injection well. It was hypothesized that the naphthalene transport was influenced by biochemical oxidation in addition to advection and dispersion. In the first stages of pumping, the naphthalene concentration increased because the bacteria were insufficient to accomplish the biochemical oxidation of naphthalene. Once these bacteria were established, the biochemical oxidation rate increased and there was a sharp decrease in naphthalene at the test well. Finally, after the biological oxidation process reached a steady state, the naphthalene concentration stabilized at less than one tenth of the concentration at the injection well.

The work by Roberts and his associates provides a basis for extending the mass balance analyses for conservative residuals to include solute transformations such as adsorption and biochemical oxidation. However, much additional research is needed before the fate of reactive substances in groundwater can be predicted with confidence. Regardless of the solute type, the absence of field data to estimate model parameters limits the widespread use of mathematical models to predict groundwater quality.

KEY CONCEPTS AND TERMS

REPRESENTATIONS OF THE HYDROLOGIC CYCLE
- Surface runoff
- Infiltration
- Percolation
- Evapotranspiration
- Hydrograph
- Hydrologic simulation model
- Calibration

INFLUENCE OF LAND USE CHANGES ON FLOOD FLOWS
- Flood frequencies
- Urbanization
- Channelization

PARAMETERS USED TO MEASURE IMPACTS ON WATER QUALITY
- Coliform bacteria
- Suspended solids
- Bottom sediments
- Inorganic and organic solutes
- Volatile and nonvolatile solutes
- Total dissolved solids
- Biochemical oxygen demand
- Dissolved oxygen
- Heavy metals

COMMON SOURCES OF WATER-BORNE RESIDUALS
- Domestic wastewater
- Industrial cooling water
- Cleaning and process water
- Nonpoint sources
- Urban runoff
- First flush effect
- Farm runoff
- Contaminated leachates
- Thermal stratification
- Hypolimnion

FORECASTING CHANGES IN SURFACE WATER QUALITY
- Mass balance analysis
- Conservative solute
- Biochemical oxidation
- First-order reaction kinetics
- Oxygen sag equation
- Sediment transport
- Adsorption and desorption
- Partition coefficient
- Pollutograph

FORECASTING CHANGES IN GROUNDWATER QUALITY
- Porous media
- Advection
- Dispersion
- Reactive solute
- Adsorption capacity

DISCUSSION QUESTIONS

16–1 Consider a proposed reservoir project that would supply water for irrigation and hydroelectric power generation. How might the project influence water flows in various components of the hydrologic cycle? Indicate which hydrologic impacts are likely to be the most significant.

16–2 When stream channels are cleared and widened to reduce flooding there is sometimes an increase in flooding downstream. Give two possible explanations for this.

16–3 Provide four examples demonstrating how adsorption and desorption phenomena affect the quality of both surface water and groundwater.

16–4 Explain the difference between advection and dispersion in surface water. How would you modify your statement to explain the difference between advection and dispersion in groundwater?

16–5 Consider a situation in which the following oxygen sag equation parameters are given:

$$k_1 = 0.1/\text{day}$$
$$k_2 = 0.2/\text{day}$$
$$D_0 = 1 \text{ mg/l}$$
$$C_s(x) = 8 \text{ mg/l for all } x$$
$$L_0 = 15 \text{ mg/l}$$

(i) Plot the oxygen sag curve for $t = 1, 2, 3, \ldots, 7$, where t, the time of travel, equals (x/u) in equation 16–3. Streamflow velocity is constant for this steady-state analysis.

(ii) Replot the sag curve for the case where wastewater treatment reduces L_0 to 7.5 mg/l. Use the same graph paper as in **(i)**.

(iii) Employ the results of **(i)** and **(ii)** to characterize the effects of wastewater treatment on stream dissolved oxygen.

REFERENCES

Beard, L. R., and S. Chang, 1979, Urbanization Impact on Streamflow. *Journal of the Hydraulics Division, Proceedings of the American Society of Civil Engineers* **105** (HY6), 647–659.

Biswas, A. K., 1970, *History of Hydrology.* North Holland, Amsterdam.

Council on Environmental Quality, 1980, *Eleventh Annual Report.* Council on Environmental Quality, Washington, D.C.

Council on Environmental Quality, 1981, *Contamination of Ground Water by Toxic Organic Chemicals.* Council on Environmental Quality, Washington, D.C.

Crawford, N.H., and R. K. Linsley, 1966, *Digital Simulation in Hydrology: The Stanford Watershed Model IV,* Technical Report No. 39, Department of Civil Engineering, Stanford University, Stanford, Calif.

Dunne, T., and L. B. Leopold, 1978, *Water in Environmental Planning.* Freeman, San Francisco.

Field, R., E. J. Struzeski, H. E. Masters, and A. N. Tafuri, 1975, Water Pollution and Associated Effects from Street Salting, *in* W. J. Jewell and R. Swan (eds.), *Water Pollution in Low Density Areas: Proceedings of a Rural Environmental Engineering Conference,* pp. 317–340. University Press of New England, Hanover, N.H.

Förstner, V., and G. T. W. Wittman, 1981, *Metal Pollution in the Aquatic Environment,* 2nd revised ed. Springer-Verlag, Berlin.

Freeze, R. A., and J. A. Cherry, 1979, *Groundwater.* Prentice–Hall, Englewood Cliffs, N.J.

James, L. D., 1965, Using a Digital Computer to Estimate the Effects of Urban Development on Flood Peaks. *Water Resources Research* **1** (2), 223–234.

Konikow, L. F., 1981, Role of Numerical Simulation Models in Analysis of Ground-Water Quality Problems, *in* W. Van Duijvenbooden, P. Glasbergen, and H. van Lelyveld (eds.), *Quality of Groundwater,* Vol. 17, pp. 823–836, Studies in Environmental Science, Elsevier, Amsterdam.

Krenkel, P. A., and V. Novotny, 1980, *Water Quality Management.* Academic Press, New York.

Linsley, R. K., 1976, Rainfall-Runoff Models, *in* A. K. Biswas (ed.), *Systems Approach to Water Management,* pp. 16–53. McGraw–Hill, New York.

Loucks, D. P., J. R. Stedinger, and D. A. Haith, 1981, *Water Resources Systems Planning and Analysis.* Prentice–Hall, Englewood Cliffs, N.J.

National Academy of Sciences, 1977, *Drinking Water and Health.* NAS, Washington, D.C.

Neely, W. B., 1980, *Chemicals in the Environment: Distribution Transport, Fate, Analysis.* Dekker, New York.

Nemerow, N. L., 1978, *Industrial Water Pollution: Origins, Characteristics, and Treatment.* Addison–Wesley, Reading, Mass.

O'Connor, D. J., R. V. Thomann, and D. M. Di Toro, 1976, Ecologic Models, *in* A. K. Biswas (ed.), *Systems Approach to Water Management,* pp. 294–334. McGraw–Hill, New York.

Ortolano, L. (ed.), 1973, *Analyzing the Environmental Impacts of Water Projects,* Report No. 73-3. U.S. Army Engineer Institute for Water Resources, Ft. Belvoir, Va.

Overton, D. E., and M. E. Meadows, 1976, *Stormwater Modeling.* Academic Press, New York.

Pirner, S. M., and L. L. Harms, 1978, Rapid City Combats the Effect of Urban Runoff on Surface Water. *Water and Sewage Works* **125** (2), 48–53.

Roberts, P. V., P. L. McCarty, M. Reinhard, and J. Schreiner, 1980 , Organic Contaminant Behavior during Groundwater Recharge. *Journal Water Pollution Control Federation* **52** (1), 161–172.

Sartor, J. D., and G. B. Boyd, 1975, Water Quality Improvement Through Control of Road Surface Runoff, *in* W. I. Jewell and R. Swan (eds.), *Water Pollution Control in Low Density Areas, Proceedings of a Rural Environmental Engineering Conference,* pp. 301–316. University Press of New England, Hanover, N.H.

Velz, C. V., 1970, *Applied Stream Sanitation,* Wiley, New York.

Wanielista, M. P., 1978, *Stormwater Management, Quantity and Quality,* Ann Arbor Science, Ann Arbor, Mich.

Whipple, W., N. S. Grigg, T. Grizzard, C. W. Randall, R. P. Shubinski, and L. S. Tucker, 1983, *Stormwater Management in Urbanizing Areas,* Prentice–Hall, Englewood Cliffs, N.J.

Wolman, M. G., and A. P. Shick, 1967, Effects of Construction on Fluvial Sediment, Urban and Suburban Areas of Maryland. *Water Resources Research* **3** (2) 451–464.

INDEX